# Springer Proceedings in Mathematics & Statistics

Volume 108



More information about this series at http://www.springer.com/series/10533

# Springer Proceedings in Mathematics & Statistics

This book series features volumes composed of select contributions from workshops and conferences in all areas of current research in mathematics and statistics, including OR and optimization. In addition to an overall evaluation of the interest, scientific quality, and timeliness of each proposal at the hands of the publisher, individual contributions are all refereed to the high quality standards of leading journals in the field. Thus, this series provides the research community with well-edited, authoritative reports on developments in the most exciting areas of mathematical and statistical research today.

Constantine Georgakis • Alexander M. Stokolos
Wilfredo Urbina
Editors

# Special Functions, Partial Differential Equations, and Harmonic Analysis

## In Honor of Calixto P. Calderón

Springer

*Editors*
Constantine Georgakis
DePaul University
Chicago, IL, USA

Alexander M. Stokolos
Georgia Southern University
Statesboro, GA, USA

Wilfredo Urbina
Roosevelt University
Chicago, IL, USA

ISSN 2194-1009 ISSN 2194-1017 (electronic)
ISBN 978-3-319-37982-1 ISBN 978-3-319-10545-1 (eBook)
DOI 10.1007/978-3-319-10545-1
Springer Cham Heidelberg New York Dordrecht London

Mathematics Subject Classification (2010): 01-06, 01A65, 42B25, 44A10, 42B05, 42C10, 46E35, 35J70, 32A55, 34H10, 37N25, 37N35, 42B37, 42B35, 92B99, 97M60



Printed on acid-free paper

Springer is part of Springer Science+Business Media (www.springer.com)

Calixto P. Calderón

**In the back row (from left to right):** Alberto P. Calderón (brother), Margarita Calderón (sister), with her daughter in arms Raquel Peña y Lillo (niece), Matilde García Gallo (mother), Matilde Calderón (sister)
**In the front row (from left to right):** Silvestre Peña y Lillo (brother in law), Silvestre Peña y Lillo, Jr. (nephew), Haydee Peña y Lillo (niece), María Teresa Calderón (sister), Dr. Pedro Calderón (father) and Calixto P. Calderón (himself).

# Preface

Calixto P. Calderon, like many Argentine mathematicians of his generation, was introduced and mentored into Harmonic Analysis and Partial Differential Equations at the University of Buenos Aires by Alberto Gonzalez Dominguez, his teacher and Ph.D. advisor, whose mentor was David Tamarkin of Brown University. Moreover, Calixto P. Calderon proved very early in his career to be an influential teacher and mentor for Luis A. Caffarelli, a world authority on Free Boundary Value Problems in Partial Differential Equations and recent recipient of the prestigious Israel's Wolf Prize.

The papers presented by his friends or colleagues and some of his former doctoral students at the conference in honor of Calixto P. Calderon at Roosevelt University were indicative to a large extent of the wide scope and depth of his research work and scholarship in pure and applied mathematics. Furthermore, the friendly and festive atmosphere among the participants at the conference from the USA and Latin America reflected very much his wonderful charismatic personality and humility. A memorable moment was the scene where the president of Roosevelt University, while addressing the participants briefly, asked those in the audience who were Calixto P. Calderon's students to raise their hands, and Luis A. Caffarelli's hand went up!

Calixto P. Calderon's numerous research publications include: Classical Fourier series and Orthogonal Expansions, Calderon-Zygmund Theory of Singular Integrals, Ordinary and Partial Differential Equations especially the Navier–Stokes p.d.e, Probability Theory, Mathematical Models Applied to Biology and Medicine, and History of Mathematics and Science. Clearly, his work in pure Mathematics was influenced by Alberto Gonzalez Dominguez, Alberto C. Calderon—his brother, Antoni Zygmund, and several other collaborators and colleagues. Whereas his considerable research work in Mathematical Biology and Medicine might have been the result of the influence he received while growing up from his father, Dr. Pedro J. Calderon, an accomplished physician trained in Medicine in Buenos Aires and Paris. One can't help but note that, likewise, the father of Aristotle, Nicomachus, who did experimental research in Biology and Botany, although he was Plato's greatest student of pure philosophy, was also a medical physician and, in fact, the personal physician of King Philip of Macedon.

Except for the appointments at the University of Cuyo, the University of Buenos Aires, and the University of Minnesota, Calixto P. Calderon spent most of his distinguished academic career at the University of Illinois at Chicago (U.I.C.). He taught a variety of courses ranging from Harmonic Analysis and Partial Differential Equations to Probability Theory and Statistics, and had several Ph.D. students. His strong interest in probability and statistics may be attributed in part to his Statistics teacher—Roque Carranza at the University of Buenos Aires. He wrote two papers on the subject including a paper on Kolmogorov's strong law of large numbers for pair-wise independent random variables. The themes of the thesis of his Ph.D. students, like his research, ranged from Harmonic Analysis to the history of Spanish mathematics and related science. During his tenure at U.I.C., Calixto P. Calderon was an active participant and invited speaker at seminars in the Chicago area, including the Calderon-Zygmund Seminar at the University of Chicago, and the analysis seminars at U.I.C., and at DePaul University. Furthermore, Calixto P. Calderon was an excellent and eloquent lecturer, and remained actively engaged in issues related to mathematics education at U.I.C.

The editors of this volume of chapters would like to thank the organizers of the conference in honor of Calixto P. Calderon and the authors of the contributed papers.

Chicago, IL, USA — Constantine Georgakis
Statesboro, GA, USA — Alexander M. Stokolos
Chicago, IL, USA — Wilfredo Urbina

# Contents

# Remembrances and Silhouettes

**Calixto P. Calderón**

My life in Mathematics began when I transferred from The University of Cuyo,[1] San Juan, to the University of Buenos Aires in 1961. My brother Alberto helped me economically and morally for the jump. Upon my arrival to Buenos Aires, Dr. Alberto González Domínguez (1904–1982) helped and oriented me with the change. The first subject I took in the Math Department, School of Exact Sciences, was Funciones Reales I, first course on Lebesgue Integration. Prof. Evelio Oklander was the instructor.

With the years, I would take Complex Analysis, Ordinary Differential Equations, Funciones Reales II, Partial Differential Equations, Topology, Functional Analysis, Projective Geometry, Differential Geometry, and other important subjects. We had a first rate faculty teaching these fundamental courses. Among them I remember: Alberto González Domínguez, Mischa Cotlar, Luis Santaló, Mario Gutiérrez Burzaco, Rafael Panzone, Agnes Benedek, Miguel Herrera, and Prof Roque Carranza (statistician). This period was replete of personal experiences. I met mathematicians like Dr. Julio Rey Pastor and Beppo Levi during the mathematical teas that were held in the OEA building of Avenida de Mayo. A full floor had been loaned to the School of Exact Sciences of the University of Buenos Aires.

---

*Introductory speech given at "Special Functions, Partial Differential Equations and Harmonic Analysis", A Conference in honor of Calixto P. Calderón, November 16–18, 2012, Department of Mathematics and Actuarial Sciences, Roosevelt University.

[1]There I had many friends interested in mathematics. Among them: Luis Matons and Julian Araoz. The latter would join me in Buenos Aires.

C.P. Calderón (✉)
Department of Mathematics, University of Illinois at Chicago, Chicago, IL 60607, USA
e-mail: cpc@uic.edu



A.M. Stokolos et al. (eds.), *Special Functions, Partial Differential Equations, and Harmonic Analysis*, Springer Proceedings in Mathematics & Statistics 108,
DOI 10.1007/978-3-319-10545-1_1

I forged new friendships with fellow students, among them: Domingo Herrero, Marta Herrero, Julio Bouillet, Guillermo Hansen, Alberto Torchinsky, Alvaro González Villalobos, Susana Trione, Constantino Unguriano, Lilian Rudin, Julio Villalba, Mrs. Horacio Porta (Piqui), and Tomás Schonbek. I met also a number of brilliant new graduates: Néstor Rivière, Horacio Porta, Carlos Segovia, Cora Sadosky, Héctor Fattorini, Ricardo Nirenberg, Fausto Toranzos, Lidia Luquet, and Beatriz Margolis. I also remember faculty that had graduated before 1961, and met personally, namely: Emilio Roxin, Juan Carlos Merlo, Vera Winitzky de Spinadel, and Alberto Galmarino.

My interest in sciences began in my childhood, perhaps inspired by my father, who always challenged my sister Matilde and me with puzzles, mental calculations and medical diagnosis thought experiments. My father, Dr. Pedro J. Calderón, had been an accomplished physician trained in Buenos Aires and later in Paris under the celebrated French urologist and surgeon Georges Marion. My father came from an old colonial family in Argentina. Among his ancestors there were "conquistadores," administrators, landowners, and soldiers. My mother, Matilde Garcia Gallo, the daughter of a Frenchman[2] with roots in the "hidalgo" nobility of Burgos, Spain, instilled in me the love for foreign languages. My father was fluent in French and Italian and my mother in French with a basic knowledge of German (instilled on her by her step-father Diplom. Ing. Franz Robert Winter).

Shortly before my graduation as " Licenciado en Ciencias Matemáticas," I took a course on "Boundary Values of Analytic Functions" that was taught by Dr. González Domínguez. That course was an introduction to "mes premiéres armes" in mathematics. Dr González Domínguez would become my thesis adviser later. I learned from him topics such as the Central Limit Theorem in Probability Theory, Introduction to Functions of Hille (Hermite Expansions), Laguerre Expansions, Theory of Approximation and Theory of Distributions. He was a personality larger than life, a prodigy child that spoke 14 languages and was well versed in Philosophy, Literature and History. He would frequently cite portions of French and Spanish classic literature. One of his favorites was Baudelaire. However, the most important part of his personality was his generosity and selflessness. He had been a student of Dr. Julio Rey Pastor, and after his graduation he went on a Guggenheim fellowship to Brown University, USA, to study under David Tamarkin. Nevertheless, the most discernible influence on him was Einer Hille's work "A Class of Reciprocal Functions." That influence was passed on to me and from me on to some of my own students. During my formative years, Dr. Rafael Panzone and his wife Dr. Agnes Benedek had a tremendous influence in developing my mathematical taste and knowledge.

I started writing my dissertation in 1967 while my adviser had a visiting appointment in Hawaii. At that time I was teaching at the University of Cuyo, San Luis. There I met three remarkable mathematicians: Wilhelm Damköhler, Ezio Marchi, and Felipe Zo. The topic of my dissertation was "Summability of Multiple

[2]Edilbert Garcia Gallo, born in Hendaye, French Basque Country, in 1881.

Hermite and Laguerre Series and Multiple Weierstrass Transform." Parts of the thesis would appear later in Studia Mathematica (see [3] and [2] in the references). One of the highlights of my appointment in San Luis was teaching a modern course on Distributions Theory and Linear Partial Differential Equations. I was honored with the attendance to the course by Wilhelm Damköhler, Felipe Zo, Pedro Egarter, Giorgio Zgrablich, José Tala, and other distinguished guests.

In August 1969, I went to the University of Minnesota on a visiting appointment. At that time I met Eugene Fabes, who would have a great influence on my career. I expanded my professional interests as to include Singular Integrals, Differentiation Theory and, much later in time, Differential Equations. This was the time when I became acquainted with the Navier Stokes Equations. The work of Fabes, Rivière, and Jones played a decisive role on my research. More than 15 years later I published a sequence of three papers on Navier Stokes Equations (see [30,31,39]). During this period, I established a personal connection with B. Frank Jones, Walter Littman, Jesús Gil de Lamadrid, Robert Cameron, and Siegfried K. Grosser, who later invited me to Vienna. Each one was a highly accomplished mathematician. At this time Julio Bouillet and Norberto Fava were graduate students working on their Ph.D. dissertation in Minnesota. They were the first two argentine students that would join the Fabes-Rivière research group.

In 1970, I returned to Buenos Aires and joined the math faculty of Ciencias Exactas at the University of Buenos Aires. I met there a number of extraordinarily talented young graduate students. Among them I remember Jorge Fernández, Néstor Aguilera, Leonor Harboure, Lolina Alvarez Alonso, and Pedro Asdeu. This was the time when I met Luis A. Caffarelli, a young man of immense talent. He became my first doctoral student. I began work with him on the study of multiple Jacobi Series. This association resulted in two important publications [7, 8]. At the time I was directing an undergraduate analysis seminar attended by highly qualified students. I remember distinctively two of them; namely: Eduardo Gatto and Cristián Gutiérrez. These two students would later become accomplished mathematicians in the USA. Throughout the years, I kept a lasting friendship with Gatto. During this period the Math Dept. had two important visitors on temporary appointments, namely: Yoram Sagher and Richard Wheeden. These two researchers would have a measurable impact on the formation of young Argentine mathematicians.

In 1971, I returned to the University of Minnesota on a regular faculty appointment. This was a very fruitful period because of my collaboration with Néstor Rivière and Eugene Fabes (see [6]). An idea that originally Fabes and Rivière had to create a research group in Harmonic Analysis in Minnesota began to give fruits. In fact, L. Caffarelli, N. Aguilera, L. Harboure, C. Unguriano, and F. Zo would join the people working with or under Fabes and Rivière. This exceptional group consisted at the time of almost exclusively Argentine mathematicians. Later on, a number of very gifted mathematicians would join the group, among them: Max Jodeit and Carlos Kenig. May I add that at a personal level, this was a time full of accomplishments, as I completed a substantial amount of mathematical work (see [4, 5, 9]). This also was the time when L. Caffarelli began his outstanding career, becoming a leader in his field.

In September of 1974, I moved to Chicago on a tenured faculty position at the University of Illinois Chicago Circle. There I began a long professorial career that would end with my retirement in 2003. At UIC I would find new friends, namely: Jeff Lewis, James Moller, Melvin Heard, Charles Lin, Herbert Alexander, S. Friedlander, and David Tartakoff.[3] This period was very important because of the diversification of my research interests. I included a number of new areas in my research, namely: Non Linear Partial Differential Equations, Ordinary Differential Equations, Probability Theory, Mathematical Biology, History of Mathematics, and Commutator Singular Integrals. During this period I collaborated with a number of mathematicians, namely: Alberto Calderón (my brother), Jeff Lewis, Yoram Sagher, E. Fabes, N. Rivière, Max Jodeit, and Mario Milman. Later, in the mid-1990s, I held a temporary appointment in Mendoza, Argentina, my birthplace. There, I had the chance to collaborate with Virginia Vera de Serio, jointly writing three papers [42, 43, 45].

This time was also very rich in collaborative work at the University of Illinois. I started there my joint work with Thor Kwembe and Evans Afenya on mathematical models applied to Biology and Medicine. Part of my work with both is included in a Review publication (see [50]). I received a great stimulation also from my other doctoral students: Marwan Abu El Rub, S. Krejca, S. Robbert, A. Gorgius, and Myrna La Rosa. During this period, I started a series of historical papers on Mathematics and allied sciences [33, 34, 52]. After my retirement, I remained in Chicago. I contacted the analysts at De Paul University: Marshall Ash, Jonathan Cohen, Eduardo Gatto, Constantine Georgakis, Alexander Stokolos, and Wilfredo Urbina-Romero. I had the chance to participate actively in their Analysis Seminar.

I cannot close these lines without mentioning the impact that Antoni Zygmund and Robert Fefferman had on my work. Both encouraged me. Part of my research stems from their work. My most recent work was completed at De Paul University and stems from collaboration with Wilfredo Urbina (see [54]). Finally, I would like to recognize Alberto Torchinsky, N. Etemadi, and Miguel de Guzmán for their work had an important impact in my research. During my short period at Rice University I came in contact with Salomon Bochner, an extraordinary scholar and a great mathematician. He illuminated me on many subjects, in particular on "Singularities of Solutions of Linear Partial Differential Equations." Here, John Polking, a professor at Rice University, was the person that led me to learn the subject and to S. Bochner in particular.

## List of Publication of Calixto P. Calderón

1. Some remarks on the pointwise convergence of sequences of multiplier operators. Revista Unión Mat. Argentina y de la Asociación Física Argentina **23**, 153–171 (1968)
2. On Abel summability of multiple Laguerre series. Stud. Math. **33**, 273–294 (1969)

[3]Aside from the analysts, I forged friendships with Viktor Guggenheim, Pete Bousfield, N. Etemadi, Emad El Newihi and Sam Hedayat.

3. Some remarks on the multiple Weierstrass transform and Abel summability of multiple fourier-Hermite series. Stud. Math. **32**, 119–148 (1969)
4. Conjugate kernels and convergence of harmonic singular integrals. Stud. Math. **39**, 39–58 (1971)
5. Differentiation through starlike sets in $\mathbb{R}^m$. Stud. Math. **48**, 1–13 (1973)
6. Maximal Smoothing operators (with E. Fabes and N.M. Rivière). Indiana Univ. Math. J. **23**, 889–898 (1974)
7. On Abel summability of multiple Jacobi series (with L.A. Caffarelli). Colloq. Math. **30**, 277–288 (1974)
8. Weak type estimates for the Hardy-Littlewood maximal functions (with L.A. Caffarelli). Stud. Math. **49**, 213–219 (1974)
9. On commutators of singular integrals. Stud. Math. **53**, 139–174 (1975)
10. Maximal smoothing operators and some Orlicz classes (with J.E. Lewis). Stud. Math. **57**, 285–296 (1976)
11. On the differentiability of functions of several real variables (with J.E. Lewis). Ill. J. Math. **20**, 535–542 (1976)
12. On parabolic Marcinkiewicz Integrals. Stud Math. **59**, 93–105 (1976)
13. Applications of the cauchy integral on Lipschitz curves (with A.P. Calderón, E. Fabes, M. Jodeit and N.M. Rivière). Bull. Am. Math. Soc. **84**, 287–290 (1978)
14. On a lemma of Marcinkiewicz. Ill. J. Math. **22**, 36–40 (1978)
15. On the fractional differentiation of the commutator of the Hilbert Transform. Trabajos de Matemáticas 19, Consejo Nacional de Investigaciones Cientícas y Técnicas, Instituto Argentino de Matemática, Buenos Aires (1978)
16. Lacunary spherical means. Ill. J. Math. **23**, 476–484 (1979)
17. On a singular integral. Stud. Math. **65**, 313–335 (1979)
18. Smooth functions and convergence of singular integral. Ill. J. Math. **23**, 497–509 (1979)
19. On a condition of Marcinkiewicz and the convergence of singular integrals. Actas de la Reunión de El Escorial. Spanish Math. Assoc. 65–85 (1980)
20. On the Fourier Series of certain smooth functions (with Y. Sagher). Ill. J.Math. **24**, 437–439 (1980)
21. On the fractional differentiation of the commutator of the Hilbert transform II. Revista Unión Mat. Argentina **29**, 131–138 (1980)
22. Smooth functions and convergence of singular integrals II. Ill. J. Math. **24**, 426–436 (1980)
23. On the Dini test and the divergence of the Fourier series. Proc. Am. Math. Soc. **83**, 382–384 (1981)
24. Existence of singular integrals in $L^1$. Indiana Univ. Math. J. **32**, 615–633 (1983)
25. Interpolation of Sobolev spaces: the real method (with M. Milman). Indiana Univ. Math. J. **32**, 794–801 (1983)
26. On Etemadi's proof of the strong law of the large numbers. Math. Notae **30**, 31–36 (1983)
27. Lacunary differentiation in $R^n$. J. Approx.Theory **40**, 148–154 (1984)
28. Approximation units and sum of independent random variables. J. Approx. Theory **45**, 133–139 (1985)
29. Diffusion and nonlinear population theory. Revista Unión Mat. Argentina **35**, 283–288 (1990)
30. Existence of weak solutions for the Navier-Stokes equations with initial data in $L^p$. Trans. Am. Math. Soc. **318**, 179–200 (1990)
31. Global solutions of the Navier-Stokes equations. Trans. Am. Math. Soc. **318**, 201–207 (1990)
32. On the classical trapping problem (with T. Kwembe). Math. Biosci. **102**, 183–190 (1990)
33. The sixteenth century Iberian calculatores. Revista Unión Mat. Argentina. **35**, 245–258 (1990)
34. Alvaro Thomas and the Iberian calculatores. Interamerican Review, Puerto Rico **21**(1, 2), 124–132 (1991)
35. Modeling dispersal (with T. Kwembe). In: Proceedings of the X ELAM, August 1991; Rev. Un. Mat. Argentina **37**, 212–229 (1991)
36. Modeling tumor growth (with T. Kwembe). Math. Biosci. **103**, 97–114 (1991)

37. Variational principles in Biology. In: Proceedings of the X ELAM, August 1991; Rev. Unión Mat. Argentina **37**, 16–23 (1991)
38. Diverse ideas in modeling tumor growth (with T. Kwembe). Acta Científica Venezolana **43**(2), 63–75 (1992)
39. On the initial values of solutions of Navier-Stokes equations. Proc. AMS **117**(3), 761–766 (1993)
40. Remark on a non linear integral equation (with E. Afenya). Revista Unión Mat. Argentina. **39**, 223–227 (1995)
41. Normal cell decline and inhibition in Acute Leukemia: A Biomathematical approach (with E. Afenya). Cancer Detect. Prev. **20**, 171–179 (1996)
42. Abel summability of Jacobi type series (with Virginia N. Vera de Serio). Ill. J. Math. **41**(2), 237–265 (1997)
43. Successive approximations and Osgood's Theorem (with Virginia Vera de Serio). Revista de la Unión Mat. Argentina **40**, 3, 4, 73–81 (1997)
44. A remark on leukemogenesis (with E. Afenya). Int. J. Math. Stat. Sci. **8**(2), 1–7 (1999)
45. Successive approximations and Osgood's Theorem II. Revista de la Unión Mat. Argentina **41**(2), 25–38 (1999)
46. A representation formula and its applications to singular integrals. Indiana J. Math. **49**, 1–5 (2000)
47. Diverse ideas on the growth of disseminated cancer cells (with E. Afenya). Bull. Math. Biol. **62**, 527–542 (2000)
48. Summability of orthonormal polynomial series. Rev. de la Unión Mat. Argentina **42**(2), 35–42 (2001)
49. Growth kinetics of cancer cells prior to detection and treatment (with E. Afenya). In: Proceedings of the WSEAS Conferences (2003)
50. Modeling disseminated cancers: a review of the mathematical models (with E. Afenya). Comments Theor. Biol. **8**(2–3), 225–253 (2003)
51. Growth kinetics of cancer cells prior to detection and treatment: an alternative view (with E. Afenya). Discrete Contin. Dyn. Syst. Ser. B **4**(1), 25–28 (2004)
52. Copernico el Mito y la Controversia. Anales de la Fundación Francisco Elías de Tejada, Madrid, vol. 11, Spain, 2005 (appeared in January 2006)
53. Métodos Reales en la Teoría de Conmutadores de Integrales Singulares. In: VII Simposio Chileno de Matemática, Conferencias, Comunicaciones, Sociedad Matemática de Chile (2007)
54. On Abel summability of Jacobi polynomials series, the Watson kernel and applications (with W. Urbina). Illinois J. Math. **57**(2), 343–371 (2013)

# The Calderón Brothers, a Happy Mathematical Relation

**Alexandra Bellow**

It is not often that one has the opportunity to observe at close range two remarkable brothers, mathematicians. I had the privilege of being Alberto Calderón's wife—second wife—which made me Calixto Calderón's sister-in-law and gave me an unusual vantage point. Allow me then to say a few words about the remarkable Calderón brothers.

Alberto and Calixto were born some 20 years apart in Argentina, in the city of Mendoza, that golden city at the foot of the Andes, the eternally snow-capped Andes. With its luxuriant vineyards and olive groves where children roamed freely, Mendoza helped shape Alberto and later Calixto, as they were growing up. For them Mendoza never lost its magic spell.

Alberto and Calixto Calderón were in fact half-brothers: same father, different mothers. Alberto's mother, Haydée, a spirited woman—reputed to be the first woman in Mendoza to drive a car—died unexpectedly, prematurely. Sometime later, the father, Dr. Pedro Calderón—a renowned surgeon in Mendoza—remarried. His second wife was Matilde, a charming, much younger woman, Calixto's mother. I met Matilde a number of years ago. Dr. Pedro Calderón had a natural affinity for arithmetic and music. He would have undoubtedly subscribed to Leibniz's famous saying that: "Music is the secret arithmetic of the soul, unaware of its act of counting." The fact is that he tried to instill in his sons, at an early age, a keen interest in mathematics and music. "At the dinner table he would challenge Alberto, a boy of six or seven, to make rapid mental calculations; or he would play classical music for Alberto and his older sister Nenacha." This scenario repeated itself, at the dinner table, some 20 years later, with Calixto and his older sister, Matilde.

A. Bellow (✉)
Mathematics Department, Northwestern University, Evanston, IL 60208, USA
e-mail: a_bellow@math.northwestern.edu



A.M. Stokolos et al. (eds.), *Special Functions, Partial Differential Equations, and Harmonic Analysis*, Springer Proceedings in Mathematics & Statistics 108,
DOI 10.1007/978-3-319-10545-1_2

Considering the large gap in age between Alberto and Calixto, it is probably fair to say that Alberto was more of a father-figure than a brother-figure for Calixto. Given Alberto's meteoric rise in the world of mathematics, his air of quiet authority but unassuming manner, his impact on his much younger, impressionable brother was unavoidable. Indeed, as a teenager, Calixto felt closer to Alberto than to his own parents and knew all life long that he could trust Alberto, count on his kindness, infinite patience, affection. As a matter of fact, generosity, loyalty and unmistakable chivalry were traits of character that the Calderón brothers shared.

Dr. Pedro Calderón was a man of authority, a stern father. It was not surprising, therefore, that Alberto and Calixto's education should follow more or less similar patterns. In elementary school both boys passed through the Colegio San José of the Maristas Brothers. Alberto developed a passion for Mathematics early on. A number of years later, Calixto became strongly attracted to Theoretical Physics. Dr. Pedro Calderón, however, was firmly convinced that you could not earn a living as a mathematician, or for that matter, a theoretical physicist. I do not know how much soul-searching went on, but the fact is that Alberto went to the School of Engineering of the University of Buenos Aires and graduated as a civil engineer, while sometime later Calixto attended the Engineering School of the University of Cuyo where he received basic training in Mathematics, Physics, Technical Drawing, and Chemistry. But the final destination, the preordained destiny was the same for both, namely, Mathematics! The Institute of Mathematics at the University of Buenos Aires was a powerful magnet that had attracted several brilliant Spanish refugees, and that was brimming over with mathematical activity. At the center of the Institute, as we had already learned this morning, was the legendary personality of Dr. Alberto González Domínguez, a man of vast humanistic culture and human wisdom, who had left behind Greek, Latin and Philology, for the sake of Mathematics. Dr. González Domínguez became Alberto's mentor, protector, devoted friend, and unsurpassed "fan." Years later he became Calixto Calderón's doctoral thesis advisor. The story of how Alberto got his doctorate at the University of Chicago is familiar to older mathematicians, so I shall not repeat it.

One of the things that make this country great is the vast influx of talent from abroad. Alberto and Calixto's academic careers zigzagged through various mathematical centers in the USA and Argentina, including important stints at MIT for Alberto and the University of Minnesota for Calixto. In the end both Alberto, and years later Calixto, settled in Chicago: Alberto at the University of Chicago, Calixto at the University of Illinois. Alberto took early retirement from the University of Chicago and returned to Buenos Aires when his wife—his first wife Mabel—became seriously ill. He returned to the University of Chicago on a post-retirement appointment in 1989. Calixto retired from the University of Illinois in 2003, and taught afterwards at Oakton College and DePaul University for a number of years. Alberto and Calixto had parallel professional lives. Their many collaborators and doctoral students made these professional lives most lively and interesting. As a matter of fact Calixto's first doctoral student, the one and only Luis Caffarelli, and one of Alberto's late doctoral students, the incomparable Carlos Kenig, both performed at this meeting, earlier today.

As far as professional interests go, Alberto roamed widely but stayed within the confines of Mathematics, with a side interest in mathematical education. Calixto ventured outside and tried related fields, such as Biomathematics, History of Calculus and Biographies of Scientists. Noteworthy are the *"Biography of Copernicus,"* the article on *"16th century Iberian Calculatores"* (these were the precursors of Galileo's Modern Mechanics and of Newton's Calculus), as well as the brief but highly interesting essay *"Dr. Pedro Calderón and Urology in Mendoza"* (in his heyday, Dr. Pedro Calderón was the preeminent urologist in Mendoza.) The *"Biography of Copernicus,"* in particular, took Calixto several years of research. There is an unforgettable quote in this biography of the great astronomer. A close friend and adviser of Martin Luther, by the name of Melanchton, refers to Copernicus, sarcastically in Latin: *"Il Sarmaticus Astronomus qui movet Terram et figit Solem"* (*"The Sarmatian Astronomer who moves the Earth but fixes, immobilizes the Sun."*) " Sarmatian" here is used disparagingly to mean "outsider," a "barbarian" for Calixto goes on to explain that Sarmatia in ancient times was a vast geographic area that stretched from the basin of the Vistula to the Caspian Sea.

Alberto admired the facility and grace with which Calixto was able to write. He also admired Calixto's unusual erudition and exceptional memory: "Where does he store all this information?" Alberto wondered. For Alberto writing a letter, a review, an essay was not exactly a pleasant task. He nevertheless had a real feeling for poetic beauty: he did a stunning translation into English of Gustavo Adolfo Becquer's classic poem *"Volverán las oscuras golondrinas"* just before we were married.

Over the years I often heard mathematicians comment on the Bernoulli brothers of Basel, Switzerland, Jacob Bernoulli, the older brother, the founder of the Bernoulli "dynasty," and Johann Bernoulli, his much younger brother. The Bernoulli brothers lived more than 3 and a half centuries ago, and were endowed with magnificient creative gifts, but their relationship was fraught with difficulties, animosity, bitter rivalry. Not so in the case of the Calderón brothers. Nothing illustrates this better, in my opinion, than "the posthumous" paper by Alberto Calderón and Calixto Calderón.

Let me backtrack a little. Alberto Calderón died in April 1998. The volume *"Selected papers of Alberto Calderón with Commentary"* or for short *Calderón Selecta* took 10 years to see the light of day, but the AMS did a fine job. The Editors were: Paul Malliavin—the great French probabilist and Alberto's oldest mathematical friend outside Argentina, Carlos Kenig—Alberto's former doctoral student, and myself—Alberto's wife (second wife). This project was a tribute to Alberto's mathematical legacy and a labor of love. I would like to single out three of the papers in this volume:

(1) Paul Malliavin, *"On the analytical side of the proof of the Index Theorem, some personal recollections."*

    This essay (Commentary) contains Malliavin's account of the history of the Index Theorem and the role played by Alberto Calderón.

(2) Yves Meyer, *"Complex analysis and operator theory in Alberto Calderón's work."*

This is one of the most substantial, beautiful and at times poetic essays (Commentary) in the volume.

And last but perhaps most important and relevant to this talk, the "posthumous" paper, the joint paper that Calixto Calderón published in the Indiana University Mathematics Journal in 2000 and that we included in the *Calderón Selecta,*

(3) A. P. Calderón and C. P. Calderón, *"A representation Formula and its Applications to Singular Integrals."*

Let me now go back to the year 1977. Alberto Calderón had just published the landmark paper on the *"Cauchy Integral on Lipchitz curves,"* the Cauchy integral here being the singular integral on the curve, i.e. the analog of the Hilbert transform on the real line. This triggered a frenzy of activity. First, mathematicians worked hard to remove the bound on the norm of the Lipschitz curve. For the next 20 years or so, first-rate analysts were busy finding new proofs of the boundedness in $L^2$ of this Cauchy operator. In the process, they discovered connections with such diverse areas of Mathematics as: the Traveling Salesman problem, Ahlfors Regular Curves, Menger curvature, to mention only a few. This last, shall we say "geometric" proof, given by Melnikov and Verdera in 1995, and using as a tool the Menger curvature, is considered by many to be the simplest and most beautiful proof of the boundedness of the Cauchy Integral on Lipschitz curves.

If we are in the complex plane $\mathbb{C}$, have an open disc $D$, a function $F(z)$ that is analytic on the closure of $D$, we know we can use the classical Cauchy kernel, integrate on the boundary of $D$, $\partial D$, to recapture the function inside $D$

$$\frac{1}{w-z} = \frac{1}{1-\frac{z}{w}} \cdot \frac{1}{w} \tag{1}$$

This Cauchy formula representation can be extended to the $n$-dimensional complex space, $\mathbb{C}^n$ and functions $F(z_1, z_2, \cdots, z_n)$ analytic in a polydisc, by using the standard "product" Cauchy kernel,

$$\prod_{i=1}^{n} \frac{1}{1-\frac{z_i}{w_i}} \cdot \frac{1}{w_1 w_2 \cdots w_n} \tag{2}$$

This is indeed the Cauchy kernel that Hormander introduced in 1966. But there is another side of the story, and this is where the Calderón brothers come in.

There is another Cauchy type kernel in $n$ dimensions, more elusive, mysterious, but gloriously beautiful, that had been known to Alberto and Calixto since the mid-seventies and that had an underground kind of existence all these years. Alberto Calderón died in April, 1998. A few months before Alberto died, I remember Calixto coming to visit us: I remember Calixto and his older brother deeply engrossed in mathematical conversation, trying to recapture their "elusive" Cauchy type kernel. But they were unable to; memory did not cooperate. This was all the more frustrating since apparently Alberto and Calixto had made use of this "kernel," without explicitly exhibiting it, in their 1978 paper, the only other paper that Alberto and Calixto wrote together: this was a paper with five authors, called *"Applications of the Cauchy integral on Lipschitz curves."* After Alberto died, Calixto was determined to complete the job. He was finally able to recover their "elusive" kernel and to write up the paper. As I said before, the paper, the "posthumous" paper, appeared in the Indiana Journal of Mathematics in 2000. We included it also in the *Calderón Selecta.* It is the supreme tribute and gesture of love that Calixto paid to his older brother.

A few words now about this "elusive" Cauchy type kernel which I shall refer to as *"the Calderóns"* kernel. Unlike *"the Hörmander"* kernel which is based on "multiplication," the Calderóns kernel is based on "addition."

$$\frac{1}{1-\sum_{i=1}^{n}\frac{z_i}{w_i}}\cdot\frac{1}{w_1w_2\cdots w_n} \tag{3}$$

This kernel too yields a Cauchy type formula, i.e. the representation of a function analytic in a polydisc, by integration on the boundary of a transformed function.[1] Futhermore, the Calderóns kernel permits to answer the important Calderón Conjecture which Yves Meyer calls the *magic key* opening new chapters in complex analysis, linear PDE and nonlinear PDE, namely, let:

$$K(x,y)=F\left(\frac{A(x)-A(y)}{x-y}\right)\frac{1}{x-y} \tag{4}$$

*where:* $A:\mathbb{R}\to\mathbb{R}^n$ *is Lipschitz and* $F$ *is analytic. Then the singular integral operator defined by the kernel* $K(x,y)$ *is continuous in* $L^2$.

What a pity that the Calderóns kernel was not publicized several decades ago. It is to be hoped, however, that in the future it will be put to further good use!

Talk given at "Special Functions, Partial Differential Equations and Harmonic Analysis," A Conference in honor of Calixto P. Calderón, November 16–18, 2012, Department of Mathematics and Actuarial Sciences, Roosevelt University.

---

[1]If $a_{k_1,\cdots,k_n}$ are the coefficients of the function, those of the "transformed" one are $C\frac{k_1!\cdots k_n!}{(k_1+\cdots+k_n)!}a_{k_1,\cdots,k_n}$

# Calixto Calderón As I Knew Him

**Luis Caffarelli**

Calixto Calderón was born in Mendoza, Argentina, by the mountains and the vineyards, and spent in Cuyo his early youth. He always talked with great love of those times and places, and undoubtedly those years were very influential in giving him the gentile, generous attitude that we all cherish in him.

As he mentions in his recollections, Calixto came to the Universidad de Buenos Aires in 1961. This was a great time for mathematics, not only in Argentina but also worldwide: science was blooming, research had become highly valued and central to university life.

He found in Buenos Aires, not only an exceptional group of dedicated teachers but also condisciples that were enthusiastic and engaged. Under the direction of González Domínguez, Calixto developed his first ideas in real and harmonic analysis, mainly in the context of summability of multiple series of special functions. Many of the fundamental ideas of real analysis: singular integrals, multipliers, decompositions were relatively recent, and Calixto had to develop new methods and ideas in his work.

I met Calixto in 1970. In those years a sudden influx of bright, young mathematicians, Argentinian, who had studied in the USA, and foreign, came to spend long periods in Argentina, often for a full year. Among the foreigners: Gene Fabes, Dick Wheeden,Yoram Sager, Bob Fernholst, Luc Tartar. Among the Argentinian: Carlos Segovia, Néstor Riviére, Horacio Porta, Héctor Fatorini, Enrique Lami Dozo, and Calixto.

L. Caffarelli (✉)
Mathematics Department, The University of Texas at Austin, RLM 8.100 2515 Speedway Stop C1200, Austin, TX 78712, USA
e-mail: caffarel@math.utexas.edu



A.M. Stokolos et al. (eds.), *Special Functions, Partial Differential Equations, and Harmonic Analysis*, Springer Proceedings in Mathematics & Statistics 108,
DOI 10.1007/978-3-319-10545-1_3

Trained in the top mathematical centers, they taught the central topics of the time: Singular Integral Theory, Martingales, Several complex variables, Interpolation. All of a sudden math had jumped ahead 20 years: This was a collection of not only bright mathematicians but also of lively generous people, exchanging ideas, and mentoring in the current sense of the word.

Shortly after, Calixto became my mentor. Beyond mathematics, Calixto taught us how to be an analyst, to seek the heart of an issue, to understand a problem. In fact, Calixto is a non-conventional mathematician and his work has expanded in diverse areas, and presented new insights. By that time Calixto had already started to think on new issues concerning harmonic analysis and partial differential equations.

Two years later, he will return to Minnesota, and I followed him shortly after. A contingent of exceptional students from Argentina (Elena Fernández, Néstor Aguilera, Eleonor (Pola) Harboure, Tino Unguriano) followed shortly after. Felipe Zo, another bright Argentinian mathematician had arrived before I did.

In my view, Minnesota was at the time, with Courant and Berkeley one of the three most exciting places in partial differential equations. Our weekly seminar included Aronson, Weinberger, Littman, McCarthy, Meyer, Fabes, Riviere, Calixto, Jodeit among the senior people and there was a congenial atmosphere and the sense of being part of a larger enterprise.

Calixto developed there new fundamental ideas related to the problem of commutators of singular integrals, that is linked to boundary data of PDE, giving an alternative approach to the good lambda inequalities through interpolation techniques. He also became interested in partial differential equations, some of which, like with his work with Alberto, and his papers on Navier Stokes bore fruit several yeas later.

In the fall of 74 Calixto went with a tenured position at the University of Illinois at Chicago. Once more, he embedded himself in new problems and ideas, some connected to his past research, others bold and new. From these times are his articles on existence of solutions to Navier Stokes Equations in a very particular norm and through the years Calixto published articles in diverse areas such as tumor growth, multiscale species competition, probability. For Calixto, doing mathematics never felt like an obligation but always an enjoyment.

He also took a strong interest at this time in the history of science and in particular, the early contributions to science of the Hispanic culture.

To meet in Chicago in November 2012 to celebrate Calixto's research and academic achievements in his long academic career has been a wonderful occasion, full of sweet remembrances and laughter.

Calixto: We wish you many more happy years of success!

Luis

# An Appraisal of Calixto Calderón's Work in Mathematical Biology

**Evans Afenya**

**Summary.** The body of investigative biomathematical work undertaken by Prof. Calixto Calderón is reviewed. The appraisal demonstrates that Prof. Calderón has not only been active in the area of harmonic analysis but has also been an active researcher in the area of mathematical biology and has been a strong proponent of the use of mathematical ideas and techniques in the broad areas of biology and medicine as a way of giving these areas firm systematic support.

## 1 Introduction

This article is written in honor of Prof. Calixto Calderón and is dedicated to his contributions in the area of mathematical biology. Aside from his primary areas of interest in harmonic analysis, it is important to mention that at some point in the career of Prof. Calderón, he started getting interested in the applications of mathematical ideas and techniques to problems in biology and medicine. It had always been the view of Prof. Calderón that biology and medicine could benefit from systematized mathematical support just as how the disciplines of physics and engineering benefitted immensely from such support many years ago. It is, therefore, not surprising that a number of doctoral candidates produced by Prof. Calderón completed their dissertations on topics related to mathematical biology. We note that the advances made to date in the prevention, detection, and treatment of various forms of cancer have been remarkable. Current developments in the areas of oncology and hematopathology, including the impact of phenotyping, cytogenetics, molecular probes, and growth factors, are noteworthy. The development of technologies such as flow cytometry and polymerase chain reaction (PCR), in which minimal residual tumors can be detected at extremely low levels, holds great promise for the future in the fight against cancer. Despite all the advances, however, many problems

E. Afenya (✉)
Department of Mathematics, Elmhurst College, 190 Prospect Ave, Elmhurst, IL 60126, USA
e-mail: evansa@elmhurst.edu



A.M. Stokolos et al. (eds.), *Special Functions, Partial Differential Equations, and Harmonic Analysis*, Springer Proceedings in Mathematics & Statistics 108,
DOI 10.1007/978-3-319-10545-1_4

still exist in cancer prevention, detection, treatment, and management. According to the American Cancer Society (ACS), about 1,529,560 new cases of cancer were expected to be diagnosed in 2010 and the occurrence of cancer comes along with enormous cost to society. Available year 2010 estimates of overall costs from the National Institutes of Health (NIH) stand at $263.8 billion. Out of this figure $102.8 billion represents direct medical costs (total of all health expenditures), $20.9 billion stood for indirect morbidity costs (cost of lost productivity due to illness), and $140.1 billion for indirect mortality costs (cost of lost productivity due to premature death). The 5-year relative survival rate for all cancers diagnosed between 1999 and 2005 is 68 %, which is up from 50 % in 1975–1977. Improvement in survival reflects progress in diagnosing certain cancers at an earlier stage and improvements in treatment. However, cancer deaths remain the second most common cause of death in the USA, exceeded only by heart disease. Cancer accounts for nearly one of every four deaths in the USA and data from the ACS shows that 569,490 Americans were expected to die of cancer in 2010. Not surprisingly, the World Health Organization (WHO) has recently issued a report warning that cancer rates worldwide would surge by 57 % in the next 20 years. The WHO report is a telling reminder about the problems this disease poses to the entire human race. Coupled with these nagging problems, one issue of great importance is that current treatment protocols regard a cancer patient to be in complete remission when the population of cancer cells falls below detectable levels. However, a latent malignant subpopulation that is possibly propelled by the purported existence of a suspected "cancer stem cell" population could pose great dangers. The existence of this malignant subpopulation coupled with the evolution of multidrug resistance may account for the many cases of relapse that occur today.

It is against such aforementioned background that Prof. Calderón gained the desired impetus to join the fight against malignant disease by investigating biomedical phenomena such as cancer from a mathematical viewpoint through active mathematical modeling, with the view to providing insights particularly in cases where it is particularly difficult to obtain relevant in vivo data because of the nature and character of disease evolution in human hosts.

## 2 Appraisal of Mathematical Modeling Work Done by Prof. Calixto Calderón

In what follows, we will delve into Prof. Calderón's contributions to the area of mathematical biology by reviewing a representative selection of his works.

## 2.1 Solid Tumor Modeling

In [8], Calderón and coworker Kwembe consider the meaning and limitations of certain mathematical models of mainly solid tumor growth. They introduce new derivations of existing models and present theoretical justification for the Gompertz law of growth for solid tumors. They then proceed to introduce an age-dependent Von Bertalanffy-type equation and interestingly address problems of existence and uniqueness of solutions.

Most notably, in their construction of the age-dependent model, the dynamics of a tumor cell population are described by means of an age density function $m(a,t)$ which specifies the size structure of the tumor cells in terms of the age $a$ of the cell contents of a tumor at time $t$. They specifically focus on solid spherical tumors. Taking into consideration that the growth of a solid tumor may be the result of a sum of anabolic and catabolic processes caused by the degradation and regeneration of biological materials, they consider the following age-dependent model that is based on Von Bertalanffy's [15] equation:

$$\frac{dm}{dt} = a_1 m^{\alpha} - a_2 m^{\beta} \tag{1}$$

This equation states that the rate of change of body weight, mass, or volume, $m$, is given by the differences between the processes of building up and breaking down. The constants $a_1, a_2$ are, respectively, those of anabolic and catabolic processes linked to the buildup and breaking down of biomass, and the exponents $\alpha, \beta$ indicate appropriate powers of $m$. Thus $a_1$ is the regenerative factor and $a_2$ is the degradation factor. Since their focus is on solid spherical tumors, they consider the model with $2/3 \leq \alpha < 1$ and $\beta = 1$. By building the resulting model around the assumptions that; (a) the change occurring in the population of cells of age $a$ at time $t$ in a small time interval is proportional to the size of the population and the small time interval and (b) the number of new cells introduced into the population within this small time interval is proportional to the surface area of the convex tumor, based on a form of allometric principle, Calderón and his coworker derive the model given by

$$\frac{\partial m(a,t)}{\partial t} + \frac{\partial m(a,t)}{\partial a} = -\mu(a,t)m(a,t) \tag{2}$$

$$m(0,t) = C\left[\int_0^{\infty} m(a,t)da\right]^{\alpha}, \; 2/3 \leq \alpha < 1 \tag{3}$$

$$m(a,0) = g(a) \tag{4}$$

where $\mu(a,t)$ is an age-specific degradation rate, $C$ is a proportionality constant, and $g$ is a smooth function that is assumed to vanish for large $a$.

With their interest mainly in existence and uniqueness of solutions, they use an approach similar to that of Hoppensteadt [12] with $m(0,t) = A(t)$ to obtain a solution of the form

$$m(a,t) = \begin{cases} g(a-t)\exp(-\int_0^t \mu(a-t+\delta,\delta)d\delta & \text{if } a \geq t, \\ A(t-a)\exp(-\int_0^t \mu(a-t+\delta,\delta)d\delta & \text{if } t > a. \end{cases} \quad (5)$$

By substituting Eq. (5) into an ensuing expression for $m(0,t) = A(t)$, they obtain a nonlinear integral equation given by

$$A(t) = C\left[f(t) + \int_0^t K(a,t)A(t-a)da\right]^{\alpha},\ 2/3 \leq \alpha < 1 \quad (6)$$

where

$$f(t) = \int_t^{\infty} g(a-t)\exp\left[-\int_0^t \mu(a-t+\delta,\delta)d\delta\right]da$$

and

$$K(a,t) = \exp\left[-\int_0^t \mu(a-t+\delta,\delta)d\delta\right]$$

Proceeding further to establish existence and uniqueness, Eq. (6) is rewritten as

$$A(t) = C\left[f(t) + \int_0^t K(t-a,t)A(a)da\right]^{\alpha},\ 2/3 \leq \alpha < 1;\ t \in [0,\beta] \quad (7)$$

By letting $\mu : [0,\beta] \times [0,\beta] \to \mathbb{R}_+$ be a continuous function, $E = L^{\infty}(0,\beta)$, and defining

$$\|f\|_{\infty} = \text{ess} \sup_{0\leq t\leq\beta} |f|$$

such that $\|f\|_{\infty} \leq M$ (a constant) almost everywhere in $0 \leq t \leq \beta$, and $L^{\infty}(0,\beta) = \{f : \|f\|_{\infty} < \infty\}$, they seek a fixed point of the nonlinear operator $T : L^{\infty}(0,\beta) \to L^{\infty}(0,\beta)$ defined by

$$(TA)(t) = C\left[f(t) + \int_0^t K(t-a,t)A(a)da\right]^{\alpha} \quad (8)$$

where $f(t)$ is continuous on $[0,\beta]$. They then obtain the equicontinuity of the family $(TA)(t)$ when $\|f\|_{\infty} \leq M$ (a constant) almost everywhere on $0 \leq t \leq \beta$ and proceed to use the Schauder–Tychonoff [13] fixed-point theorem to show that there is a ball in $C[0,\beta]$ that is mapped into itself by $T(A)$, thus establishing existence. Calderón and Kwembe show uniqueness by pointing out the fact that $f(t) > 0$ whenever $0 < t < \beta$, and the initial values of $m(a,t)$ are nonzero implying that $T$ satisfies a Lipschitz condition.

The approach of establishing existence and uniqueness of solutions in the area of mathematical modeling of biomedical phenomena is an essential one since such an approach points towards the fact that solutions are achievable and this makes such an endeavor significant.

## 2.2 *Disseminated Cancer Modeling*

The work of Prof. Calderón has not only been limited to studies of solid tumor dynamics, but it also encompasses investigations of disseminated cancers. We must mention here that cancer can be divided into two broad groups. These are solid tumor cancers and dispersed or disseminated cancers. Solid tumor cancers include those of the breast, lungs, liver, pancreas, prostate, and colon, to mention a few. The leukemias, myelomas, and lymphomas make up the disseminated ones. In solid tumors, an abnormal clone of cells forms as a tumor at a specific location in the human body. If the tumor happens to be malignant or cancerous, then it proliferates uncontrollably and metastasizes to other parts of the body. On the other hand, benign tumors do not possess metastatic properties and are rarely life-threatening. In disseminated cancers such as the leukemias, a malignant cell may emerge in the bone marrow, proliferate, and replace normal cells in the marrow, peripheral blood, and other body and blood-forming organs and, this may lead to fatalities. Having worked to find justification for solid tumors obeying the Gompertz law of growth, Prof. Calderón teamed up with another coworker to study disseminated cancer kinetics with the aim of finding out whether the conclusions drawn about Gompertz growth kinetics for solid tumors could be extended to disseminated tumors. In doing this, Afenya and Calderón [3, 4] consider observations based on the histological structure and function of the bone marrow, an organ that supports the massive activity of hematopoiesis in the human body and is said to be surprisingly uniform, and develop a number of assumptions that lead to the following two-dimensional steady state version of the diffusion equation:

$$-\nabla^2 C = \delta(x) - \kappa \text{ and } x = (x_1, x_2), \tag{9}$$

with

$$C(x_1, x_2) = 0 \text{ for } (x_1 + x_2)^{1/2} = R \tag{10}$$

where $C$ represents the concentration of metabolite, $\kappa$, assumed to be constant, represents the average consumption of metabolite, and $R$ is a constant that represents the maximum reach of metabolite. The symbols $\nabla^2$ and $\delta$ denote the Laplacian operator and the Dirac delta function, respectively. The quantity $\delta(x)$ is a source located on a representative blood vessel, which for the sake of simplicity is assumed to be a point at the origin of coordinates. The variable $x = (x_1, x_2)$, as stated above, represents the coordinate system of measurement which in this case is chosen to be the Cartesian coordinate system with $x_1^2 + x_2^2 = r^2$ (a radius). A solution of Eq. (9) with boundary condition (10) yields

$$C = C_1 + C_2 + C_3 \tag{11}$$

where

$$C_1 = h_1 \ln(1/r) \text{ with } r = \sqrt{x_1^2 + x_2^2}$$

is the contribution from the source on a representative vessel,

$$C_2 = h_2 \int \int_{|y| \leq R} -\kappa \ln \left( \frac{1}{|x - y|} \right) dy$$

is a convolution representing the contribution from metabolite consumption, and $C_3$ is an harmonic function which for $r = R$ takes on values equal to those of $-C_1 - C_2$.

The malignant mass at time $t$ around a representative blood vessel is taken to be $m(t)$ with the hypothesis indicating that $(1/m)dm/dt$ should be proportional to the concentration of metabolite $C(r)$ with $r$, which lies in the interval [0, $R$], being proportional to the square root of $m(t)$ where $R$ represents the maximum reach of the malignant mass at time $t$. That is,

$$r^2 = \alpha m(t) \text{ and } R^2 = \alpha m_{\max}$$

where $m_{\max}$ is the absolute maximum level to which the malignant mass could get around a blood vessel. It is assumed for the sake of fixing ideas that the maximum reach of the malignant mass at time $t$ is circular in shape and the following result is obtained from further analysis:

$$C_1 >> |C_2 + C_3|.$$

Thus, putting together all the contributions results in the following equation:

$$\frac{dm}{dt} = hm(t) \ln \left[ \frac{m_{\max}}{m(t)} \right] + \lambda(m(t))m(t) \tag{12}$$

where $h$ is a constant and $\lambda(m(t)) << h \ln[m_{\max} m(t)]$ for small values of $m(t)$. It is noted that $\lambda$ (an error term) would be small even for moderate values of $m(t)$ and these arguments lead to an inequality of the type;

$$(1 - \epsilon)h \leq \frac{m'}{m(t) \ln \left[ \frac{m_{\max}}{m(t)} \right]} \leq (1 + \epsilon)h \tag{13}$$

Here, the quantity $\epsilon$ is a small positive constant given by

$$\epsilon = \lambda(m(t))/h \ln[m_{\max} m(t)].$$

Since $\epsilon$ is small and could be neglected in expression (13) above, the analysis shows that $m(t)$ could be chosen, to satisfy a differential equation of the Gompertz type;

$$\frac{dm}{dt} = hm(t) \ln\left[\frac{m_{\max}}{m(t)}\right]. \tag{14}$$

An empirical validation of the analytical conclusions is carried out with the use of the leukemia data of Skipper and Perry [14]. Their work with the data revealed that the best fitting curve is the Gompertz curve in comparison with fits of the logistic, exponential, and cubic curves.

Expression (13) basically suggests that malignant growth of disseminated cancer cells can be sandwiched between two Gompertzians. Added to this, the histological structure and function of the bone marrow where such malignant cells owe their origins leads naturally to the diffusion equation that engenders a solution that is logarithmic and hence Gompertzian. Afenya and Calderón [4] have also employed entropy considerations to arrive at Gompertz kinetics of disseminated cancer cells. Therefore, from an analytical and empirical standpoint, it is justifiable to say that the law that governs the growth of disseminated cancer cells is essentially Gompertzian. The work of Prof. Calderón and his coworker in this case strongly suggests that it is appropriate to use Gompertzian representations to model solid and disseminated tumor dynamics without necessarily blurring the differences inherent in the physiology of these cell systems.

## *2.3 Other Biomathematical Works of Prof. Calderón*

Undoubtedly, it is practically impossible for us to exhaustively review all the other contributions of Prof. Calderón within the limited space available. As such, we herein provide brief summaries of a few more and point the reader to the relevant cited sources.

In [7], Calderón and Kwembe consider the problem of finding the probability density for the location of an untrapped pest satisfying a diffusion equation with the scaled Laplacian. By taking the initial data to the diffusion process to be in some $L^p$ spaces, they establish the existence and uniqueness of local solutions in these spaces and the existence and uniqueness of weak global solutions in $L^{p,q}$, with $p, q > 3$. The approach used here is of particular interest because a method of successive approximations is introduced and this makes it amenable to numerical treatment. In [5], Afenya and Calderón interestingly use analytical arguments to enhance findings related to the Gompertzian growth kinetics of disseminated cancer cells and end up showing that such cells could also obey kinetics described by a modified Gompertz (hyper-Gompertz) representation arising from looking at the bone marrow as a porous medium. In [2], Afenya and Calderón show that the elusive phenomenon of leukemogenesis could be addressed through the use of probabilistic models and in 2006 Prof. Calderón's knowledge of biological processes enabled him to play the lead role among a number of his coworkers in studying the dynamic features of the deterministic (drift) component when the output of an atomistic simulation is approximated as a diffusion process. They performed traditional scientific computing tasks such as integration, steady state and closed orbit computation, and stability analysis on the drift component using a stochastic simulation algorithm (SSA) of the cyclic Lotka–Volterra biological system as an illustrative example [11]. Additional biomathematical contributions of Prof. Calixto Calderón can be found in [1, 6, 9, 10].

## 3 Concluding Remarks

All told, Prof. Calderón's ability to do research in multiple areas is a clear testament to his intellectual prowess and versatility. His body of work exemplifies that of an individual who always pursues knowledge with passion and with the utmost aim of satisfying his scientific curiosities. His work will remain as a source of inspiration to younger up-and-coming researchers who aspire to investigate various fields of research endeavor.

## References

1. Afenya, E.K., Calderón, C.P.: A brief look at normal cell decline and inhibition in acute leukemia. J. Can. Det. Prev. **20**(3), 171– 179 (1996)
2. Afenya, E.K., Calderón, C.P.: A remark on leukemogenesis. Int. J. Math. Stat. Sci. **8**(2), 199–205 (1999)
3. Afenya, E.K., Calderón, C.P.: Diverse ideas on the growth kinetics of disseminated cancer cells. Bull. Math. Biol. **62**, 527–542 (2000)
4. Afenya, E.K., Calderón, C.P.: Modeling disseminated cancers-a review of mathematical models. Commun. Theor. Biol. **8**(2), 225–253 (2003)
5. Afenya, E.K., Calderón, C.P.: Growth kinetics of cancer cells prior to detection and treatment: an alternative view. Discrete Continuous Dyn. Syst. Ser. B **4**(1), 25–28 (2004)
6. Calderón, C.P.: Diffusion and nonlinear population theory. Rev. Un. Mat. Argentina **35** 283–288 (1991)
7. Calderón, C.P., Kwembe, T.A.: On the classical trapping problem. Math. Biosci. **102**(2), 183–190 (1990)
8. Calderón, C.P., Kwembe, T.A.: Modeling tumor growth. Math. Biosci. **103**, 97–114 (1991)
9. Calderón, C.P., Kwembe, T.A.: Diverse ideas in modeling tumor growth. Acta Cient. Venez. **43**(2), 64–75 (1992)
10. Calderón, C.P., Kwembe, T.A.: Dispersal models. Rev. Un. Mat. Argentina **37**(3–4), 212–229 (1992)

11. Calderón, C.P., Tsekouras, G.A., Provata, A., Kevrikidis, I.G.: Coarse-graining the cyclic Lotka-Volterra model: SSA and local maximum likelihood estimation. In: Model Reduction and Coarse-graining Approaches for Multiscale Phenomena, pp. 247–267. Springer, Berlin (2006)
12. Hoppensteadt, F.C.: Mathematical Theories of Populations: Demographics, Genetics and Epidemics. Regional Conference Series in Applied Mathematics, vol. 20. SIAM, Philadelphia (1975)
13. Miller, R.K.: Nonlinear Volterra Integral Equations. W. A. Benjamin, Menlo Park, (1971)
14. Skipper, H.E., Perry, S.: Kinetics of normal and leukemic leukocyte populations and relevance to chemotherapy. Cancer Res. **30**, 1883–1897 (1970)
15. Von Bertallanfy, L.: Principles and theorem of growth. In: Nowinsky, W.W. (ed.) Fundamental Aspects of Normal and Malignant Growth. Elsevier, Amsterdam (1960)

# Remarks on Various Generalized Derivatives

**J. Marshall Ash**

**Summary.** Various generalized derivatives are defined and related. Some of these are the Peano derivatives, the symmetric (Peano) derivatives, the symmetric Riemann derivatives, a generalized derivative from numerical analysis, the very large family of $A$ derivatives, symmetric quantum derivatives, and quantum symmetric Riemann derivatives. Additionally, $L^p, 1 \leq p < \infty$ versions of many of these derivatives are considered. Relations between some of these derivatives are mentioned. Some counterexamples showing that other relations are not true are also given.

## 1 Generalized $n$th Derivatives

The ordinary $n$th derivative $f^{(n)}(x)$ is defined by an $n$ step recursive process. For $f^{(n)}$ to exist at a real number $x$, it is necessary that $f^{(n-1)}$ exist in a neighborhood of $x$. We will look at two different levels of generalizations of the $n$th-derivative $f^{(n)}$.

The first level of generalization will involve viewing the $n$th-derivative of a function $f$ as the coefficient in a polynomial approximation to $f$. In this category are the Peano derivative $f_{(n)}(x)$ and the symmetric (Peano) derivative $f^s_{(n)}$. The most fundamental of these, and in a sense to be made precise below, the most important generalization is the Peano derivative. We say the function $f$ has a Peano derivative of order $n$ at $x$ and write $f \in t_n(x)$, if there are constants $f_{(1)}(x), f_{(2)}(x), \ldots, f_{(n)}(x)$ such that

$$f(x+h) = f(x) + f_{(1)}(x)h + \cdots + \frac{f_{(n)}(x)}{n!}h^n + o(h^n) \text{ as } h \to 0. \tag{1}$$

J.M. Ash (✉)
Department of Mathematics, DePaul University, Chicago, IL 60614, USA
e-mail: mash@math.depaul.edu



A.M. Stokolos et al. (eds.), *Special Functions, Partial Differential Equations, and Harmonic Analysis*, Springer Proceedings in Mathematics & Statistics 108,
DOI 10.1007/978-3-319-10545-1_5

Notice that this definition implies that $f_{(1)}(x) = f'(x)$. A function $f$ having one derivative at $x$ means that $f$ is well approximated by a line at $x$. A function $f$ having an $n$th Peano derivative at $x$ means that $f$ is well approximated by an $n$th degree polynomial at $x$.

A version of Taylor's theorem that is slightly stronger than the usual ones found in calculus texts was proved by de la Vallée-Poussin in the late nineteenth century. It can be rephrased as asserting that if $f$ has an $n$th derivative at a point $x$, then $f$ has an $n$th Peano derivative at $x$ and, furthermore, $f_{(n)}(x) = f^{(n)}(x)$.

The first derivative and the first Peano derivative have the same definition. But the second Peano derivative is defined by

$$f_{(2)}(x) = \lim_{h \to 0} \frac{2!}{h^2} \{ f(x+h) - f(x) - f_{(1)}(x) h \}.$$

The other generalized derivative involving polynomial approximation is the symmetric (Peano) derivative. If $n$ is odd, say the function $f$ has a symmetric derivative of order $n$ at $x$ if there are $\frac{n+1}{2}$ constants $f^s_{(1)}(x), f^s_{(3)}(x), f^s_{(5)}, \dots, f^s_{(n)}(x)$ such that

$$\frac{f(x+h) - f(x-h)}{2} = f^s_{(1)}(x) h + \frac{f^s_{(3)}(x)}{3!} h^3 \cdots + \frac{f^s_{(n)}(x)}{n!} h^n + o(h^n) \text{ as } h \to 0.$$

Similarly, if $n$ is even, only even indices arise and the analogous relation is

$$\frac{f(x+h) + f(x-h)}{2} = f(x) + \frac{f^s_{(2)}(x)}{2!} h^2 + \frac{f^s_{(4)}(x)}{4!} h^4 \ldots + \frac{f^s_{(n)}(x)}{n!} h^n + o(h^n) \text{ as } h \to 0.$$

The existence of the $n$th Peano derivative $f_{(n)}$ at $x$ immediately implies that the $n$th symmetric derivative exists at $x$ and $f^s_{(n)}(x) = f_{(n)}(x)$.

The second level of generalization consists of higher order derivatives that are defined directly as difference quotients. One such generalization is the $n$th symmetric Riemann derivative. If $\Delta_n$ is the $n$th symmetric difference (examples:

$$\begin{aligned}
\Delta_1 f(x) &= f\left(x + \frac{h}{2}\right) - f\left(x - \frac{h}{2}\right) \\
\Delta_2 f(x) &= \Delta_1 (\Delta_1 f(x)) = \Delta_1 f\left(x + \frac{h}{2}\right) - \Delta_1 f\left(x - \frac{h}{2}\right) \\
&= \left(f\left(x + \frac{h}{2} + \frac{h}{2}\right) - f\left(x + \frac{h}{2} - \frac{h}{2}\right)\right) \\
&\quad - \left(f\left(x - \frac{h}{2} + \frac{h}{2}\right) - f\left(x - \frac{h}{2} - \frac{h}{2}\right)\right) \\
&= f(x+h) - 2f(x) + f(x-h)
\end{aligned}$$

$$\Delta_3 f(x) = \Delta_1(\Delta_2 f(x)) = f\left(x + 3\frac{h}{2}\right) - 3f\left(x + \frac{h}{2}\right) + 3f\left(x - \frac{h}{2}\right)$$
$$- f\left(x - 3\frac{h}{2}\right)$$
$$\Delta_4 f(x) = f(x+2h) - 4f(x+h) + 6f(x) - 4f(x-h) + f(x-2h)$$

then the $n$th Riemann derivative $D_n f$ is defined by

$$D_n f(x) = \lim_{h \to 0} \frac{\Delta_n f(x)}{h^n}.$$

This probably motivated Leibniz's notation $\frac{d^n f}{dx^n}$. The first and second symmetric Riemann derivatives are the same as the first and second symmetric derivatives. The second symmetric Riemann derivative is often called the Schwarz derivative. If the Peano derivative $f_{(2)}(x)$ exists, then

$$\begin{aligned}
\Delta_2 f(x) &= f(x+h) - 2f(x) + f(x-h) \\
&= \left\{ f + f_{(1)}h + \frac{f_{(2)}}{2!}h^2 + o\left(h^2\right) \right\} - 2\{f\} \\
&\quad + \left\{ f + f_{(1)}(-h) + \frac{f_{(2)}}{2!}(-h)^2 + o\left(h^2\right) \right\} \\
&= \{f - 2f + f\} + \{f_{(1)} - f_{(1)}\} h + \left\{ \frac{f_{(2)}}{2!}h^2 + \frac{f_{(2)}}{2!}(-h)^2 \right\} + o\left(h^2\right) \\
&= f_{(2)}h^2 + o\left(h^2\right).
\end{aligned}$$

Divide by $h^2$ and then let $h \to 0$ to see that $D_2 f(x)$ exists and is equal to $f_{(2)}(x)$. This simple argument easily generalizes to all higher $n$. A very similar argument also shows that even the existence of only $f_{(n)}^s(x)$ implies that $D_n f(x)$ also exists and that $D_n f(x) = f_{(n)}^s(x)$.

The symmetric Riemann derivative is one special case of the following general derivative. Let

$$A = \{a_0, a_1, \ldots, a_{n+e}; b_0, b_1, \ldots, b_{n+e}\}$$

be a set of real numbers with the $b_i$ being distinct and satisfying

$$\sum_{i=0}^{n+e} a_i b_i^j = \begin{cases} 0 & \text{if } j = 0, 1, \ldots, n-1 \\ n! & \text{if } j = n. \end{cases} \tag{2}$$

We say that $f$ has an $n$th $A$ derivative at $x$ if there is a number $A_n f(x)$ such that

$$\lim_{h\to 0}\sum_{i=0}^{n+e} a_i f(x+b_i h) = A_n f(x)\, h^n + o(h^n). \tag{3}$$

Equivalently,

$$A_n f(x) = \lim_{h\to 0}\frac{\sum_{i=0}^{n+e} a_i f(x+b_i h)}{h^n}.$$

When $\{b_i\} = \left\{-\frac{n}{2}, -\frac{n}{2}+1, -\frac{n}{2}+2, \ldots, \frac{n}{2}\right\}$, $A_n$ coincides with the $n$th symmetric derivative $D_n$. Every $A_n$ derivative must be based on at least $n+1$ $b_i$s. If the excess $e=0$, the Eq. (2) uniquely determine the coefficients $\{a_i\}$. For $D_n$, we find that each $a_i = (-1)^{n-i}\binom{n}{i}$. An example from numerical analysis of a first generalized derivative with positive excess is

$$\lim_{h\to 0}\frac{\begin{array}{c}\left(3-2\sqrt{3}\right) f\left(x+\left(\frac{1}{\sqrt{3}}+1\right)h\right)+4\sqrt{3}f\left(x+\frac{1}{\sqrt{3}}h\right)\\ -\left(3+2\sqrt{3}\right) f\left(x+\left(\frac{1}{\sqrt{3}}-1\right)h\right)\end{array}}{6h}.$$

Of course there are infinitely many choices of $\{a_0, a_1, a_2\}$ satisfying the associated system

$$\begin{cases} a_0+a_1+a_2=0 \\ a_0\left(\frac{1}{\sqrt{3}}+1\right)+a_1\left(\frac{1}{\sqrt{3}}\right)+a_2\left(\frac{1}{\sqrt{3}}-1\right)=1 \end{cases}.$$

The particular choice made here is the only one that makes this difference quotient interesting for someone wanting to efficiently numerically approximate the derivative of $f$ at $x$. See [AJ, AJJ] for details.

A simple way to generate new generalized derivatives from old ones is to translate all the base points by a single number. For example, adding $\frac{3}{2}$ to each $b_i$ changes the third symmetric Riemann difference

$$f\left(x+3\frac{h}{2}\right)-3f\left(x+\frac{h}{2}\right)+3f\left(x-\frac{h}{2}\right)-f\left(x-3\frac{h}{2}\right)$$

to the third forward Riemann difference

$$f(x+3h)-3f(x+2h)+3f(x+h)-f(x).$$

It is obvious from system (2) that any translation of an $A_n$ derivative is still an $A_n$ derivative.

Here are four basic pointwise implications relating the ordinary $n$th derivative $f^{(n)}$, the $n$th Peano derivative $f_{(n)}$, the $n$th symmetric derivative $f^{s}_{(n)}$, the symmetric Riemann derivative $D_n(x)$, and any $A_n$derivative. For each point $x$,

$$\begin{aligned} \text{there exists } f^{(n)}(x) &\Longrightarrow \text{there exists } f_{(n)}(x) \\ \text{there exists } f_{(n)}(x) &\Longrightarrow \text{there exists } f^{s}_{(n)}(x) \\ \text{there exists } f^{s}_{(n)}(x) &\Longrightarrow \text{there exists } D_n(x) \\ \text{there exists } f_{(n)}(x) &\Longrightarrow \text{there exists } A_n(x). \end{aligned}$$

The first implication is trivially reversible if $n = 1$, because $f'$ and $f_{(1)}$ have the same definition. The second implication is never reversible. The third implication is trivially reversible if $n = 1$ or $n = 2$. The fourth implication is irreversible when $n \geq 2$, is trivially reversible when $n = 1$ and $A_1$ is chosen to be the ordinary first derivative, and is irreversible for most, but not all, other choices of $A_1$.[1] The function

$$f(x) = \begin{cases} x^3 \cos x^{-1} & \text{if } x \neq 0 \\ 0 & \text{if } x = 0. \end{cases}$$

at the point $x = 0$ shows the first implication to be irreversible when $n = 2$. In fact $f(0+h) = 0 + 0h + 0h^2 + o(h^2)$ means that $f'(0) = f_{(1)}(0) = f_{(2)}(0) = 0$, while

$$\begin{aligned} f''(0) &= \lim_{h\to 0} \frac{\{f'(h)\} - \{f'(0)\}}{h} = \lim_{h\to 0} \frac{\{3h^2 \cos h^{-1} + h^3 (h^{-2} \sin h^{-1})\} - \{0\}}{h} \\ &= \lim_{h\to 0} \sin \frac{1}{h} \end{aligned}$$

does not exist.

The second implication is not reversible for any $n$. To see this when $n = 1$, consider the function $|x|$ at the point $x = 0$. At $x = 0$, $|x|$ has first symmetric Riemann derivative equal to 0, but $|x|$ is not differentiable at $x = 0$. When $n = 2$, consider the function $\operatorname{sgn}(x)$ at the point $x = 0$.

Next look at the third implication. As soon as $n \geq 3$, it is easy to give a function having $D_n f(0)$ existing, but not having $f^{s}_{(n)}(0)$ existing. If $n$ is 3, let $f(x) = 3^{-k}$ on $[3^{-k}, 3 \cdot 3^{-k})$, $k = 0, 1, 2, \ldots$. This defines $f$ on $(0, 3)$. Extend $f$ to be odd on $(-3, 3)$. On the one hand,

---

[1]When I gave this talk in the fall of 2012, I knew that many, many cases of $A_1$ had manifested irreversibility. This led me to speculate to the audience that, except for the trivial case, irreversibility of the fourth implication when $n = 1$ always occurred. My black swan moment (see [Ta]) arrived with my discovery in [ACC] of a small family of cases of $A_1$ that *do* imply ordinary differentiability. All such cases are classified in [ACC].

$$\frac{f\left(0+3^{-k}\right)-f\left(0-3^{-k}\right)}{2}=f\left(3^{-k}\right)=3^{-k}$$

so that $\limsup_{h\to 0}\frac{f(0+h)-f(0-h)}{2h}\geq 1$, while if $h$ is just slightly less than $3\cdot 3^{-k}$,

$$\frac{f\left(0+h\right)-f\left(0-h\right)}{2}=3^{-k}\approx\frac{1}{3}h$$

so that $\liminf_{h\to 0}\frac{f(0+h)-f(0-h)}{2h}\leq\frac{1}{3}$ and $f$ does not even have a first symmetric derivative at $x=0$. On the other hand, for small $h$,

$$\Delta_3 f\left(h\right)=2\left(f\left(\frac{3}{2}h\right)-3f\left(\frac{1}{2}h\right)\right)$$

is zero so $D_3 f\left(0\right)=0$; furthermore, for any odd $n\geq 5$, $\Delta_n f\left(h\right)=\Delta_{n-3}\left(\Delta_3\left(h\right)\right)=0$ and $D_n f\left(0\right)=0$.

Similarly, if $n$ is 4, let $f\left(x\right)=4^{-k}$ on $\left[4^{-k},4\cdot 4^{-k}\right)$, $k=0,1,2,\ldots$. This defines $f$ on $(0,4)$. Let $f\left(0\right)=0$. Extend $f$ to be even on $(-4,4)$. On the one hand,

$$\frac{f\left(0+4^{-k}\right)+f\left(0-4^{-k}\right)}{2}-f\left(0\right)=f\left(4^{-k}\right)=4^{-k}$$

so that $\limsup_{h\to 0}\frac{f(0+h)+f(0-h)}{2h}-f\left(0\right)\geq 1$, while if $h$ is just slightly less than $4\cdot 4^{-k}$,

$$\frac{f\left(0+h\right)+f\left(0-h\right)}{2}-f\left(0\right)=4^{-k}\approx\frac{1}{4}h$$

so that $\liminf_{h\to 0}\frac{f(0+h)-f(0-h)}{2h}-f\left(0\right)\leq\frac{1}{4}$ and $f$ does not even have a second symmetric derivative at $x=0$. On the other hand, $\Delta_4 f\left(h\right)=2\left(f\left(2h\right)-4f\left(h\right)\right)$ is zero so $D_4 f\left(0\right)=0$; furthermore, for any even $n\geq 6$, $\Delta_n f\left(h\right)=\Delta_{n-4}\left(\Delta_4\left(h\right)\right)=0$ and $D_n f\left(0\right)=0$.

Now look at each of the above four implications generically. In other words, ask whether the converse of an implication can fail at each point of a set of positive measure. The first implication is still irreversible; but the other three all have a converse at almost every point. More specifically, in 1954 H. W. Oliver gave an example of a function $f$ having a second Peano derivative at every point of a certain set $E$ of positive measure although $f''$ fails to exist at every point of $E$ [Ol]. On the other hand, in 1936 Marcinkiewicz and Zygmund proved

**Theorem 1.** *If $f\left(x\right)$ has an nth Riemann symmetric derivative $D_n$ at every $x\in E\subset\mathbb{R}$, then $f$ has an nth Peano derivative at a.e. $x\in E$ [MZ].*

We will abbreviate the statement of this theorem, by saying that it establishes the generic implication

$$\text{there exists } D_n(x) \Longrightarrow \text{there exists } f_{(n)}(x) \ a.e.$$

This theorem obviously generically reverses both implication two and implication three. In 1967 I extended this theorem from $D_n$ to every $A_n$,

$$\text{there exists } A_n(x) \Longrightarrow \text{there exists } f_{(n)}(x) \ a.e., \tag{4}$$

thereby generically reversing the fourth implication [As].

## 2 Generalized Differentiation in the $L^p$ Sense

Let $1 \leq p < \infty$ and let $f \in L^p[x-\epsilon, x+\epsilon]$ for some $\epsilon > 0$. We may extend all of the definitions of generalized derivatives given in Sect. 1 to definitions in $L^p$, $1 < p < \infty$. For example, a function $f$ is said to have at $x$ an $n$th Peano derivative in $L^p$ if there are numbers $f_{(0)p}(x), \ldots, f_{(n)p}(x)$ such that

$$\left(\frac{1}{h}\int_0^h \left| f(x+t) - \left\{ f_{(0)p}(x) + f_{(1)p}(x)\, t + \cdots + \frac{f_{(n)p}(x)}{n!} t^n \right\} \right|^p dt\right)^{1/p}$$
$$= o(h^n) \text{ as } h \to 0.$$

There are obvious corresponding definitions for the symmetric $L^p$ derivative $f^s_{(n)p}$, the symmetric Riemann $L^p$ derivative $D_{np}$, and the generalized $L^p$ derivatives $A_{np}$.

If for $1 \leq p < \infty$, we define

$$\|g(h)\|_p = \left(\frac{1}{h}\int_0^h |g(t)|^p \, dt\right)^{1/p}$$

and also make the slightly unusual definition of

$$\|g(h)\|_\infty = \sup_{|t| \leq h} |g(t)|,$$

then Sect. 1 may be called the $L^\infty$ theory and a completely parallel $L^p$ theory has also been developed. All four of the direct implications are true pointwise and just as easy to prove (by the same methods) as in the $L^\infty$ case. The study of generalized $L^p$ differentiation began to become important in the 1950s in connection with the study of finding $L^p$ solutions for partial differential equations. At that time there was only one generic reverse implication known, the landmark theorem of Marcinkiewicz and Zygmund,

$$\text{there exists } D_n(x) \Longrightarrow \text{there exists } f_{(n)}(x) \text{ a.e.}$$

So the Zygmund school of analysis formulated the natural conjecture

$$\text{there exists } D_{np}(x) \Longrightarrow \text{there exists } f_{(n)p}(x) \ a.e. \tag{5}$$

This was indeed true, but the path to its solution was a little bit convoluted. The first step was taken by Mary Weiss, who proved

$$\text{there exists } f^{s}_{(n)p}(x) \Longrightarrow \text{there exists } f_{(n)p}(x) \ a.e. \text{ [We]}$$

Her work was deep and added important methodology for use on conjecture (5), but resolved that conjecture only for $n = 1$ and $n = 2$, where the symmetric and symmetric Riemann derivatives coincide. Notice that the introduction of the $L^\infty$ symmetric derivative into Sect. 1 was a little artificial since the main generic result of Marcinkiewicz and Zygmund went directly all the way back from the symmetric Riemann derivative to the Peano derivative, leapfrogging the symmetric derivative entirely. The reason I put it in was to help clarify the meaning of Mary Weiss' result here.

In 1963, Prof. Antoni Zygmund proposed conjecture (5) as my thesis problem. I will indicate how the $A_n$ generalized derivatives arose in my investigations. As a simple example, suppose $D_{4,1}(x)$ exists. This implies that we have

$$\frac{1}{h}\int_0^h \{f(x+2t)-4f(x+t)+6f(x)-4f(x-t)+f(x-2t)\}\,dt=O\left(h^4\right).$$

Denote an antiderivative of $f$ by $F$, multiply by $h$, and integrate to get

$$\frac{1}{2}F(x+2h)-4F(x+h)+6f(x)h+4F(x-h)-\frac{1}{2}F(x-2h)=O\left(h^5\right).$$

If the $6f(x)h$ term can be removed, we will have essentially a generalized $L^\infty$ fifth derivative of $F$. So, on the one hand, substitute $h \to 2h$, on the other hand, just multiply the equation by 2, and then subtract the second equation from the first.

$$\frac{1}{2}F(x+4h)-4F(x+2h)+12f(x)h+4F(x-2h)-\frac{1}{2}F(x-4h)=O\left(h^5\right)$$

$$F(x+2h)-8F(x+h)+12f(x)h+8F(x-h)-F(x-2h)=O\left(h^5\right)$$

$$\frac{1}{2}F(x+4h)-5F(x+2h)+8F(x+h)-8F(x-h)+5F(x-2h)$$
$$-\frac{1}{2}F(x-4h)=O\left(h^5\right)$$

We have more or less arrived at a fifth generalized $L^{\infty}$ derivative with no excess, but the $b_i$ are $\{-4,-2,-1,1,2,4\}$ and so that derivative is not a $D_n$. If we could establish that $F$ has five $L^{\infty}$ Peano derivatives, the methodology of Mary Weiss would allow us to conclude that $f$ has four $L^p$ Peano derivatives. So we are motivated to prove the $L^{\infty}$ result

$$\text{there exists } A_n(x) \Longrightarrow \text{there exists } f_{(n)}(x)\ a.e., \tag{6}$$

at least for generalized derivatives of zero excess. If we had started with a function $f$ having derivative $D_{6,1}(x)$ and applied exactly the same procedure, the associated antiderivative $F$ would have an $A_7$ antiderivative with the $b_i$ being $\{-6,-4,-3,-2,-1,1,2,3,4,6\}$. This should have 8 points, but it has 10. So we need an $A_n$ theory that allows excess. This is the reason that I formulated and proved the $L^{\infty}$ result (6), even for the case of positive excess.

So the conjecture (5) was reached as a corollary of the (quite difficult) lemma (6). Of course the result

$$\text{there exists } D_{np}(x) \Longrightarrow \text{there exists } f^{s}_{(n)p}(x)\ a.e.$$

follows immediately from (5) and the trivial implication [there exists $f_{(n)p}(x) \Longrightarrow$ there exists $f^{s}_{(n)p}(x)\ a.e.$]

In my thesis, I also prove this.

$$\text{there exists } A_{np}(x) \Longrightarrow \text{there exists } f_{(n)p}(x)\ a.e.. \tag{7}$$

I give three reasons for proving this. It completes the process of bringing all three of the major $L^{\infty}$ results over to $L^p$; its proof is exactly the same as the proof of conjecture (5); and the $A_n$ derivatives, at least in the case of zero excess, were defined by A. Denjoy in 1935 and may be of some interest in their own right.

## 3 Quantum $L^{\infty}$ and $L^p$ Derivatives

The three most common derivatives are the ordinary derivative $f'(x) = \lim_{h\to 0}\frac{f(x+h)-f(x)}{h}$, the symmetric derivative $f_1^s(x) = \lim_{h\to 0}\frac{f(x+h)-f(x-h)}{2h}$, and the Schwarz derivative $f_2^s(x) = \lim_{h\to 0}\frac{f(x+h)-2f(x)+f(x-h)}{h^2}$. The quantum analogues of the first of these is

$$\lim_{q\to 1}\frac{f(qx)-f(x)}{qx-x}.$$

In fact,this is nothing but a rewriting of $f'(x)$ by making the substitution $h = qx - x = (q-1)x$.

It is slightly annoying that the difference quotients for ordinary derivatives can be defined for all real $x$, whereas those for the quantum derivatives can only be defined when $x \neq 0$.

Generalizing the next two is not as simple. Here are two possible generalizations. One generalization for the symmetric derivative is

$$S_1^0 f(x) = \lim_{q \to 1} \frac{f(qx) - f(q^{-1}x)}{qx - q^{-1}x}.$$

One generalization for the Schwarz derivative is the number $S_2^1 f(x)$ satisfying

$$\frac{q^{-1/2} f(qx) + q^{1/2} f(q^{-1}x)}{2} = f(x) + S_2^1 f(x) \frac{\left(q^{1/2}x - q^{-1/2}x\right)^2}{2} + o\left(\left(q^{1/2}x - q^{-1/2}x\right)^2\right).$$

The reason for the notation $S_n^a$ follows. First we equivalently redefine $S_1^0 f(x)$ as the number satisfying

$$\frac{q^{-0/2} f(qx) - q^{0/2} f(q^{-1}x)}{q^{1/2} + q^{-1/2}} = S_1^0 f(x) \left(q^{1/2}x - q^{-1/2}x\right)^1 + o\left(\left(q^{1/2}x - q^{-1/2}x\right)^1\right).$$

We now fit these two examples into an infinite family, indexed by the arbitrary integer $a$, of symmetric quantum $n$th-derivatives. For each integer $a$, the symmetric derivatives $S_n^a f = S_n^a f(x)$ are given inductively by

$$\frac{q^{-\frac{a}{2}} f(qx) + (-1)^n q^{\frac{a}{2}} f(q^{-1}x)}{2_{an}} = \sum_{\substack{k \equiv n \bmod 2 \\ 0 \le k \le n}} S_k^a f(x) \frac{\Delta^k}{k!} + o(\Delta^n), \tag{8}$$

where $\Delta = \left(q^{1/2} - q^{-1/2}\right) x$,

$$[k]_q = \frac{q^{k/2} - q^{-k/2}}{q^{1/2} - q^{-1/2}}, \text{ for } k = 0, \pm 1, \pm 2, \ldots, \tag{9}$$

and

$$2_{an} = \begin{cases} 2, & \text{if } n \equiv a \bmod 2, \\ [2] = [2]_q = q^{\frac{1}{2}} + q^{-\frac{1}{2}} & \text{otherwise.} \end{cases}$$

(If $q$ is close to 1, $[k]_q \approx k$; in particular, $[2]_q \approx 2$.)

Note that Taylor expanding the left-hand sides about $x$ shows that whenever $f_{(n)}(x)$ exists,

$$S_n^a f(x) = \frac{1}{2_{an}} \left( [2]_{q^{n-a}} f_{(n)}(x) + \frac{n [n-a-1]_q}{x} f_{(n-1)}(x) \right).$$

Unfortunately, this does not look as pretty as the corresponding formula for the ordinary symmetric derivative, namely $f^s_{(n)}(x) = f_{(n)}(x)$ except for the following cases:

$$S_j^0 f(x) = f_{(j)}(x),\ j = 0, 1$$
$$S_j^1 f(x) = f_{(j)}(x),\ j = 0, 2$$
$$S_j^{-1} f(x) = f_{(j)}(x),\ j = 0.$$

The relation $S_j^a f(x) = f_{(j)}(x)$ holds in general for no other pairs $(a, j)$.

The naturality of these families of symmetry derivatives is discussed in [AC]. It is fairly easy to prove that at any point $x$,

$$\text{there exists } f_{(n)}(x) \Longrightarrow \text{there exists } S_n^a f(x),$$

and it is also true that for each $n$ and $a$

$$\text{there exists } S_n^a f(x) \Longrightarrow \text{there exists } f_{(n)}(x)\ a.e.$$

There is also true to the corresponding $L^p$ result, that for every $n$, $p$, and $a$,

$$\text{there exists } S_{np}^a f(x) \Longrightarrow \text{there exists } f_{(n)p}(x)\ a.e.\text{[AC]}$$

The $n$th symmetric Riemann derivative is defined by

$$\lim_{h \to 0} \frac{\sum_{k=0}^{n} (-1)^k \binom{n}{k} f\left(x + \left(\frac{n}{2} - k\right) h\right)}{h^n}.$$

There are several natural quantum versions of this. One of these is defined by

$$\mathcal{D}_n^S f(x) = \lim_{q \to 1} \frac{\sum_{k=0}^{n} (-1)^k \begin{bmatrix} n \\ k \end{bmatrix}_q q^{(k-1)k/2} f\left(q^{n/2-k} x\right)}{q^{(n-1)n/2} (q-1)^n x^n},$$

where the $q$-binomial coefficients are defined by setting $[n]_q = 1 + q + q^2 + \cdots + q^{n-1} = \frac{q^n - 1}{q-1}$ if $n = 1, 2, \ldots,$

$$[n]_q! = \begin{cases} 1 & \text{if } n = 0 \\ [1]_q [2]_q \dots [n]_q & \text{if } n = 1, 2, \dots \end{cases},$$

and

$$\begin{bmatrix} n \\ k \end{bmatrix}_q = \frac{[n]_q!}{[k]_q! [n-k]_q!} \text{ for } n \geq k \geq 0.$$

(Notice that $[2]_q$ is defined one way when defining the symmetric quantum derivatives and a different way here.) One reason for the greater complexity in the quantum case is that the coefficients no longer remain constant as $q$ approaches 1 in the quantum case. Compare, for example, the limits

$$\lim_{h \to 0} \frac{f(x+h) - 2f(x) + f(x-h)}{h^2}$$

and

$$\lim_{q \to 1} \frac{f(q^2 x) - (1+q) f(qx) + qf(x)}{q(qx - x)^2}$$

To see other, quite possibly more natural versions of an $n$th quantum Riemann derivative, see the discussion in my 2008 paper with Stefan Catoiu [AC]. One hint towards more symmetry is to note that a more symmetric definition of $[n]_q$ might be

$$[n]_q = \frac{q^{n/2} - q^{-n/2}}{q^{1/2} - q^{-1/2}}.$$

The derivative $\mathcal{D}_n^S f(x)$ is also known to be generically equivalent to the Peano derivative in $L^\infty$ [ACR]. But this question is still open in the $L^p$ case. Thus the quantum $L^p$ theory is now at the same point of development as was the ordinary $L^p$ theory during the period 1964–1967 when Marry Weiss's result for the ordinary symmetric $L^p$ derivative had been published, but the solution for the ordinary $L^p$ Riemann derivative contained in my thesis had not. We recapitulate with this conjecture:

*Conjecture 1.* If the $n$th $L^p$quantum symmetric Riemann derivative exists at every point of a set $E$, then the $n$th $L^p$ Peano derivative exists at almost every point of $E$.

When I was confronted with my thesis problem of trying to prove that the existence of the $L^p$ Riemann derivative on $E$ implies the existence of the $L^p$ Peano derivative a.e. on $E$, the most direct and simple way to proceed seemed to just change every step of Marcinkiewicz and Zygmund's proof for the $L^{\infty}$ case to an $L^{p}$ analogue. This was easy for every step but one. That one asserts that you can a.e. slide the $L^{\infty}$ estimate

$$\sum a_i f\left(x + b_i h\right) = O\left(h^n\right)$$

to

$$\sum a_i f\left(x + \left(b_i + b\right) h\right) = O\left(h^n\right).$$

I did not know then, and I still do not know now, how to directly slide

$$\left\| \sum a_i f\left(x + b_i h\right) \right\|_p = O\left(h^n\right) \text{ for all } x \in E$$

to

$$\left\| \sum a_i f\left(x + \left(b_i + b\right) h\right) \right\|_p = O\left(h^n\right) \text{ for a.e. } x \in E.$$

Similarly, proving directly that the a.e. sliding of an estimate for $\|\sum a_i\,(q)\, f\left(q^{b_i} x\right)\|_p$ to the same estimate for $\|\sum a_i\,(q)\, f\left(q^{b_i+b} x\right)\|_p$ would allow an easy resolution to the conjecture as well as to the more general open problem of proving the quantum version of (7).

## 4 Irreversibility and Generic Reversibility of Implications

We have looked at many different kinds of generalized derivatives and found many implications that are true at every point. I will try to draw a few general conclusions that ought to be true in most cases. First, in a class by itself, is the implication that the existence of the ordinary $n$th derivative at a point implies the existence of the $n$th Peano derivative at that point. As I mentioned earlier, this relation is not even generically reversible [Ol]. Otherwise, just about every implication is generically reversible. This has been proved in many cases and, for me, is the right conjecture whenever a new family of generalized derivatives arises.

On the other hand, reversibility at a single point almost never happens [ACC].

Here I will just give three examples of irreversibility at a point: an $A_1$ example, an $A_2$ example, and a second quantum symmetric example.

*Example 1.* The function $|x|$ has a first symmetric derivative at $x = 0$ but does not have at that point a first Peano derivative, i.e. a first derivative.

*Example 2.* The function $\operatorname{sgn} x$ has a second symmetric derivative at $x = 0$, but does not have a second Peano derivative at that point.

*Example 3.* Let

$$f(x)=\begin{cases}x^{3/2} & \text{if } 1\le x<2\\ 0 & \text{if } x=1\\ -x^{-1/2} & \text{if } \frac{1}{2}\le x<1\end{cases}.$$

Then $f(x)$ has second quantum symmetric derivative $S_2^1 f(1)$, but is not even continuous at $x = 1$ and thus separates the second quantum symmetric derivative from the second Peano derivative.

## 5 Higher Dimensions

We use the notations $x = (x_1, \ldots, x_d)$, $h = (h_1, \ldots, h_d)$,$|h| = \sqrt{\sum h_i^2}$; for exponents $j = (j_1, \ldots, j_d) \in \{0, 1, \ldots,\}^d$, $|j| = j_1 + \cdots + j_d$; and $h^j = h^{j_1} \cdots h^{j_d}$. Here are some definitions and a conjecture.

**Definition 1.** If $f(x+h) = \sum_{|j|\le k-1} f_{(j)}(x)\, h^j + O\left(|h|^k\right)$ at $x$, say that $f$ is Lipschitz of order $k$ at $x$.

**Definition 2.** If $f(x+h) = \sum_{|j|\le k} f_{(j)}(x)\, h^j + o\left(|h|^k\right)$ at $x$, say that $f$ has a $k$th Peano derivative at $x$.

What Buczolich proved is this.

**Theorem 2.** *Let $k \ge 1$. If $f$ is Lipschitz of order $k$ at every $x \in E$, $E$ closed, then $f$ has a $k$th Peano derivative at almost every $x \in E$ [Bu].*

**Problem 1 (Small Problem).** What is the issue of $E$ being closed? This is stated by Buczolich. This appears neither in the Marcinkiewicz and Zygmund proof of Theorem 1, nor in my proof of 4. Is it necessary?

The regularity theorem for Lebesgue measure states that if $E$ is (Lebesgue) measurable and if $\epsilon > 0$ is given, then there is a closed set $C$ and an open set $U$ such that $C \subseteq E \subseteq U$ and $|U \setminus C| < \epsilon$, and in particular, $|E \setminus C| < \epsilon$. Here is a simple meta-theorem that allows us to replace "closed" by "measurable" in Buczolich's theorem.

**Theorem 3.** *Suppose we have related two properties $p$ and $q$ by a theorem which asserts that if $C$ is any closed set with $p$ being true at each point of $C$, then $q$ holds at a.e. point of $C$. Then the following statement is also true. If $E$ is any measurable set with $p$ being true at each point of $E$, then $q$ holds at a.e. point of $E$.*

*Proof.* Fix properties $p$ and $q$ such that the first implication is valid for every closed set $C$. Assume the second implication fails for a measurable set $E_0$. Then there is a measurable subset $F_0$ of $E_0$ with property $q$ being false at each point of $F_0$ and with $|F_0| = \epsilon_0 > 0$. By the regularity theorem for Lebesgue measurable functions, there is a closed subset $C_0 \subseteq E_0$ so that $|E_0 \setminus C_0| < \epsilon_0$. Since $C_0 \subseteq E_0$, property $p$ holds at each point of $C_0$. By the first implication, there is a set $E_1 \subseteq C_0$ such that $|E_1| = |C_0|$ and such that property $q$ holds at every point of $E_1$. This is a contradiction, since $F_0 \subseteq E_0 \setminus E_1$ implies

$$\epsilon_0 = |F_0| \leq |E_0 \setminus E_1| = |E_0 \setminus C_0| < \epsilon_0.$$

Here is an unpublished fact whose proof does not require much more than bringing the one dimensional Marcinkiewicz and Zygmund proof along, lemma by lemma.

*Claim.* If $f(x)$ satisfies

$$\sum_{j=0}^{k} (-1)^j \binom{k}{j} f\left(x + \left(\frac{k}{2} - j\right) h\right) = O\left(|h|^k\right)$$

for every $x \in E \subseteq \mathbb{R}^d$, then $f(x)$ has a $k$th Peano derivative at a.e. $x \in E$.

It might be fruitful to try to transplant all of the results previously mentioned to this higher dimensional context.

**Acknowledgements** The author's research was partially supported by a Summer Research Grant, College of Science and Health, DePaul University.

This paper is dedicated to Calixto Calderón.

## References

[As] Ash, J.M.: Generalizations of the Riemann derivative. Trans. Am. Math. Soc. **126**, 181–199 (1967)

[AC] Ash, J.M., Catoiu, S.: Quantum symmetric $L^p$ derivatives. Trans. Am. Math. Soc. **360**, 959–987 (2008)

[ACC] Ash, J.M., Catoiu, S., Csörnyei, M.: Generalized vs. ordinary differentiation. preprint

[ACR] Ash, J.M., Catoiu, S., Rios, R.: On the nth quantum derivative. J. Lond. Math. Soc. **66**, 114–130 (2002)

[AJ] Ash, J.M., Jones, R.: Optimal numerical differentiation using three function evaluations. Math. Comput. **37**, 159–167 (1981)

[AJJ] Ash, J.M., Janson, S., Jones, R.: Optimal numerical differentiation using n function evaluations. Estratto da Calcolo **21**, 151–169 (1984)

[Bu] Buczolich, Z.: An existence theorem for higher Peano derivatives in $R^m$. Real Anal. Exch. **13**, 245–252 (1987/88)

[MZ] Marcinkiewicz, J., Zygmund, A.: On the differentiability of functions and the summability of trigonometric series. Fund. Math. **26**, 1–43 (1936)

[Ol] Oliver, H.W.: The exact Peano derivative. Trans. Am. Math. Soc. **76**, 444–456 (1954)

[Ta] Taleb, N.N., The Black Swan: The Impact of the Highly Improbable, Random House (2007)

[We] Weiss, M.: On symmetric derivatives in $L^p$. Stud. Math. **24**, 89–100 (1964)

# Some Non Standard Applications of the Laplace Method

**Calixto P. Calderón and Wilfredo O. Urbina**

**Summary.** In this paper we consider two nonstandard applications of the Laplace method. The first one is referred to the Inversion Formula of D.V. Widder and E.L. Post, from which a maximal theorem is proved. The second one is a special Calderón–Zygmund partition that gives us a genuine generalization of Natanson's lemma in this context.

## 1 The Inversion Formula of D.V. Widder and E.L. Post

Given a function $g \in L^1(0, \infty)$ then, the Laplace transform of the function $g$, $\mathcal{L}(g, \lambda)$, is given by

$$\mathcal{L}(g, \lambda) = \int_0^\infty e^{-\lambda\alpha} g(\alpha)\, d\alpha, \quad \text{a.e. } \mathrm{Re}(\lambda) > \lambda_0. \tag{1}$$

The Riemann–Mellin formula is a well-known inversion formula for the Laplace transform [2]. A different type of inversion formula has been studied by E.L. Post and D.V. Widder, see [4, 7]. Let

$$L_{k,\xi}[f] = \frac{(-1)^k}{k!} (\frac{k}{\xi})^{k+1} \mathcal{L}(g, \lambda)^{(k)} (\frac{k}{\xi}), \tag{2}$$

C.P. Calderón
Department of Mathematics, University of Illinois at Chicago, Chicago, IL 60607, USA
e-mail: cpc@uic.edu

W.O. Urbina (✉)
Department of Mathematical and Actuarial Sciences, Roosevelt University Chicago, Chicago, IL 60605, USA
e-mail: wurbinaromero@roosevelt.edu

 
A.M. Stokolos et al. (eds.), *Special Functions, Partial Differential Equations, and Harmonic Analysis*, Springer Proceedings in Mathematics & Statistics 108,
DOI 10.1007/978-3-319-10545-1_6

which is defined for values of $k \in \mathbb{N}$ for each $\xi > 0$. Then, it can be proved, using Laplace's method, see [7, Chapter VII], that

$$L_{k,\xi}[f] \to g(\xi), \text{ a.e. as } k \to \infty. \tag{3}$$

An alternative proof can de derived from the maximal theorem below. Taking into account the formula

$$(\mathcal{L}(g,\lambda))^{(k)}(\lambda) = \int_0^\infty (-1)^k \alpha^k e^{-\lambda\alpha} g(\alpha)\, d\alpha,$$

it follows

$$\begin{aligned} L_{k,\xi}[f] &= \frac{1}{\xi}\int_0^\infty \frac{k^{k+1}}{k!}\Big(\frac{\alpha}{\xi}\Big)^k e^{-k\frac{\alpha}{\xi}} g(\alpha) d\alpha \\ &= \int_0^\infty W_0\Big(\frac{\alpha}{\xi};k\Big) g(\alpha)\, d\alpha/\alpha \end{aligned} \tag{4}$$

where

$$W_0(\beta;k) = \frac{k^{k+1}}{k!}\beta^{k+1}e^{-k\beta}.$$

Let $\mu(dx) = \frac{d\alpha}{\alpha}$ be the Haar measure in $(0,\infty)$. We will prove that $W_0(\beta;k)$ can be bounded by a Natanson's type of kernel on $((0,\infty),\mu)$. First, let us study the integral,

$$\int_0^\infty \frac{k^{k+1}}{k!}\Big(\frac{\alpha}{\xi}\Big)^{k+1} e^{-k\frac{\alpha}{\xi}} \frac{d\alpha}{\alpha}.$$

Taking $\xi = 1$, the multiplicative identity in $(0,\infty)$, then

$$\begin{aligned} \int_0^\infty \frac{k^{k+1}}{k!}\alpha^{k+1}e^{-k\alpha}\frac{d\alpha}{\alpha} &= \int_0^\infty \frac{k^{k+1}}{k!}\alpha^k e^{-k\alpha} d\alpha \\ &= \int_0^\infty \frac{k^{k+1}}{k!}\Big(\frac{\beta}{k}\Big)^k e^{-\beta}\frac{d\beta}{k} \\ &= \frac{1}{k!}\int_0^\infty \beta^k e^{-\beta} d\beta = 1 \end{aligned}$$

Then, if $g \in L^\infty(0,\infty)$, we have

$$\|L_{k,\xi}[g]\|_\infty = \Big\|\int_0^\infty \frac{k^{k+1}}{k!}\Big(\frac{\alpha}{\xi}\Big)^{k+1} e^{-k\frac{\alpha}{\xi}} g(\alpha)\frac{d\alpha}{\alpha}\Big\|_\infty \le \|g\|_\infty.$$

Now, let us consider the case that $g \in L^p((0,\infty),\mu)$, and consider also the kernel

$$\left[(\frac{\alpha}{\xi})^{1+\frac{1}{k}}e^{-\frac{\alpha}{\xi}}\right]^k.$$

Observe that

$$\begin{aligned}\frac{d}{ds}(s^{1+\frac{1}{k}}e^{-s}) &= (1+\frac{1}{k})s^{\frac{1}{k}}e^{-s} - s^{1+\frac{1}{k}}e^{-s}\\ &= (s^{\frac{1}{k}}e^{-s})\left(\frac{k+1}{k}-s\right)=0,\end{aligned}$$

for $s=\frac{k+1}{k}=1+\frac{1}{k}$. $s=1+\frac{1}{k}$ is a maximum. The following integral, on $[1,\frac{k+1}{k}]$, can be estimated as follows:

$$\begin{aligned}\int_1^{\frac{k+1}{k}} \frac{k^{k+1}}{k!}\left(\frac{k+1}{k}\right)^{k+1} e^{-k(\frac{k+1}{k})}\frac{d\alpha}{\alpha} &= \log\frac{k+1}{k}e^{-(k+1)}\frac{(k+1)^{k+1}}{k!}\\ &\sim \log\frac{k+1}{k}e^{-(k+1)}\frac{(k+1)^{k+1}}{k^k e^{-k}\sqrt{2\pi k}}\\ &\sim \log\frac{k+1}{k}e^{-(k+1)}(k+1)\left(\frac{k+1}{k}\right)^k\frac{1}{e^{-k}\sqrt{2\pi k}}\\ &\sim \frac{1}{k}e^{-1}(k+1)e\frac{1}{\sqrt{2\pi k}}\sim\frac{1}{\sqrt{2\pi k}}.\end{aligned}$$

The kernel $W_0(s,k)$ is monotone increasing for $0<s<\frac{k+1}{k}$ and monotone decreasing for $\frac{k+1}{k}<s<\infty$. Consider the mayorant kernel $W_0^*(s,k)$ defined as

$$W_0^*(s,k)=W_0(s,k),\quad s\le 1,$$

$$W_0^*(s,k)=W_0\left(\frac{k+1}{k},k\right),\quad 1<s<\frac{k+1}{k},$$

$$W_0^*(s,k)=W_0(s,k),\quad \frac{k+1}{k}<s<\infty.$$

Trivially,

$$W_0(s,k)\le W_0^*(s,k).$$

Moreover, the kernel $W_0^*(s,k)$ is monotone increasing for $0<s<1$ and monotone decreasing for $1<s<\infty$. Therefore, from what has been said

$$\int_0^\infty W_0^*(s,k)\frac{ds}{s}=\int_0^1 W_0(s,k)\frac{ds}{s}+\int_1^{\frac{k+1}{k}}W_0\left(\frac{k+1}{k},k\right)\frac{ds}{s}+\int_{\frac{k+1}{k}}^\infty W_0(s,k)\frac{ds}{s}.$$

Altogether, we get a bound of the type, w

$$(1+\varepsilon)\left(1+\frac{1}{\sqrt{2\pi k}}\right), \quad k > k_0, \quad \text{for } k_0 \text{ sufficiently large.}$$

Then, the kernel is a Natanson kernel with respect to the measure $d\mu = \frac{dt}{t}$. By using Natanson's lemma[1] on $((0,\infty),\mu)$, see [3] and also [1], we get

$$m_g = \sup_k \int_0^\infty W_0^*(s,k)g(s)\,\frac{ds}{s} \leq Cg^*(\xi),$$

where $g^*$ is the Hardy–Littlewood maximal function on non-symmetric intervals, with respect to the measure $\mu$, see [5]. It follows a *Maximal Theorem:*

$$\|g^*\|_{L^p(\mu)} \leq C\ \|g\|_{L^p(\mu)}, \quad 1 < p \leq \infty \tag{5}$$

and

$$\mu(\{g^* > \lambda\}) \leq \frac{C}{\lambda}\ \|g\|_{L^1(\mu)}. \tag{6}$$

The same is valid for $m_g$.

Using this maximal theorem, we can provide an elegant alternative proof of (3). In fact, given $g \in L^1(0,\infty)$ it can be decomposed as $g = g_1 + g_2$, where $g_1$ is supported in $[0,\varepsilon]$, $\varepsilon > 0$ arbitrarily small and $g_2 \in L^1(\mu)$, $L_{k,\xi}[g_1] \to 0$, $\xi > \varepsilon$ and $f_1(\lambda) = \mathcal{L}(g_1,\lambda)$.

## 2 Another Application: Calderón–Zygmund Decomposition

In this second part, we will consider Weierstrass type kernels and a Calderón–Zygmund decomposition using Whitney decomposition for open sets (see Whitney decomposition Stein [6, Propositions 2 and 3, p. 169].)

A Weierstrass type kernel $k(x)$ is a kernel that satisfies

(i) $k(x) \geq 0$,
(ii) $k(x)$ has an absolute maximum at $x = 0$, that is $k(0) > k(x)$ if $x \neq 0$,
(iii) $k(0) = 1$,

[1]Natanson's lemma: Given $- \quad \leq a < b \leq \infty$, a Borel measure $\mu$ with support in $(a,b)$ and $K(r,x,y)$ be a Natanson kernel (i.e., $K$ it is monotone increasing in $y$, for $a < y < x$, monotone decreasing in $y$, for $b > y > x$ and $\int_a^b K(r,x,y)\mu(dy) \leq M$, for some $M$ is independent of $x$ and $r$. Then, for $f \in L^1(\mu)$, we have $|\int_a^b K(r,x,y)f(y)\mu(dy)| \leq Mf_\mu^*(x)$.

(iv) $k(x)$ is continuous,
(v) $k \in L^1(\mathbb{R})$, that is $\int_{-\infty}^{\infty} k(x)\,dx < \infty$.
(vi) $k$ satisfies the *Maximum Principle*, see below.

## 2.1 Construction of the Kernel Function

Let us consider a Weierstrass type kernel $k(x)$ and consider $\frac{1}{\Gamma_m} k^m(x)$ with $\Gamma_m = \int_{-\infty}^{\infty} k^m(x)\,dx$. Therefore

$$\frac{1}{\Gamma_m} \int_{-\infty}^{\infty} k^m(x)\,dx = 1.$$

Let be $\delta > 0$ so that $|x| \geq \delta$ then $k(x) < 1 - \varepsilon$, and $\delta' < \delta$ chosen so that if $|x| \leq \delta'$ then $k(x) \geq 1 - \varepsilon/2$. We have the following properties:

$$C\delta'(1 - \varepsilon/2)^m \leq \int_{|x|<\delta'} k^m(x) \leq \Gamma_m, \tag{7}$$

hence

$$\frac{1}{\Gamma_m} k^m(x) \leq \frac{1}{C(\delta')^m} \frac{(1-\varepsilon)^m}{(1-\varepsilon/2)^m}, \quad |x| \geq \delta. \tag{8}$$

This estimate ensures that

(j) $\frac{1}{\Gamma_m} k^m(x) \to 0, \quad m \to \infty$, uniformly on $|x| \geq \delta$,
(jj) $\int_{|x|<\delta} \frac{1}{\Gamma_m} k^m(x) \to 1, \quad m \to \infty$.

This, together with the fact that, $\frac{1}{\Gamma_m} \int_{-\infty}^{\infty} k^m(x)\,dx = 1$, gives

$$\frac{1}{\Gamma_m} k^m * f \to f, \text{ in } \; L^1.$$

## 2.2 A Calderón–Zygmund Type of Decomposition

Consider $G$ an open set with finite Lebesgue measure, and a Whitney type decomposition of G, see [6]:

$$G = \bigcup_{k=1}^{\infty} Q_k, \; F = G^c,$$

such that $Q_k^o \cap Q_j^o = \phi$, for $k \neq j$ and there exists constants $C_1, C_2$ such that

$$C_1 \operatorname{diam}(Q_k) \leq d(Q_k, F) \leq C_2 \operatorname{diam}(Q_k).$$

Now, given a $f \geq 0$ and $\lambda > 0$, let us consider the open set $G$ defined as

$$G = \left\{ x : \sup_{x \in Q} \frac{1}{|Q|} \int_Q f(x)\, dx > \lambda \right\}. \tag{9}$$

We claim that there exists a constant $C_3 = C_3(C_1, \epsilon, n)$ such that

$$\frac{1}{|Q_k|} \int_{Q_k} f(x)\, dx \leq C_3 \lambda.$$

There exists a constant $\gamma > 1$, depending on $C_2$ (but not $k$) such that

$$\gamma Q_k \cap F \neq \phi.$$

$$\frac{1}{|Q_k|} \int_{Q_k} f(x)\, dx \leq \frac{1}{|Q_k|} \int_{\gamma Q_k} f(x)\, dx \leq \frac{C}{|\gamma Q_k|} \int_{\gamma Q_k} f(x)\, dx \leq C\lambda.$$

## 2.3 Maximum Principle

Regarding condition (vi) above, we have the following *Maximum Principle*: if $Q$ is a cube, with edges parallel to the coordinate axis, $d(Q, 0) \geq 5 \operatorname{diam}(Q)$ then, there exists a cube $\overline{Q}$ adjacent to $Q$ with $\operatorname{diam}(\overline{Q}) = \frac{1}{8} \operatorname{diam}(Q)$ such that

$$\max_{x \in Q} k(x) \leq \frac{1}{|\overline{Q}|} \int_{\overline{Q}} k(x)\, dx. \tag{10}$$

From (10), as a consequence of Jensen inequality, we have also a maximum principle for $k^m$,

$$\max_{x \in Q} k^m(x) \leq (\max_{x \in Q} k(x))^m \leq \left( \frac{1}{|\overline{Q}|} \int_{\overline{Q}} k(x)\, dx \right)^m \leq \frac{1}{|\overline{Q}|} \int_{\overline{Q}} k^m(x)\, dx.$$

Thus,

$$\max_{x \in Q} k^m(x) \leq \frac{1}{|\overline{Q}|} \int_{\overline{Q}} k^m(x)\, dx. \tag{11}$$

In particular, this principle applies if $k$ is a monotone non-increasing function of the distance $d(x, 0)$ (for the case $n > 1$ minor changes give us the same result).

Hence, a Calderón–Zygmund partition can be done using Whitney type cubes $\{Q_k\}$, decomposing the given function $f > 0$, and $G$ given as in (9), as

$$f = f_1 + f_2,$$

where $f_1$ is the "good part," $f_1 \leq \lambda$, a. e. and the "bad part" $f_2$ satisfies

$$\frac{1}{|Q_k|} \int_{Q_k} f_2(x)\, dx \leq C\lambda.$$

Let $x \in \mathbb{R}^n$ such that $d(x, \overline{G}) > 0$, where $\overline{G} = \bigcup_{k=1}^{\infty} 5Q_k$, ($5Q$ is the five times dilation of $Q$ from its center). Then, $f_2 = \sum_k f\chi_{Q_k}$ and by (11) we get

$$\begin{aligned}
\frac{1}{\Gamma_m} \int k^m(x-y) f_2(y)dy &= \frac{1}{\Gamma_m} \sum_k \int_{Q_k} k^m(x-y) f(y)dy \\
&\leq \frac{1}{\Gamma_m} \sum_k (\max_{y \in Q_k} k^m(x-y)) \int_{Q_k} f(y)dy \\
&\leq \frac{1}{\Gamma_m} \sum_k \frac{1}{|\overline{Q_k}|} \int_{\overline{Q_k}} k^m(x-y)\, dx \times C\lambda |Q_k| \\
&= \frac{C}{\Gamma_m} \sum_k \lambda \int_{\widehat{Q_k}} k^m(x-y)\, dx \leq C\lambda,
\end{aligned}$$

where $\widehat{Q_k} = (1 + \frac{1}{4})Q_k$, $Q_k \cup \overline{Q_k} \subset \widehat{Q_k}$, and each $x \in \mathbb{R}^n$ belongs to at most $C = C(n)$ different $\widehat{Q_k}$.

This leads to a genuine generalization of the Natanson lemma in this context.

We would like to thank the referee for his comments leading to important improvements in the presentation of these results.

## References

1. Calderón, C.P., Urbina, W.: On Abel summability of Jacobi polynomials series, the Watson kernel and applications. Illinois J. Math. **57**(2), 343–371 (2013)
2. Hille, E., Phillips, R.: Functional Analysis and Semigroups. AMS Colloquium, vol. 31. American Mathematical Society, New York (1957)
3. Natanson, I.P.: Theory of Functions of Real Variable, vol. 2. Ungar, New York (1960)
4. Post, E.L.: Generalized differentiation. Trans. Am. Soc. **32**, 723–781 (1930)
5. Robert, S.: On Classical Functions on Real Analysis. Ph.D. Thesis. University of Illinois at Chicago (1993)
6. Stein, E.M.: Singular Integrals and Differentiability Properties of Functions. Princeton University Press, Princeton (1970)
7. Widder, D.V.: Laplace Transform. Dover, New York (2012)

# Fejér Polynomials and Chaos

**Dmitriy Dmitrishin, Anna Khamitova, and Alexander M. Stokolos**

**Abstract** We show that given any $\mu > 1$, an equilibrium $x$ of a dynamic system

$$x_{n+1} = f(x_n) \tag{1}$$

can be robustly stabilized by a nonlinear control

$$u = -\sum_{j=1}^{N-1} \varepsilon_j \left(f\left(x_{n-j+1}\right) - f\left(x_{n-j}\right)\right), \ |\varepsilon_j| < 1, \ j = 1, \dots, N-1, \tag{2}$$

for $f'(x) \in (-\mu, 1)$. The magnitude of the minimal value $N$ is of order $\sqrt{\mu}$. The optimal explicit strength coefficients are found using extremal nonnegative Fejér polynomials. The case of a cycle as well as numeric examples and applications to mathematical biology are considered.

## 1 Introduction

Nonnegative trigonometric polynomials appear in many problems of harmonic analysis, univalent mappings, approximation theory, orthogonal polynomials on the torus, and number theory and in other branches of mathematics. Some of the most natural and beautiful examples of application of the properties of nonnegative trigonometric polynomials occur when solving extremal problems.

For instance, for the nonnegative trigonometric polynomials $\sum_{j=0}^{N} a_j \cos jt$, $a_0 = 1$, the following inequality established by L. Fejér [8] is valid:

$$|a_1| \le 2 \cos \frac{\pi}{N+2}.$$

D. Dmitrishin
Odessa National Polytechnic University, 1 Shevchenko Ave., Odessa 65044, Ukraine
e-mail: dmitrishin@opu.ua

A. Khamitova • A. Stokolos (✉)
Georgia Southern University, Statesboro, GA 30458, USA
e-mail: anna_khamitova@georgiasouthern.edu; astokolos@georgiasouthern.edu



A.M. Stokolos et al. (eds.), *Special Functions, Partial Differential Equations, and Harmonic Analysis*, Springer Proceedings in Mathematics & Statistics 108,
DOI 10.1007/978-3-319-10545-1_7

The inequality is sharp, and the extremal polynomial $\frac{2}{N+2} \cdot \sin^2 \frac{\pi}{N+2} \cdot \Phi_N^{(1)}(t)$ is unique, where

$$\Phi_N^{(1)}(t) = \left( \frac{\cos \frac{N+2}{2} t}{\cos t - \cos \frac{\pi}{N+2}} \right)^2 .$$

Another example is a classical Fejér kernel; namely, in 1900, Fejér [9] demonstrated that the trigonometric polynomial

$$1 + 2 \sum_{j=1}^{N} \left( 1 - \frac{j}{N+1} \right) \cos jt$$

is nonnegative when he proved the multiplicative representation

$$1 + 2 \sum_{j=1}^{N} \left( 1 - \frac{j}{N+1} \right) \cos jt = \frac{1}{N+1} \cdot \Phi_N^{(2)}(t),$$

where

$$\Phi_N^{(2)}(t) = \left( \frac{\sin \frac{N+1}{2} t}{\sin \frac{t}{2}} \right)^2 .$$

He established [10] the extremal property of the function $\Phi_N^{(2)}(t)$: *The maximal value of the nonnegative trigonometric polynomial* $\sum_{j=0}^{N} a_j \cos jt, a_0 = 1$, *does not exceed* $N + 1$. Moreover, the equality occurs only for the polynomial $\frac{1}{N+1} \cdot \Phi_N^{(2)}(t)$ and only at the points $2\pi k, k = 0, \pm 1, \pm 2, \ldots$.

The functions $\Phi_N^{(1)}(t)$ and $\Phi_N^{(2)}(t)$ are called Fejér kernels and possess a variety of extremal properties. Many papers and book chapters have been published on this subject, and we will not reproduce well-known facts; rather we refer to the nice expository article by D. Dimitrov [5].

The problem of optimal impact on a chaotic regime is a fundamental of nonlinear dynamics. The aim is either to synchronize chaotic motion or, conversely, to make a regular motion chaotic. Moreover, the admissible controls are only small ones that, however, totally change the regime of the motion.

In this chapter, we focus on the simplest case of the dynamic system (1). The solutions of the system typically exhibit chaotic behavior called "deterministic chaos," or just "chaos"—a term coined by J. A. Yorke and T. Y. Li in their paper "Period Three Implies Chaos" [19]. They proved that any one-dimensional system that exhibits a regular cycle of period three will also display regular cycles of every other length as well as completely chaotic cycles. Note that the famous Sharkovsky theorem [30] includes this result as a special case (cf. [4, p. 79])

This field has various aspects comprising stabilization of unstable periodic orbits embedded in a deterministic chaotic attractor, which is generally referred to as chaos control, stabilization of unstable fixed points (steady states). According to [22], controlling chaos consists of perturbing a chaotic system in order to stabilize a given unstable periodic orbit embedded in the chaotic attractor.

In a formal way, we want to change $f(\cdot)$ to $F(\cdot, u)$, where $f(x) = F(x, u)$ for the equilibrium point $x$ and the control $u$ is a function of the prehistory $x_n, \dots, x_0$. If $u$ is a linear function, the control is called *linear*; otherwise, it is said to be *nonlinear*. An intrinsic form of the control is an additive; that is, the closed-loop system looks like $x_{n+1} = f(x_n) + u$.

## 2 Linear Control

Various methods of control have been developed since the ground-breaking work of E. Ott, C. Grebogi, and J. A. Yorke [22] in 1990, who suggested a method to stabilize the chaos by making only small time-dependent perturbations of an available system parameter. They demonstrated that small time-dependent changes in the control parameters of a nonlinear system can turn a previously chaotic trajectory into a stable periodic motion.

The next step forward was made by K. Pyragas [26] in 1992. He suggested a very simple linear scheme where the control force is constructed from time-delayed signals of the type $u = K(x_n - x_{n-1})$. It is noninvasive since the control forces vanish if the target state is reached. Thus, the unstable states of the uncontrolled system are not changed themselves, but only their neighborhood is adjusted such that neighboring trajectories converge to it; that is, the control forces act only if the system deviates from the state to be stabilized. Let's mention that ideas of the delayed control were discussed several decades earlier by V. I. Zubov [40]. Perhaps it was too early for an extensive development of the subject.

Pyragas's method has turned out to be very robust and universal to apply as well as easy to implement experimentally. It has been used in a large variety of systems in physics, chemistry, biology, medicine, and engineering (cf. [13, 27, 29]). The method has the acronym DFC, for delayed feedback control. Some authors use the abbreviation TDAS, for time-delay autosynchronization.

In 1996, T. Ushio [35] showed that the DFC method has several serious limitations. In particular, the admissible region of the multiplier is (−3,1), and so it does not serve the whole range of negative values, as is desired. Thus, some modifications of the linear DFC methods were proposed, including *prediction-based control* (PBC) [33], *dynamic delayed feedback control* (DDFC) [37], and *recursive delayed feedback control* (RDFC) [34]. An extension to multiple time delays has been proposed by Socolar et al. [31], who considered multiple delays in the form of an infinite series (ETDAS), or an average of $N$ past iterates (N time delay autosynchronization, or NTDAS) [32]. The proposed controls became more and more sophisticated. Analytical insight into those schemes can be found in

[1, 3, 6, 12, 14–18, 25, 28, 38]. A mysterious story involved the often-invoked *odd-number limitation*, which had been believed to impose serious restrictions for a long time but finally has been refuted [11].

## 3 Nonlinear Control

In the early 1990s, the dynamic community started to realize that linear DFC methods even in quite sophisticated forms still had serious limitations in applications. A natural step was to consider nonlinear delays. One of the earliest nonlinear controls was introduced in 1994 by S. Bielawski, D. Derozier, and P. Glorieux [2]. The controlled logistic map in this case is given by $x_{n+1} = 4(h + \epsilon_n)x_n(1 - x_n)$, with $\epsilon_n = K(x_n - x_{n+1})$. The method has been generalized in [31], with the controlled logistic map given by the preceding equation, where $\epsilon_n = K(x_n - x_{n+1}) + R\epsilon_{n-1}$. Obviously, this form includes memory of all past states of the system.

In 1996, M. de Sousa Vieira and A. J. Lichtenberg suggested a more robust version of nonlinear DFC $u = K(f(x_n) - f(x_{n-1}))$ (see [36, equation (6)]) and compared it with the linear one. In their "Conclusion and discussion" section, they wrote, "The generalization consists of feeding back the nonlinear mapping signal rather than a signal linearized around the fixed point. This increases the basin of attraction of the controlled signal and thus decreases the sensitivity to noise. *However, the range of parameters for which control can be achieved is limited.*"

Further development of the nonlinear control procedure took place in 2005. In [39], a version of nonlinear DFC close to one in [36] was considered. In [24], an applied predictive control method $u = \epsilon(f_{(p+1)s+1}(x_n) - f_{ps+1}(x_n))$ was introduced. Here $f_j$ means the $j$ folded iteration of the map $f$. These controls are more complicated, but they allow one to bypass many restrictions of the linear DFC methods. On the other hand, the approach has strong limitations because it is nonrobust with respect to system dynamics.

Recent work has focussed on the one hand on basic aspects such as developing novel control schemes and gaining analytical insights, and, on the other hand, on applications to optical and electronic systems, to chemical and electrochemical reactions, and to biological and medical systems.

The most recent development to the method was made by Ö. Morgul [20] in 2012. In particular, he extensively studied a control of the type $\frac{K}{K+1}(x_n - f(x_n))$, which allows the consideration of a broader range of the system multipliers. However, in some important cases it stabilizes to the value of zero, which is different from stabilization to the equilibrium.

The majority of modifications to the DFC methods rely on linear algebra analysis, namely, on validating numerically Schur–Cohn or Routh–Hurwitz stability tests. Our approach, in contrast, is based on analysis rather than on algebra. This allows us to construct a robust and optimal control for the basic cases of an equilibrium and a cycle.

## 4 Statement of the Problem

The characteristic polynomial for the linear part of the closed-loop system (1) and (2) is

$$Q(\lambda) = \lambda^N + k\left(a_1\lambda^{N-1} + \cdots + a_N\right), \tag{3}$$

where $a_1 = 1 - \varepsilon_1, a_j = \varepsilon_{j-1} - \varepsilon_j, j = 2, \ldots, N-1, a_N = \varepsilon_{N-1}, k = -f'(x)$.

It is clear that $\sum_{j=1}^{N} a_j = 1$ and therefore $Q(1) < 0$ if $k < -1$. On the other hand, if $\lambda$ is positive and large enough, then $Q(\lambda) > 0$. Therefore, it should be a positive root outside a unit disk. Thus, the doable case is $k \geq -1$ or, equivalently, $f'(x) \leq 1$.

A change in parameter $k$ could produce a series of bifurcations in the system leading to the emergence of a chaotic attractor. The first bifurcation value of the parameter relates to the momentum of the system's losing stability. This value is related to the region of Schur stability of polynomial (3) in the space of parameters $k$. For $k = 0$, all polynomials are stable. The zeros of polynomials are continuously dependent upon parameters. Therefore, there are $k_1 > 0$ and $k_2 > 0$ such that for $k \in (-k_1, k_2)$, the family (3) is still stable, while for $k = k_2 + \varepsilon$ or $k = -k_1 - \varepsilon$, the stability fails.

It is a point of interest to find a direction determined by the vector of coefficients $(a_1, \ldots, a_N)$ for which the length of the connected component will be maximal in the space of parameter $k$. In other words, we try to maximize the length $k_1 + k_2$ of robust stability and to find the precise values of $k_1$ and $k_2$. The problem of finding a minimal value for $N^*$ is dual to the problem of finding a maximal value of $k_1 + k_2$. Both problems can be reduced to determining the minimal power of a polynomial and its coefficients in a way to keep all zeros in the unit disk $\{z : |z| < 1\}$ of the complex plane for any admissible values of $k$.

We need to maximize the length of the robust stability segment, that is, the function

$$\Phi\left(a_1, \ldots, a_N\right) = k_1\left(a_1, \ldots, a_N\right) + k_2\left(a_1, \ldots, a_N\right). \tag{4}$$

The function (4) has a simple geometrical meaning. Since

$$\frac{Q\left(e^{it}\right)}{k \cdot e^{iNt}} = \frac{1}{k} + \sum_{j=1}^{N} a_j \cos jt - i \cdot \sum_{j=1}^{N} a_j \sin jt,$$

the points of intersection of the curve $\left\{x = \sum\limits_{j=1}^{N} a_j \cos jt,\ y = -\sum\limits_{j=1}^{N} a_j \sin jt\right\}$ on the $OXY$-plane with $OX$-axis correspond to those values of parameter $k$ that the polynomial $Q$ has zeros on the unit circle. The length of the longest segment defined by these points of intersection is $\frac{1}{k_1} + \frac{1}{k_2}$.

Since $Q(1) = 1 + k$, then $\max\limits_{a_1,\ldots,a_N} \{k_1(a_1, \ldots, a_N)\} \leq 1$.

Now, if a root appears on the unit circle, then

$$1 + k\sum_{j=1}^{N} a_j e^{-ijt} = 0. \tag{5}$$

Because $k$ is positive, we can conclude that $\sum_{j=1}^{N} a_j e^{-ijt}$ is real and negative or

$$Arg\left(\sum_{j=1}^{N} a_j e^{-ijt}\right) = \pi.$$

So if we introduce a quantity

$$I_N = \inf_{\sum_{j=1}^{N} a_j = 1}\left[\max_{t\in[0,\pi]}\left\{\left|\sum_{j=1}^{N} a_j e^{-ijt}\right| : Arg\left(\sum_{j=1}^{N} a_j e^{-ijt}\right) = \pi\right\}\right]$$

then the condition

$$I_N \cdot k < 1 \tag{6}$$

guarantees the existence of $a_1, \ldots, a_N$ such that (5) never occurs for the given value of $k$ and any smaller nonnegative one.

This observation immediately implies that given $|k| < 1$, a choice of $a_1 = 1$ and $a_j = 0, j = 2, \ldots, N$, yields (6) for all $t \in [0, 2\pi]$ and therefore implies stability. Thus, $\max\limits_{a_1,\ldots,a_N} \{k_1(a_1, \ldots, a_N)\} = 1$. Hence, the interesting case that remains is $f'(x) \leq -1$.

Denote

$$C(t) = \Re\left(\sum_{j=1}^{N} a_j e^{-ijt}\right) = \sum_{j=1}^{N} a_j \cos jt; \; S(t) = \Im\left(\sum_{j=1}^{N} a_j e^{-ijt}\right) = \sum_{j=1}^{N} a_j \sin jt.$$

**Lemma 1.** *The quantity*

$$\rho(a_1, \ldots, a_N) = \min_t \left\{C(t) : S(t) = 0, \sum_{j=1}^{N} a_j = 1\right\} \tag{7}$$

*is negative.*

*Proof.* Let $F(z) = \sum_{j=1}^{N} a_j z^j$, where $z$ is a complex variable. It is clear that $z_0 = 0$ is zero of $F(z)$. Because of the continuous dependence of polynomial zeros on the coefficients, the function

$$F_\epsilon(z) = \epsilon + \sum_{j=1}^{N} a_j z^j$$

has zero $z_\epsilon$ such that $|z_\epsilon| < 1$ if $\epsilon$ is small enough.

Now, assume that $\rho$ is nonnegative. Then, for any positive $\epsilon$, the graph of the function $x + iy = F_\epsilon(e^{it})$ does not intersect the negative part of the real axis in the $(x, y)$-plane, that is, does not enclose the origin for $t \in [0, 2\pi]$. By the argument principle, the function $F_\epsilon(z)$ does not have zeros inside the unit disk, which is not the case. The lemma has been proved.

Since the polynomials $C(t)$ and $S(t)$ are periodic, it is enough to consider the minimum in the definition of $\rho(a_1, \ldots, a_N)$ on the interval $[0, \pi]$.

Denote

$$J_N = \sup_{\sum_{j=1}^{N} a_j = 1} \rho(a_1, \ldots, a_N).$$

Lemma 1 implies that $J_N$ is negative and, moreover, that $J_N = -I_N$. Therefore, the criteria of stability for the family of the polynomials (3) for all $k \in (0, \ |\mu^*|)$ can be written as

$$|J_N| \cdot k < 1.$$

The following theorem is announced in [7].

**Theorem 1.** *The following formulas are valid:*

$$-J_N = \tan^2 \frac{\pi}{2(N+1)} \sim \frac{1}{N^2},$$

*and the optimal control coefficients are uniquely defined by*

$$\varepsilon_j = \sum_{k=j+1}^{N} a_k^0, \ j = 1, \ldots, N-1,$$

*where*

$$a_j^0 = 2 \cdot \tan \frac{\pi}{2(N+1)} \cdot (1 - \frac{j}{N+1}) \cdot \sin \frac{\pi j}{N+1}, \ j = 1, \ \ldots \ , N.$$

*If $x$ is an equilibrium point and $f'(x) \in (-\mu^*, -1]$, then*

$$N^* = \left\lfloor \frac{\pi}{2 \cot^{-1} \sqrt{\mu^*}} \right\rfloor - 1 \sim \sqrt{|\mu^*|}.$$

Let's provide the proof of the most important part of Theorem 1—the inequality

$$J_N \geq -\tan^2 \frac{\pi}{2(N+1)}. \tag{8}$$

Recall that

$$J_N = \sup_{a_1+\cdots+a_N=1} \min_{0\leq t\leq \pi} \{C(t) : S(t) = 0\}. \tag{9}$$

Let

$$\rho_1 \equiv \sup_{\sum a_j=1;\; S(t)>0, 0<t<\pi} \min_{0\leq t\leq \pi} \{C(t) : S(t) = 0\}.$$

It is clear that $J_N \geq \rho_1$. Since $C(0) = a_1 + \cdots + a_N = 1$, then

$$\rho_1 = \sup_{\sum a_j=1;\; S(t)>0, 0<t<\pi} C(\pi).$$

Let's increase $\rho_1$ a little bit to allow $S(t)$ to take zero values; that is, let

$$\tilde{\rho}_1 = \sup_{\sum a_j=1;\; S(t)\geq 0, 0<t<\pi} C(\pi).$$

It is clear that

$$\rho_1 \leq \tilde{\rho}_1. \tag{10}$$

Further, let's use the following presentation of $S(t)$, which is crucial in this setting:

$$S(t) = \sin t \cdot (\gamma_1 + 2\gamma_2 \cos t + \cdots + 2\gamma_N \cos(N-1)t).$$

Here $\gamma_s = \sum a_j$, and summation runs over the indices $s \leq j \leq N$, the same parity with $s$. This implies

$$\gamma_1 + \gamma_2 = \sum_{j=1}^{N} a_j = 1 \quad \text{and} \quad C(\pi) = \gamma_2 - \gamma_1,$$

so

$$\tilde{\rho}_1 = \sup_{\gamma_1,\ldots,\gamma_N} \{-\gamma_1 + \gamma_2 :\ \gamma_1 + \gamma_2 = 1,\ S(t)/\sin t \geq 0, 0 < t < \pi\}.$$

Since $S(t)/\sin t = \gamma_1 + 2\gamma_2 \cos t + \cdots + 2\gamma_N \cos(N-1)t$ is a nonnegative even trigonometric polynomial, then we can use the previously mentioned Fejér inequality [8] (see also [23, Sect. 6.7, Problem 52] or [5, (4.8)])

$$|\gamma_2| \leq \cos \frac{\pi}{N+1} \cdot |\gamma_1| \tag{11}$$

to obtain

$$\tilde{\rho}_1 \leq \rho_2 = \max_{\gamma_1,\gamma_2} \left\{ -\gamma_1 + \gamma_2 : \gamma_1 + \gamma_2 = 1, \, |\gamma_2| \leq \cos \frac{\pi}{N+1} \cdot |\gamma_1| \right\}. \tag{12}$$

The maximum in $\rho_2$ is achieved for

$$\gamma_1^0 = \frac{1}{1 + \cos \frac{\pi}{N+1}}, \quad \gamma_2^0 = \frac{\cos \frac{\pi}{N+1}}{1 + \cos \frac{\pi}{N+1}} \tag{13}$$

and is equal to

$$\rho_2 = -\frac{1 - \cos \frac{\pi}{N+1}}{1 + \cos \frac{\pi}{N+1}} = -\tan^2 \frac{\pi}{2(N+1)}.$$

Since $\gamma_1^0$ and $\gamma_2^0$ turn the Fejér inequality (11) into equality, there is a (unique) trigonometric polynomial $S^0(t)$ with the coefficients determined by $\gamma_1^0$ and $\gamma_2^0$ such that $S^0(t)/\sin t \geq 0, 0 < t < \pi$. Therefore,

$$\tilde{\rho}_1 \geq C^{(0)}(\pi) = -\tan^2 \frac{\pi}{2(N+1)} = \rho_2,$$

which together with (12) implies

$$\tilde{\rho}_1 = \rho_2 = -\tan^2 \frac{\pi}{2(N+1)}.$$

Now, let

$$a_1^\varepsilon = \frac{a_1^0 + \varepsilon}{1 + \varepsilon}, \quad a_j^\varepsilon = \frac{a_j^0}{1 + \varepsilon}, \; j \geq 2,$$

and

$$S^\varepsilon(t) = \sum_{j=1}^{N} a_j^\varepsilon \sin(jt), \quad C^\varepsilon(t) = \sum_{j=1}^{N} a_j^\varepsilon \cos(jt).$$

It is easy to verify that $a_1^\varepsilon + \cdots + a_N^\varepsilon = 1$ and that

$$S^\varepsilon(t) = \frac{S^0(t)}{1+\varepsilon} + \frac{\varepsilon}{1+\varepsilon} \sin t, \quad C^\varepsilon(t) = \frac{C^0(t)}{1+\varepsilon} + \frac{\varepsilon}{1+\varepsilon} \cos t.$$

Hence, for all $t \in (0, \pi)$ and $\varepsilon > 0$, we have $S^\varepsilon(t) > 0$. Since $S^\varepsilon(\pi) = 0$, we have

$$\rho_1 \geq \min_{0 \leq t \leq \pi} \{C^\varepsilon(t) : S^\varepsilon(t) = 0\} = C^\varepsilon(\pi) = \frac{C^0(\pi)}{1+\varepsilon} - \frac{\varepsilon}{1+\varepsilon}.$$

Therefore,

$$\rho_1 \geq \frac{C^0(\pi)}{1+\varepsilon} - \frac{\varepsilon}{1+\varepsilon},$$

and letting $\varepsilon \to 0^+$, we get

$$\rho_1 \geq C^0(\pi) = -\tan^2 \frac{\pi}{2(N+1)} = \tilde{\rho}_1.$$

This, together with (10), implies

$$\rho_1 = -\tan^2 \frac{\pi}{2(N+1)}.$$

Since $J_N \geq \rho_1$, we get (8).

*Example 1.* Let $f : [0, 1] \to [0, 1]$ be a one-parameter logistic map

$$f(x) = h \cdot x \cdot (1-x), \ 0 \leq h \leq 4.$$

For $h \in (3, 4]$, the equilibrium point $x^* = 1 - \frac{1}{h}$ is unstable and the multiplier $\mu \in [-2, -1)$. Therefore, $\frac{\pi}{2 \cdot \cot^{-1} \sqrt{2}} \approx 2,55$ and the minimal depths of prehistory with the delayed feedback is $N^* = 1$. The optimal strength coefficient $\varepsilon_1^0 = \frac{1}{3}$, and the optimal control is $u = -\frac{1}{3}\left(f(x_n) - f(x_{n-1})\right)$; that is, the closed system $x_{n+1} = f(x_n) + u$ has stable equilibrium points for $h \in (3, 4]$. Figure 1 contains several solutions to the system. All are chaotic. Figure 2 displays the solutions stabilized by the control $u$.

*Example 2.* Now, let's consider a modification of the logistic map $f(x) = 4x^{1.7}(1 - x^{1.7})$. It is clear from Fig. 3 that one-term control $u = 0.3333(f(x_n) - f(x_{n-1}))$ does not stabilize the equilibrium for some initial values, while two-term control $u = 0.5606(f(x_n) - f(x_{n-1})) + 0.1464(f(x_{n-1}) - f(x_{n-2}))$ does (Fig. 4).

Figures 2 and 4 demonstrate how amazingly quickly the stabilization can happen for any initial value. It can be seen as an iterative way to solve the equation $x = f(x)$ with a very general function $f$. The rate of convergence requires more study; however, there is strong numeric evidence that the schedule works faster than Newton's method. Thus, we change the equation, making it nicer while preserving the roots values.

Another interesting observation is that the normalized coefficients $a_j^0/a_1^0 < j$ and $a_j^0/a_1^0 \to j$ as $N \to \infty$. That might indicate a connection between found extremal polynomials and univalent function theory.

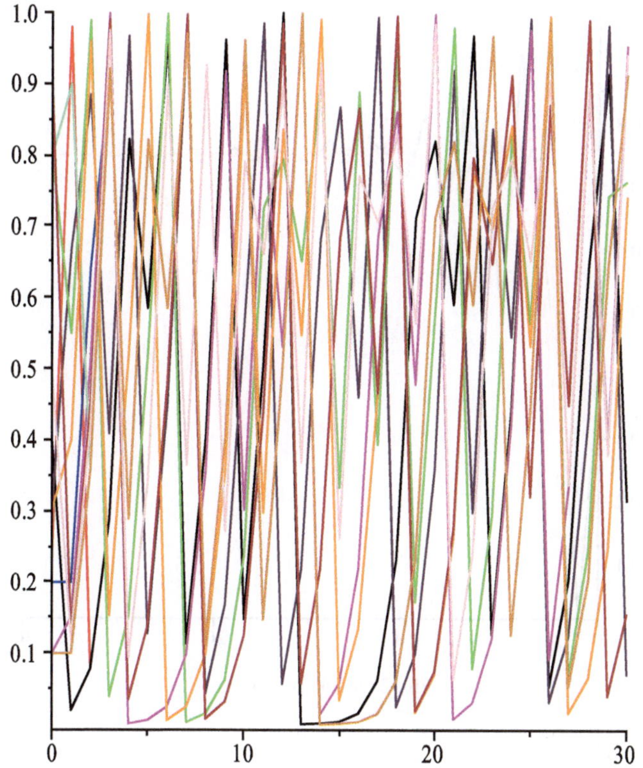


**Fig. 1** Chaotic solutions

## 5 A Cycle

Let's again consider an open scalar nonlinear discrete system (1) with a nonstable cycle $(\eta_1, \eta_2)$, which means that $\eta_1, \eta_2$ are distinct and that $\eta_2 = f(\eta_1)$, while $\eta_1 = f(\eta_2)$.

It is assumed that a differentiable function $f$ depends on a vector of parameters $h$ and that for every admissible set of these parameters the function is defined on a certain bounded interval and maps it into itself. Hence, the cycle $(\eta_1, \eta_2)$, and the cycle multiplier $\mu = f'(\eta_1) \cdot f'(\eta_2)$ will depend on these parameters.

It is assumed that $\mu \in (-\mu^*, -1)$, $\mu^* > 1$, and that for some $\mu \in (-\mu^*, -1)$, the phenomenon of quasidynamical chaos is observed. We do not know the values of the cycle and would like to detect its existence. A natural way to find a cycle is to stabilize the iterative map $f(f(x))$. Figure 5 displays the graph of the iterated standard logistic map, while Fig. 6 plots the solution of the closed-loop system. One can see that the equilibrium points with positive multipliers do not appear; only those with negative multipliers appear, as it should be.

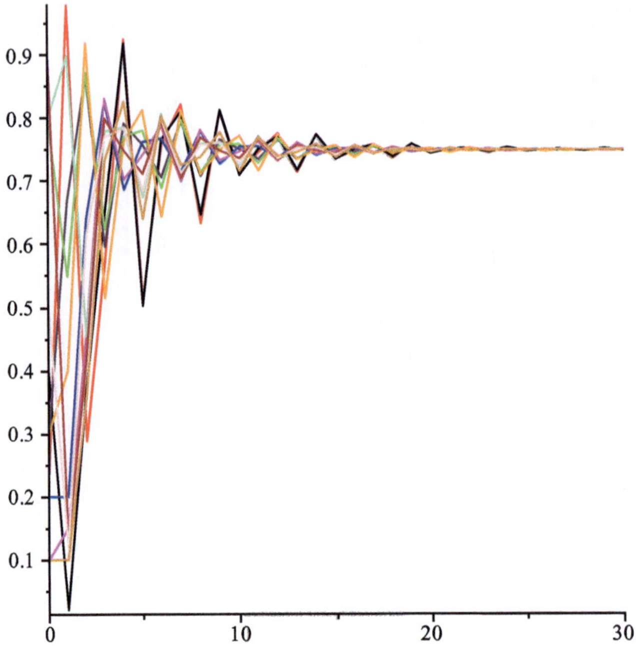


**Fig. 2** Stable solutions

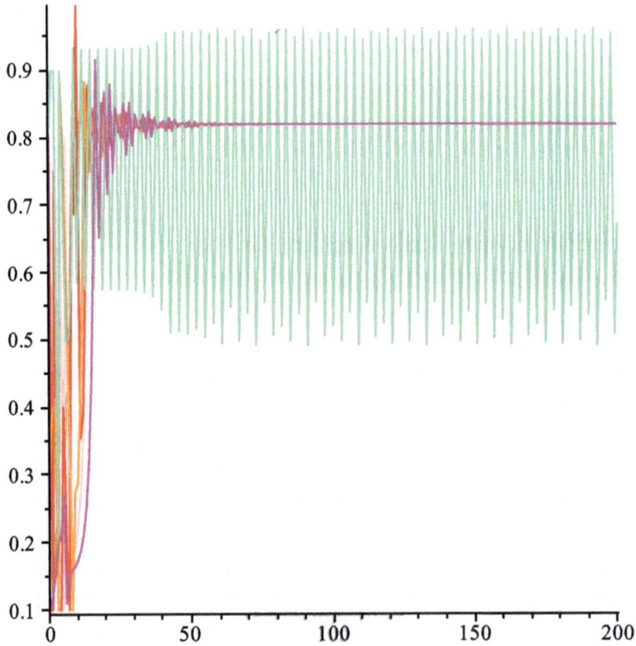


**Fig. 3** One-term control

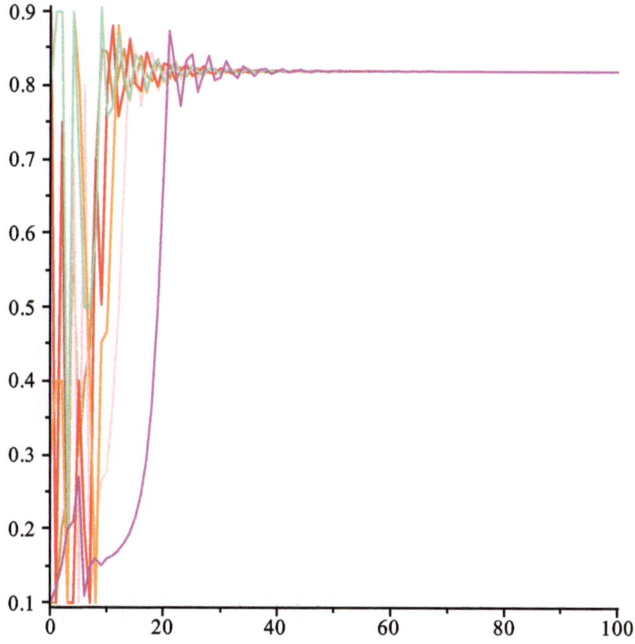


**Fig. 4** Two-term control

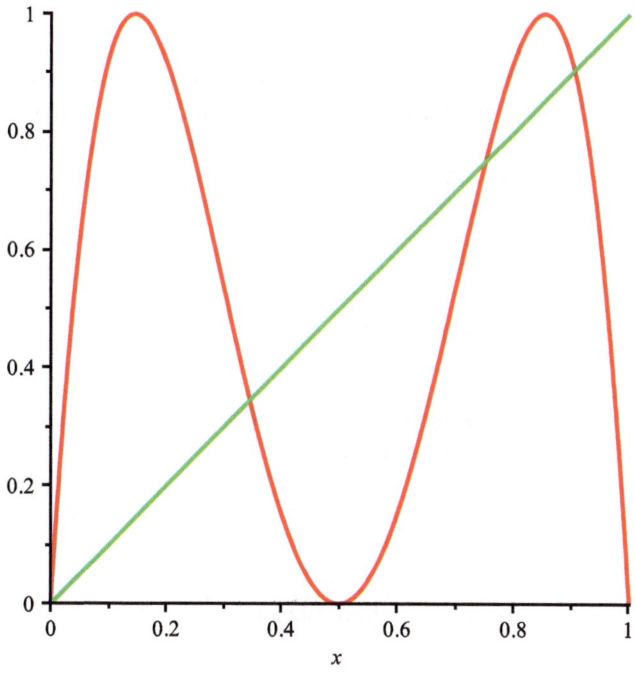


**Fig. 5** 2-Iterated logistic map

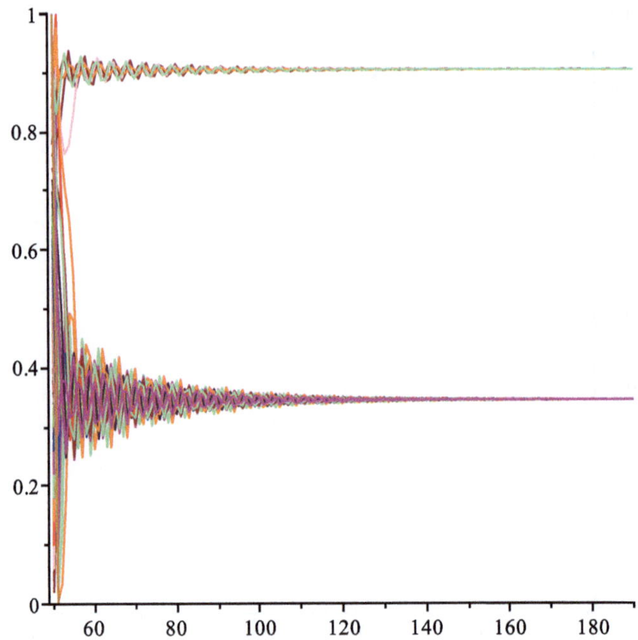


**Fig. 6** Equilibria

To detect the most interesting case of a three-cycle, one can make a triple iterative map $f(f(f(x)))$. In this case, the equilibrium for the original map $x = 0.75$ is detected (see Fig. 7) because the multiplier at this point now is negative (see Fig. 8).

Note that the iterated map contains more and more spikes, which become narrower and steeper. As a result, we see shrinkage of the basin of attraction, and we need to add more and more terms to the control.

To avoid this complication, we suggest a control of the following form:

$$u = -\sum_{j=1}^{N-1} \varepsilon_j \left(f\left(x_{n-2j+2}\right) - f\left(x_{n-2j}\right)\right), |\varepsilon_j| < 1, j = 1, \ldots, N-1. \qquad (14)$$

The standard approach to the investigation of cycle stability leads to verification of Schur stability for the family of the characteristic polynomials

$$\lambda^{2(N-1)+1} + k\,(a_1\lambda^{2(N-1)} + a_2\lambda^{2(N-1)-2} \ldots + a_N), \qquad (15)$$

where $a_1 = 1 - \varepsilon_1, a_j = \varepsilon_{j-1} - \varepsilon_j, j = 2, \ldots, N-1, a_N = \varepsilon_{N-1}$, so $a_1 + \cdots + a_N = 1$ and $k = \pm i\sqrt{|\mu|}$ (here we have taken into account that the cycle multiplier is negative).

For a sufficiently small $|\mu|$, all roots of the characteristic equation (15) are in the unit disk $|z| < 1$. With $|\mu|$ increasing, the roots could appear on the boundary

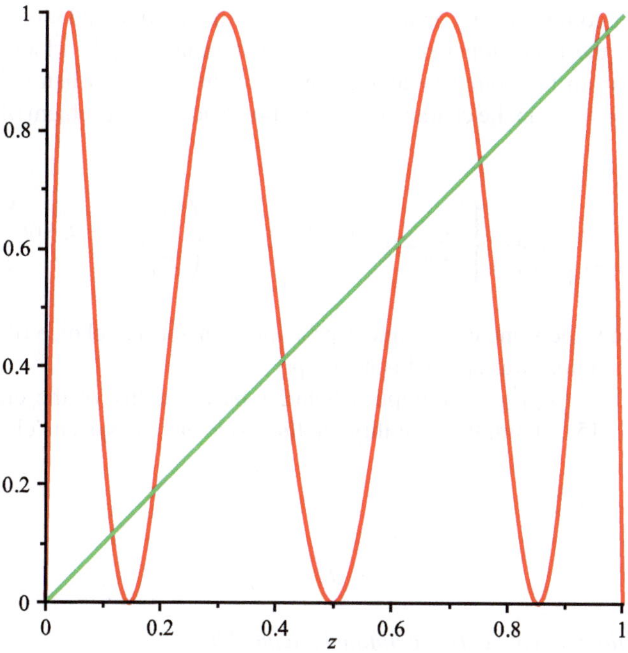


**Fig. 7** 3-Iterated logistic map

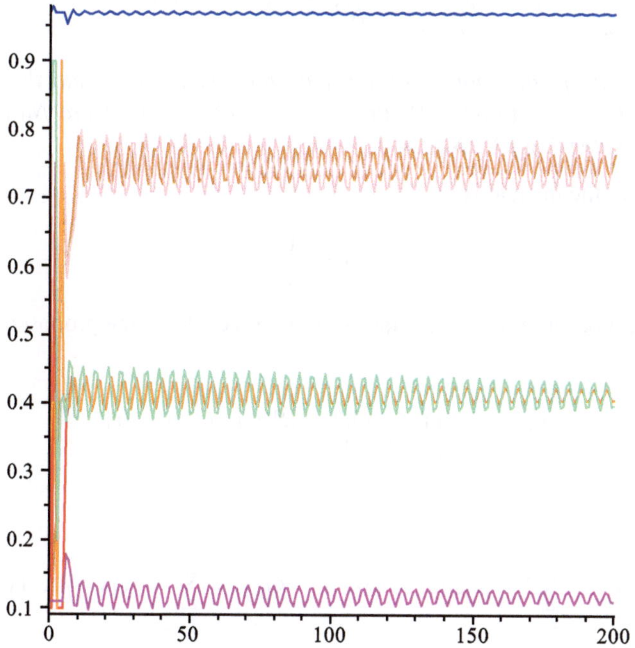


**Fig. 8** Equilibria

$|z| = 1$, then go outside and again came back to the interior, and so on. For chaos stabilization, the exceptional role plays the first value of $|\mu|$, which allows the root to appear on the boundary and move off to the exterior when $|\mu|$ increases. Substitution $\lambda = e^{it}$ in the characteristic equation leads to the problem of evaluating the quantity

$$J_N = \inf_{\sum_{j=1}^N a_j = 1} \left[ \max_{t \in [0, \frac{\pi}{2}]} \left\{ \left| \sum_{j=1}^N a_j e^{-i(2j-1)t} \right| : Arg \left( \sum_{j=1}^N a_j e^{-i(2j-1)t} \right) = \frac{\pi}{2} \right\} \right].$$

It follows from the standard argument principle that for any choice of $a_j$, the set where the maximum was applied is nonempty.

Because $|\mu \cdot (J_N)^2| < 1$ implies Schur *robust* stability of the characteristic polynomials (15), it implies stability of the cycle in the system closed by the DFC (14).

**Theorem 2.** *(i)*

$$J_N = \frac{1}{N}.$$

*(ii) The optimal coefficients are uniquely defined by*

$$\epsilon_j = \sum_{k=j+1}^{N} \frac{2(N-k)+1}{N^2}, \quad j = 1, \ldots, N-1.$$

*(iii) If $\eta_1$ and $\eta_2$ are points of an unstable cycle such that the multiplier $f'(\eta_1)f'(\eta_2) \in (-\mu^*, -1]$, then $N^* = 2(N_0 - 1)$, where $N_0$ is the least integer satisfying $\mu^* \cdot N_0^{-2} < 1$.*

Let's prove the inequality

$$J_N \leq \frac{1}{N}. \tag{16}$$

It is important because it implies stabilization of a cycle by the proposed control $u$.

Note that

$$J_N = \inf_{\sum_{j=1}^N a_j = 1} \max_{t \in [0, \frac{\pi}{2}]} \{ |S(t)| : C(t) = 0 \},$$

where

$$C(t) := \sum_{j=1}^N a_j \cos(2j-1)t, \qquad S(t) := \sum_{j=1}^N a_j \sin(2j-1)t.$$

Since we want to solve the equation $C(t) = 0$, we can try to factor it out. The best we can do with arbitrary $a_j$ is to factor out one cosine, that is, to write

$$C(t) = \cos t \cdot p(t), \quad \text{where} \quad p(t) = \gamma_1 + 2\gamma_2 \cos 2t + .. + 2\gamma_N \cos 2(N-1)t, \tag{17}$$

$$\gamma_s = \sum_{j=s}^{N} (-1)^{s+j} a_j \ , s = 1, \ldots, N. \tag{18}$$

Note that $a_1, \ldots, a_N$ and $\gamma_1, \ldots, , \gamma_N$ are in one-to-one correspondence and it is easy to check that

$$\gamma_1 + 2\sum_{j=2}^{N} \gamma_j = \sum_{j=1}^{n} a_j = 1 \quad \text{and} \quad \left|S(\frac{\pi}{2})\right| = \left|\sum_{j=1}^{N} (-1)^{j+1} \gamma_j\right| = |\gamma_1|. \tag{19}$$

By (17), we need to solve the equation $\cos t \cdot p(t) = 0$. It is unclear how to solve $p(t) = 0$ in its generality unless $p(t)$ is positive and thus the solution set is empty.

So assume that $p(t) > 0, 0 \le t < \pi/2$. Then the only root is $\pi/2$, and we need to minimize the value $|S(\pi/2)| = |\gamma_1|$. Thus, we arrive at the problem of minimizing $|\gamma_1|$ while keeping $p(t)$ positive.

It is intuitively clear that the expected minimum of $|S(\pi/2)|$ should be the same if we allow $p(t)$ to be nonnegative.

The famous Fejér inequality (cf. [23, 7, Problem 50]) implies that $|\gamma_1| \le 1/N$. Moreover, there exists an extremal nonnegative polynomial yielding equality. It is a well-known Fejér kernel of the variable $2t$. After its substitution in (17), we get a polynomial $C^0(t)$ and the conjugate polynomial $S^0(t)$ with some coefficients $a_j^0$ such that $|S(\pi/2)| = 1/N$.

Remarkably enough, $C^0(t)$ can easily be turned into a positive polynomial by a small perturbation. Following the proof of Theorem 1, consider a one-parameter family of trigonometric polynomials

$$C^\varepsilon(t) = \sum_{j=1}^{N} a_j^\varepsilon \cos(2j-1)t, \quad S^\varepsilon(t) = \sum_{j=1}^{N} a_j^\varepsilon \sin(2j-1)t,$$

where

$$a_1^\varepsilon = \frac{a_1^0 + \varepsilon}{1+\varepsilon}, \quad a_j^\varepsilon = \frac{a_j^0}{1+\varepsilon}, \ j = 2, \ \ldots, \ N, \ \varepsilon > 0, \quad \sum_{j=1}^{N} a_j^\varepsilon = 1.$$

Since $C^\varepsilon(t) > 0$ for all $t \in [0, \frac{\pi}{2})$ and for all $\varepsilon > 0$, and since $C^\varepsilon(\pi/2) = 0$, then

$$\inf_{\sum_{j=1}^{N} a_j = 1} \max_{t \in [0, \frac{\pi}{2}]} \{ |S(t)| : C(t) = 0 \} \le \max_{t \in [0, \frac{\pi}{2}]} \{ |S^\varepsilon(t)| : C^\varepsilon(t) = 0 \} = |S^\varepsilon(\frac{\pi}{2})|.$$

Now

$$\left|S^{\varepsilon}(\frac{\pi}{2})\right| \leq \frac{\left|S^{0}(\frac{\pi}{2})\right|}{1+\varepsilon} + \frac{\varepsilon}{1+\varepsilon}.$$

Thus, having in mind (19) and taking into account that $\left|S^{0}(\frac{\pi}{2})\right| < 1$, we get

$$\inf_{\sum_{j=1}^{N} a_j = 1} \max_{t\in[0,\frac{\pi}{2}]} \{\,|S(t)| : C(t) = 0\} \leq \inf_{\varepsilon} |S^{\varepsilon}(\frac{\pi}{2})| = |S^{0}(\frac{\pi}{2})| = \frac{1}{N}.$$

The coefficients $\gamma_j^0$ are derived from Fejér coefficients $(N-j)/N^2$ normalized by (19). The presentation (18) recursively defines $a_j$. This proves (16). The relation $\varepsilon_j = a_{j+1} + \cdots + a_N$, $j = 1, \ldots, N-1$, leads to the formulas in $(ii)$. Then $(iii)$ is straightforward.

This completes the proof.

An important fact has been established: It is possible to stabilize a cycle in a chaos by DFC methods *for all negative multipliers*, and the depth of the used prehistory is of the order square root of the magnitude of a multiplier.

The optimal stabilizing coefficients for an equilibrium are related to the extremal nonnegative Fejér polynomials with the maximal second coefficient compactly written as the kernel $F_N^1$, while the optimal stabilizing coefficients for two-cycle are related to the classical Fejér kernel $F_N^2$:

$$F_N^1 = \frac{2\left(1-\cos\frac{\pi}{N+1}\right)}{N+1}\left(\frac{\cos\frac{(N+1)t}{2}}{\cos t - \cos\frac{\pi}{N+1}}\right)^2, \qquad F_N^2 = \frac{1}{N}\left(\frac{\sin Nt}{\sin t}\right)^2.$$

*Example 3.* For the one-parameter logistic map $f(x) = hx(1-x)$, $1+\sqrt{6} < h \leq 4$, $x \in [0,1]$, the two-cycle

$$\eta_1 = \frac{1+h-\sqrt{h^2-2h-3}}{2h}, \; \eta_2 = \frac{1+h+\sqrt{h^2-2h-3}}{2h}$$

is unstable and the multiplier $\mu \in [-4,-1)$. Therefore, the strength coefficients are $\epsilon_1 = \frac{4}{9}$ and $\epsilon_2 = \frac{1}{9}$. Figure 9 displays the solution to the system (1). Figure 10 demonstrates that each chaotic solution in Fig. 9 is stabilizing to the cycle.

Compare the above nonlinear approach with the most popular OGY method [22]. We should mention that to apply the OGY method, one needs to know the cycle, while our method does not require that.

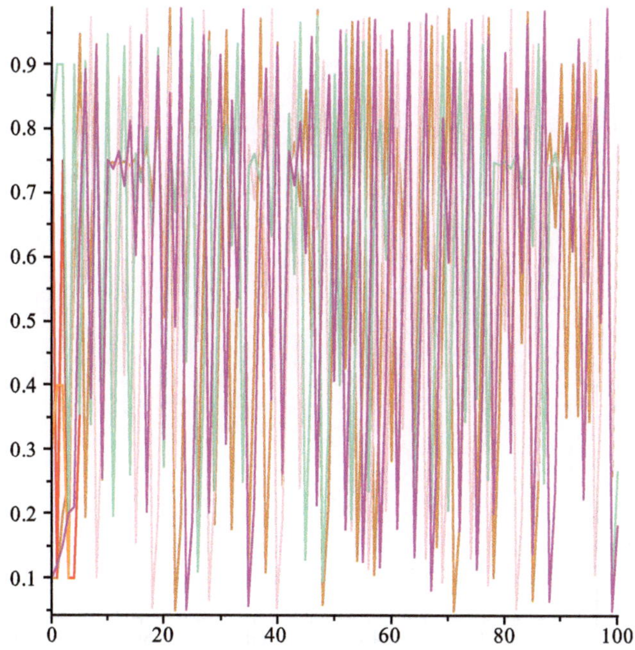


**Fig. 9** Chaotic solutions

# 6 Application to the Modeling and Optimization of Population Dynamics with Age Structure

The development and study of mathematical models for population dynamics with an age structure constitute an important applied problem.

The development of an optimal strategy for population growth is linked to the study of population species structure. The influence of the regularity of population structure on the dynamics and stability of a population has been surveyed in many publications, where several optimal strategies have been suggested.

Assume that population represents one age class and that adjacent generations are not overlapped. Then the following difference equation can be taken as a mathematical model for the dynamics of such a population:

$$x_n = f(x_{n-T}), x_n \in \mathbb{R}, f : \mathbb{R} \to \mathbb{R}. \quad (20)$$

Here $f$ is a given map or sets of maps that are typically finite parametric, and $T$ is a positive integer that characterizes the life cycle of the population.

This situation occurs, for example, for populations with nonoverlapping generations without a long diapause in the life cycle. In particular, it occurs for many kinds of insects. Their adults live for a short period of time that is sufficient for

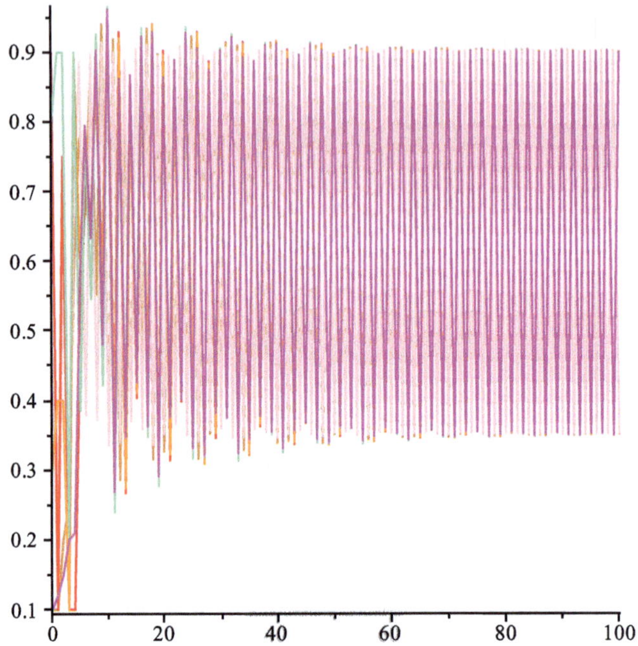


**Fig. 10** Stable solutions

laying eggs, so the previous generation ceases to exist when the new generation is born. The postulate of nonoverlapping generations is accepted implicitly for those populations, while for other populations it can be taken approximately.

Studying multiple-stage populations requires considering the fact that the population is naturally divided into discrete age classes (or stage of development), whose number depends on the number of previous (or all the other) age classes. In this case, the problem of describing the dynamics of the age structure leads to the analysis of a discrete model.

If the population is a set of discrete nonoverlapping age classes, then the population size depends on the number of those classes, which leads to the discrete model with lag

$$x_n = F(x_{n-\tau}, x_{n-2\tau}, \ldots, x_{n-N\tau}),$$

where $N$ is a number of age classes, and $\tau = \frac{T}{N}$ is an integer that characterizes the age range of the age class.

The case $N = 2$ is the most frequently considered. It is characterized by two age classes: immature and sexually mature. In this case, the model of population development corresponds to the scheme shown in Fig.11.

The corresponding model may be written as

$$\begin{cases} X_{n+1} = aY_n \\ Y_{n+1} = s_1(X_n, Y_n)X_n + s_2(X_n, Y_n)Y_n \end{cases}, \tag{21}$$

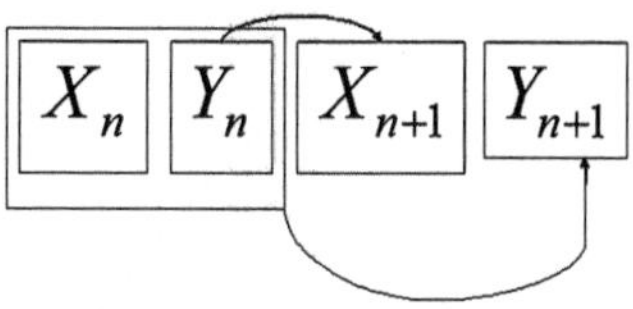


**Fig. 11** Development scheme of two-age population

where $n$ is the number of the breeding season, $X > 0$ and $Y > 0$ are the relative densities of younger and older age classes, respectively, $a > 0$ is the reproductive potential of the population, and $s_1(X_n, Y_n)$ and $s_2(X_n, Y_n)$ are the survival functions of immature and mature specimens, which depend on the density of youth and adults. This function is often considered to be linear (discrete analog of Verhulst's model for population): $s_1(X, Y) = \alpha(1 - X), s_2(X, Y) = \beta(1 - Y)$, where $\alpha > 0, \beta > 0$ are the coefficients that characterize the relative contribution to the population development of immature and mature individuals.

The system (21) in this case is reduced to the second-order difference equation

$$Y_{n+2} = a_1 s(Y_{n+1})Y_{n+1} + a_2 s(Y_n)Y_n,$$

where $s(Y) = (1 - Y)$. The model is generalized the same way for a population with $N$ age classes.

Here we assumed that the function $F$ has the form

$$F(x_{n-\tau}, \ldots, x_{n-N\tau}) = \sum_{j=1}^{N} a_j f(x_{n-j\tau}), \sum_{j=1}^{N} a_j = 1, a_j > 0, j = 1, \ldots, N,$$

where $a_j$ is the value of the $j$th age class contribution to the population growth. Also, it is assumed that the parameters $a_j$ can be controlled. In other words, the possibility exists to choose the age group distribution.

Let $N$ be given. We set the following optimization problem of population dynamics: *Choose the distribution of the age group contributions* $a_j$, $j = 1, \ldots, N$, *to maximize the margin of robustness* $\bar{\mu} = \max\limits_{a_1+\ldots+a_N=1} \{\mu^*\}$, *assuming that* $f'(x) \in (-\mu^*, 0)$, *where* $x$ *is an equilibrium.*

Theorem 1 implies that such an optimal distribution does exist and provides formulas for $a_j$.

To investigate the dependence $a_j^0$ on $j$, assume that $N$ is large enough and make a substitution $x = \frac{j}{N+1}$, $x \in [0, 1]$. Then the continuous version of this distribution is $a^0(x) = K(1 - x)\sin \pi x$. The coefficient $K$ equal to $\pi$ is determined from $\int a^0(x)dx = 1$. Therefore, the optimal population-age distribution (Figs. 12 and 13) density is

$$a^0(x) = \pi(1 - x)\sin \pi x. \tag{22}$$

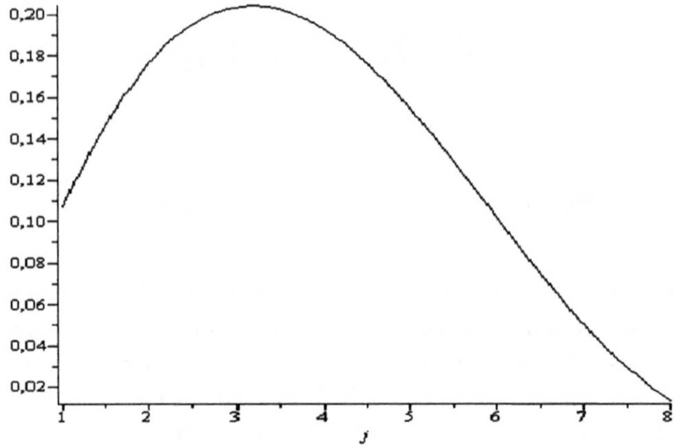


**Fig. 12** Optimal coefficients

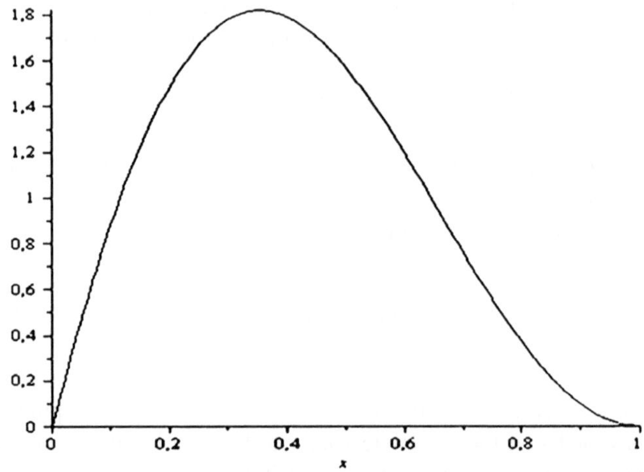


**Fig. 13** Optimal population age density

To find $\max\limits_{x\in[0,1]}\{a^0(x)\}$, apply the derivative $(a^0(x))' = -\pi(x\sin\pi x + \pi(1-x)\cos\pi x)$. Consider the equation $x\sin\pi x + \pi(1-x)\cos\pi x = 0$ and denote the root $x_0 \in (0,1)$. Let $z_0$ be the smallest positive root of the equation $\tan z + z = 0$. Then $x_0 = 1 - \frac{z_0}{\pi} \approx 0.3542, a^0(x_0) \approx 1.82$. Consequently, the maximum value of the function $a_j^0$ will be $a_{j_0}^0$, where $j_0$ is an integer closest to $(N+1)x_0$. It approximately equals $\frac{1.82}{N+1}$. Therefore, *the age $T_0$, which has the greatest influence on the population development, can be determined by the formula*

$$T_0 = 0.3542 \cdot T \cdot \frac{N+1}{N}.$$

Projecting to a human population, assume the average cycle of complete population reproduction to be $T = 80$ and the number of age groups to be $N = 4$. Then the most loaded age is $T_0 = 35 - 36$. Projecting to demographically active regions, assume the extreme case $N = 8$. Then the most important age group should be 32–33 years old.

The obtained $a^0(x)$ is a function of load distribution within one population, provided optimal development of the population and $xT$ is the age at which the load $\frac{1}{N}a^0(x)$ occurs. The function $a^0(x)$ depends neither on parameters $T$ and $N$, nor on the function $f$, so *the optimal distribution of population–age structure* (22) *is a universal law*.

Classical population models, which are described by equations of the form (21), imply that the rate of population growth follows close to a linear formula when the population density grows. In other words, the smaller population density produces a larger rate of growth. So we have $f(x) = hx(1-x)$ for Verhulst's law of population development. The rate of population growth is dependent on population density and is equal to $f'(x) = h(1-2x)$. Since this rate is decreasing, then $f'' < 0$ (for Verhulst's law $f'' = -2h$).

Now, let $f : [0,1] \to [0,1]$ in (1) and let there be three stationary points $\{0, \bar{x}, \bar{\bar{x}}\}$; that is, $f(0) = 0, f(\bar{x}) = \bar{x}, f(\bar{\bar{x}}) = \bar{\bar{x}}$. Furthermore, let $0 < f'(0) < 1, f''(0) > 0, f'(\bar{x}) = \bar{\bar{x}}, f'(\bar{\bar{x}}) < -1$. In this case, the equilibrium $x \equiv 0$ is stable while the other two equilibrium positions are unstable. Consequently, the population dies out regardless of the current density values, which is a phenomenon known as a global *Allee effect* (cf. [21, p.59–60]).

Assume that the population can be partitioned into significant age classes. Let's consider a problem to redistribute the relative contribution of each class in the development of the population so that the population can be preserved under certain current densities.

This problem always has a solution if there are enough significant age classes in the population. In this case, the unstable equilibrium $x \equiv \bar{\bar{x}}$ can be stable (equilibrium position $x \equiv \bar{x}$ is always unstable). A basin of attraction of the point $x \equiv \bar{\bar{x}}$ allows us to determine the upper and lower Allee thresholds. Note that for sufficiently broad assumptions about the function $f$, the values of lower and upper thresholds can be found by the formulas $\text{TL} = \bar{x}$, $f(\text{TU}) = \text{TL}$.

To consider a numerical example, let $f(x) = -\frac{e^{-5(2x-1)^2} - e^{-5}}{-1+e^{-5}}$ (see Fig. 14). The global Allee effect is observable for such a population (see Fig. 15).

Define $\bar{x} \approx 0.2316, \bar{\bar{x}} \approx 0.6469, \text{TL} \approx 0.2316, \text{TU} \approx 0.7684$ and split the population into two age classes, so that its dynamics is described by the equation

$$x_n = \frac{2}{3}f(x_{n-1}) + \frac{1}{3}f(x_{n-2}). \tag{23}$$

In this case, the equilibrium position $x \equiv \bar{\bar{x}}$ is unstable and the population will die out if $x_n \notin (\text{TL}, \text{TU})$, $x_{n+1} \notin (\text{TL}, \text{TU})$. However, the population does not die out for certain values of the current density; that is, if $x_0 \in (\text{TL}, \text{TU})$, then the sequence $\{x_n\}$ is separated from zero (see Fig. 16).

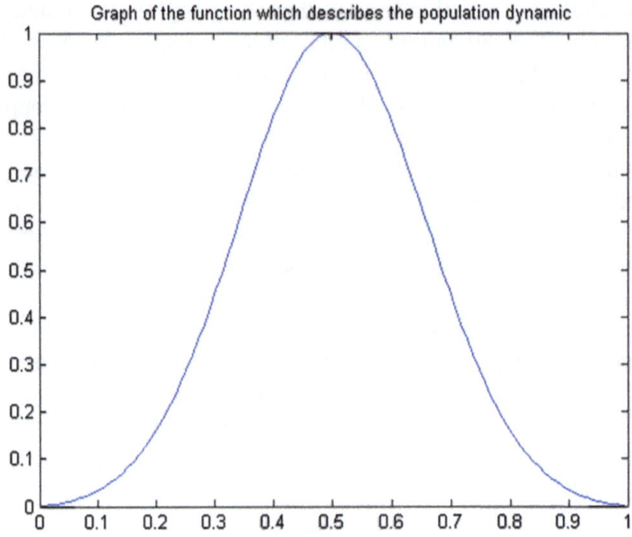


**Fig. 14** Graph of $f(x)$

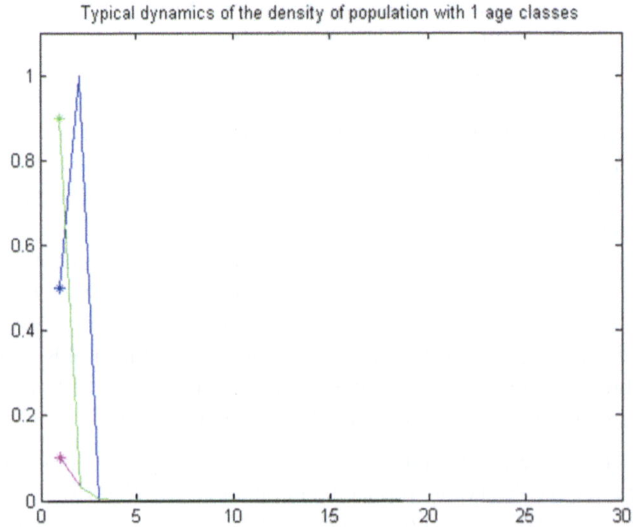


**Fig. 15** Allee effect

The population can be partitioned into three age classes and its dynamics described by the equation

$$x_n = 0.44f(x_{n-1}) + 0.41f(x_{n-2}) + 0.15f(x_{n-3}). \tag{24}$$

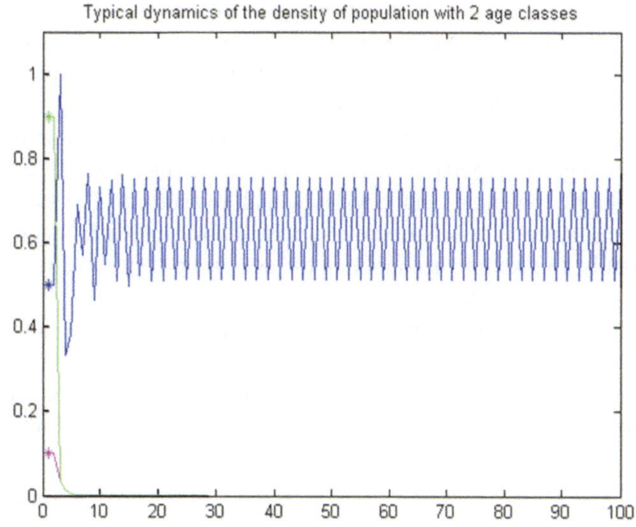


**Fig. 16** Two age classes

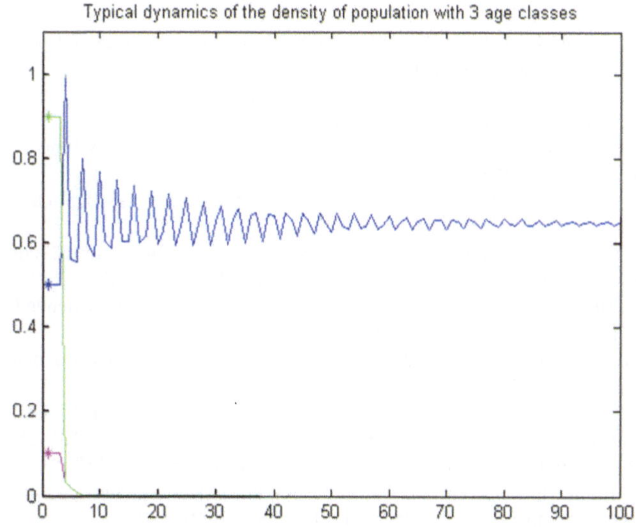


**Fig. 17** Three age classes

In this case, the equilibrium position will be stable and the population will not die out if $x_n \in (\mathrm{TL}, \mathrm{TU})$. Moreover, the population density will approach equilibrium position (homeostasis) (see Fig. 17).

Note that the lower and upper Allee thresholds depend on the function $f$ but not on the number of significant age groups. Consequently, the above control allows us to stabilize the dynamics of a dying population but unfortunately doesn't allow us to change the Allee thresholds.

We conclude that in modeling the population dynamics, it is important to take into account the age structure of the population. Such structures can be determined by the number of age groups and the distribution function of the groups that contribute to the population development.

Each population approaches a dynamic equilibrium position or homeostasis. A dramatic change in population number (density) is the result of an imbalance between its biotic potential and environmental resistance. The easiest way to maintain this balance is to distribute the contribution of each major age group optimally, that is, to take into account that the age group contribution to the relative age density is defined by (22).

Due to the intrapopulation processes, such a distribution law (or close to it) should be retained. If a population is close to extinction and it can be controlled outside, such control can be organized by redistributing the load on each population age class. However, no load transfer will lead to the desired results if the population size is below the critical value.

## References

1. Amann, A., Scholl, E., Just, W.: Some basic remarks on eigenmode expansions of time-delay dynamics. Physica A **373**, 191 (2007)
2. Bielawski, S., Derozier, D., Glorieux, P.: Controlling unstable periodic orbits by a delayed continuous feedback. Phys. Rev. E **49**, R971 (1994)
3. Bleich, M.E., Socolar, J.E.S.: Stability of periodic orbits controlled by time-delay feedback. Phys. Lett. A **210**, 87 (1996)
4. Borwein, J., Bailey, D.: Mathematics by Experiment: Plausible Reasoning in the 21st Century. A K Peters, Wellesley, MA (2003)
5. Dimitrov, D.: Extremal positive trigonometric polynomials. In: Bojoanov, B. (ed.) Approximation Theory: A Volume Dedicated to Blagovest Sendov. DARBA, Sofia (2002)
6. Dahms, T., Hovel, P., Scholl, E.: Control of unstable steady states by extended time-delayed feedback. Phys. Rev. E **76**, 056201 (2007)
7. Dmitrishin, D., Khamitova, A.: Methods of harmonic analysis in nonlinear dynamics. Comptes Rendus Mathematique **351**(9–10), 367–370 (2013)
8. Fejér, L.: Ueber trigonometrische polynome. J. Reine Angew. Math. **146**, 53–82 (1915)
9. Fejér, L.: Sur les fonctions bornées et intégrables. C. R. Acad. Sci. Paris **131**, 984–987 (1900)
10. Fejér, L.: Sur les polynomes trigonométriques. C. R. Acad. Sci. Paris **157**, 571–574 (1913)
11. Fiedler, B., Flunkert, V., Georgi, M., Hövel, P., Schöll, E.: Refuting the odd number limitation of time-delayed feedback control. Phys. Rev. Lett. **98**, 114101 (2007)
12. Fiedler, B., Yanchuk, S., Flunkert, V., Hovel, P., Wunsche, H.J., Scholl, E.: Delay stabilization of rotating waves near fold bifurcation and application toall-optical control of a semiconductor laser. Phys. Rev. E **77**, 066207 (2008)
13. Gauthier, D.J.: Resource letter: Controlling chaos. Am. J. Phys. **71**, 750 (2003)
14. Hovel, P., Socolar, J.E.S.: Stability domains for time-delay feedback control with latency. Phys. Rev. E **68**, 036206 (2003)
15. Hovel, P., Scholl, E.: Control of unstable steady states by time-delayed feedback methods. Phys. Rev. E **72**, 046203 (2005)
16. Just, W., Benner, H., Scholl, E.: Control of chaos by time-delayed feedback: a survey of theoretical and experimental aspects. In: Kramer, B. (ed.), Advances in Solid State Phyics, vol. 43, pp. 589–603. Springer, Berlin (2003)

17. Just, W., Bernard, T., Ostheimer, M., Reibold, E., Benner, H.: Mechanism of time-delayed feedback control. Phys. Rev. Lett. **78**, 203 (1997).
18. Just, W., Fiedler, B., Flunkert, V., Georgi, M., Hovel, A., Scholl, E.: Beyond odd number limitation: a bifurcation analysis of time-delayed feedback control. Phys. Rev. E **76**, 026210 (2007)
19. Li, T.-Y., Yorke, J.A.: Period three implies chaos. Am. Math. Mon. **82**(10), 985–992 (1975)
20. Morgul, Ö.: Further stability results for a generalization of delayed feedback control. Nonlinear Dyn. **70**(2), 1255–1262 (2012)
21. Murray, I.D.: Mathematical Biology: An Introduction, 3rd edn., Springer, New York (2002)
22. Ott, E., Grebogi, C., Yorke, J.A.: Controlling chaos. Phys. Rev. Lett. **64**(11), 1196–1199 (1990)
23. Polya, G., Szegö, G.: Problems and Theorems in Analysis II: Theory of Functions, Zeros, Polynomials, Determinants, Number Theory, Geometry. Springer, Berlin (1998)
24. Polyak, B.T.: Stabilizing chaos with predictive control. Autom. Rem. Control. **66**(11), 1791–1804 (2005)
25. Pyragas, K.: Analytical properties and optimization of time-delayed feedback control. Phys. Rev. E **66**, 26207 (2002)
26. Pyragas, K.: Continuous control of chaos by self-controlling feedback. Phys. Lett. **170A**, 421–428 (1992)
27. Pyragas, K.: Delayed feedback control of chaos. Phil. Trans. R. Soc. A **364**(1846), 2309–2334 (2006)
28. Scholl, E., Hovel, P., Flunkert, V., Dahlem, M.A., Nakajima, H.: On analytical properties of delayed feedback control of chaos. Phys. Lett. A **232**, 207 (1997)
29. Scholl, E., Schuster, H.G. (eds.), Handbook of Chaos Control. Wiley, Weinheim (2008), second completely revised and enlarged edition
30. Sharkovsky, A.M.: Co-existence of cycles of a continuous mapping of a line onto itself. Ukrainian Math. Z. **16**, 61–71 (1964)
31. Socolar, J.E.S., Sukow, D.W., Gauthier, D.J.: Stabilizing unstable periodic orbits in fast dynamical systems. Phys. Rev. E **50**, 3245 (1994)
32. Socolar, J.E.S., Gauthier, D.J.: Analysis and comparison of multiple-delay schemes for controlling unstable fixed points of discrete maps. Phys. Rev. E **57**, 6589 (1998)
33. Ushio, T., Yamamoto, S.: Prediction-based control of chaos. Phys. Lett. A **264**, 30–35 (1999)
34. Ushio, T., Yamamoto, S.: Prediction-based control in chaotic discrete-time systems decision and control, 1999. In: Proceedings of the 38th IEEE Conference on Decision and Control, vol. 2, pp. 2053–2054 (1999)
35. Ushio, T.: Limitation of delayed feedback control in nonlinear discrete-time systems. IEEE Trans. Circ. Syst. **43** 815–816 (1996)
36. Vieira, D.S.M., Lichtenberg, A.J.: Controlling chaos using nonlinear feedback with delay. Phys. Rev. E **54**, 1200–1207 (1996)
37. Yamamoto, S., Hino, T., Ushio, T.: Dynamic delayed feedback controllers for chaotic discrete-time systems. IEEE Trans. Circ. Syst. I **48**(6), 785–789 (2001)
38. Yanchuk, S., Wolfrum, M., Hovel, P., Scholl, E.: Control of unstable steady states by long delay feedback. Phys. Rev. E **74**, 026201 (2006)
39. Zhu, J., Tian, Y.-P.: Necessary and sufficient condition for stabiilizability of discrete-time systems via delayed feedback control. Phys. Lett. A **343**, 95–107 (2005)
40. Zubov, V.I.: Vibrations in non-linear and controlled systems. Gos. Sojuz. Izdat. Sudostroitel. Promyshl. Leningrad, 631 pp, MR0155073 (1962)

# A Note on Widder's Inequality

**Constantine Georgakis**

**Summary.** The purpose of this note is to provide an extension of Widder's inequality, which is a refinement of Hilbert's inequality, from the Hilbert sequence space $l^2$ to the Banach sequence space $l^p$, and to point out that this extension holds for a wider class of generalized Hilbert matrix transformations. These results are based on some of the sharp inequalities concerning the continuity properties for the sequence-to-function Hausdorff transformation, including power series and moment sequences studied by Hardy and Littlewood (J. Math. **157**, 141–158, 1927), which were obtained by the author in Georgakis (Proc. Am. Math. Soc. **103**, 531–542, 1988), and also some sharp inequalities for a class of generalized Hilbert matrix transformations found in Georgakis (An Extension of Hilbert's Inequality, a manuscript).

## 1 Introduction

There are three major well-known forms of Hilbert's celebrated double series theorem for positive sequences. The inequality below is the original version, it is due to Hilbert, and was published in [14].

$$\sum_{n=0}^{\infty}\sum_{m=0}^{\infty}\frac{a_m a_n}{m+n+1} \leq \pi \sum_{m=0}^{\infty} a_m^2 , \tag{1}$$

where the constant is best possible and equality is possible only if the sequence $a = \{a_n : n \geq 0\}$ is null. That is, the quadratic form corresponding to the Hilbert matrix $H(n.m) = 1/(m+n+1)$, for $m \geq 0, n \geq 0$ is bounded on $l^2$ and its norm is equal to $\pi$ or, equivalently, the Hilbert matrix transformation,

C. Georgakis (✉)
Department of Mathematics, DePaul University, Chicago, IL 60614, USA
e-mail: cgeorgak@depaul.edu



A.M. Stokolos et al. (eds.), *Special Functions, Partial Differential Equations, and Harmonic Analysis*, Springer Proceedings in Mathematics & Statistics 108,
DOI 10.1007/978-3-319-10545-1_8

$$(Ha)_n = \sum_{m=0}^{\infty} \frac{a_m}{m+n+1},$$

is bounded on $l^2$ and $||Ha||_2 \leq \pi \, ||Ha||_2$ ,where the constant is best possible and equality is possible only if the sequence $a$ is null. A second form of Hilbert's inequality is the following.

$$\sum_{n=0}^{\infty}\sum_{m=0}^{\infty} \frac{a_m b_n}{m+n+1} \leq \left(\pi \csc \frac{\pi}{p}\right) \left(\sum_{m=0}^{\infty} a_m^p\right)^{1/p} \left(\sum_{n=0}^{\infty} b_n^{p'}\right)^{1/p'}, \tag{2}$$

where the constant $\pi \csc(\pi/p)$ is best possible and equality is possible only if the sequence $a$ or $b$ is null, assuming the indices $p, p'$ are conjugate, i.e., $1/p + 1/p' = 1$, for $1 < p < \infty$. This is known as Hilbert's double series for conjugate indices. That is, the Hilbert matrix transformation $H$ is bounded on $l^p$, for $1 < p < \infty$, and

$$||Ha||_p \leq (\pi \csc \frac{\pi}{p}) \, ||a||_p \, ,$$

where the constant is best possible and equality is possible only if the sequence $a$ is null. There is also a third version of Hilbert's inequality with non-conjugate parameters, namely,

$$\sum_{n=1}^{\infty}\sum_{m=0}^{\infty} \frac{a_m b_n}{(m+n+1)^{\lambda}} \leq (\pi \csc \frac{\pi}{\lambda q'})(\sum_{m=0}^{\infty} a_m^p)^{1/p}(\sum_{n=0}^{\infty} b_n^q)^{1/q}, \tag{3}$$

where $p$, $q$ are non-conjugate indices, i.e., $\lambda = 2 - \frac{1}{p} - \frac{1}{q}, 0 < \lambda \leq 1$, for $1 < p$, $q < \infty$.

Hardy, Littlewood, and Polya in [12, Chapter IX] provide an extensive account of the early research related to Hilbert's double series inequality. Here, we are only interested in (1) and (2). In the case of the Hilbert space $l^2$, Widder in [13] found a second quadratic form, which dominates Hilbert's quadratic form and has identical norm, i.e.,

$$\sum_{n=0}^{\infty}\sum_{m=0}^{\infty} \frac{a_m a_n}{m+n+1} \leq \pi \sum_{n=0}^{\infty}\sum_{m=0}^{\infty} \binom{m+n}{m} \frac{a_m a_n}{2^{m+n+1}} \leq \pi \sum_{m=0}^{\infty} a_m^2, \tag{4}$$

where the constants are best possible and equality is possible only if the sequence $a$ is null. Widder's original proof was based on complex variables. But Hardy in [9] gave a real variable proof and an extension to more general transformations.

## 2 Main Results

In the sequel, it is convenient to use the following notation for the Hilbert and Widder matrix transformations $H$ and $W$, the Borel transform $B$ and the transformation $\Psi$ of the Abel means of a sequence $a = \{a_n; n \geq 0\}, i.e.,$

$$(Ha)_n = \sum_{m=0}^{\infty} \frac{a_m}{m+n+1}, \quad (Wa)_n = \sum_{m=0}^{\infty} \binom{m+n}{m} \frac{a_m}{2^{m+n+1}}, \quad for\ n \geq 0$$

$$(Ba)(y) = \sum_{m=0}^{\infty} a_m e^{-y} \frac{y^m}{m!}, \quad (\Psi a)(x) = (1-x) \sum_{m=0}^{\infty} a_m x^m, \text{for } y \geq 0, 0 \leq x < 1.$$

Thus, the inequality in (1) together with that in (4), and the inequality in (2) can be restated in more general form for real or complex valued sequences as follows:

i) Both $H$ and $W$ are bounded self-adjoint positive definite transformations on $l^2$ with norm equal to $\pi$, they have no extremals, and the quadratic form for $H$ is dominated by that for $W$;
ii) $H$ is a bounded transformation on $l^p$ with norm equal to $\pi \csc(\pi/p)$ and has no extremals, for $1 < p < \infty$.

We need the following results from [1].

**Theorem 2.1.** *(a) For a in $l^p$ and $1 < p < \infty$,*

$$\left[ \int_0^1 |(\Psi a)(x)|^p (1-x)^{-2} \right]^{1/p} \leq \Gamma(\frac{1}{p'}) \, ||a||_p ,$$

*(b)*

$$||Ba||_p \leq ||a||_p \quad for\ 1 \leq p \leq \infty,$$

*Moreover, the constants are best possible and equality is possible only if the sequence a is null.*

The assertion in $(a)$ is simply $(a)$ in Corollary 2 in [1] with $\nu = 1$ and $s = 0$, and the assertion in $(b)$ is Corollary 1 in [1] with $s = 0$.

**Lemma 2.2.** *For a in $l^p$, and $1 < p < \infty$, we have:*

*(a)*

$$\left( \int_0^1 |(\Psi a)(x)|^p (1-x)^{-2} dx \right)^{1/p} = \left( \int_0^{\infty} \left| \int_0^{\infty} (Ba)(yt) e^{-t} dt \right|^p dy \right)^{1/p}$$
$$\leq \Gamma(\frac{1}{p'}) \, ||Ba||_p ,$$

*and equality is possible only if the sequence a is null.*

*For a in* $l^2$, *we have:*

(b) $(Ha,\overline{a}) = \sum\limits_{n=0}^{\infty}\sum\limits_{m=0}^{\infty} \frac{a_m\overline{a_n}}{m+n+1} = \int_0^1 \left|\sum\limits_{m=0}^{\infty} a_m x^m\right|^2 dx,.$

(c) $(Wa,\overline{a}) = \sum\limits_{n=0}^{\infty}\sum\limits_{m=0}^{\infty} \binom{n+m}{m} \frac{a_m\overline{a_n}}{2^{m+n+1}} = ||Ba||_2,$

*and both* $H$ *and* $W$ *are self-adjoint and positive definite.*

*Proof.* Clearly, $(b)$ and $(c)$ are true because

$$\int_0^1 \left|\sum_{m=0}^{\infty} a_m x^m\right|^2 dx = \sum_{m=0}^{\infty} a_m\overline{a_n} \int_0^1 x^{m+n} dx = (Ha,\overline{a}), \text{ and}$$

$$||Ba||_2^2 = \int_0^{\infty} \left|\sum_{m=0}^{\infty} a_m e^{-y} \frac{y^m}{m!}\right|^2 dy = \sum_{n=0}^{\infty}\sum_{m=0}^{\infty} a_m\overline{a_n} \int_0^{\infty} \frac{y^{m+n}}{m!n!} e^{-2y} dy$$

$$= \sum_{n=0}^{\infty}\sum_{m=0}^{\infty} \binom{n+m}{m} \frac{a_m\overline{a}_{nm}}{2^{m+n+1}} = (Wa,\overline{a}).$$

Furthermore,

$$(Ha,\overline{a}) = 0 \Rightarrow \int_0^1 \left|\sum_{m=0}^{\infty} a_m x^m\right|^2 dx = 0 \Rightarrow \sum_{m=0}^{\infty} a_m x^m = 0, \text{ for } x \text{ in } [0,1)$$

$$(Wa,\overline{a}) = 0 \Rightarrow ||Ba||_2^2 = 0 \Rightarrow e^y(Ba)(y) = \sum_{m=0}^{\infty} a_m \frac{y^m}{m!} = 0, \text{for } y \geq 0,$$

and either one implies that the sequence $a$ is null.

$(a)$ The change of variable $x = y/(1+y)$ maps $[0,\infty)$ onto $[0,1)$ so that $dy = (1-x)^{-2}dx$, and

$$(\Phi a)(y) = \int_0^{\infty} (Ba)(yt)e^{-t} dt = \sum_{m=0}^{\infty} a_m e^{-(1+y)t} \frac{(yt)^m}{m!}$$

$$= \frac{1}{1+y}\sum_{m=0}^{\infty} a_m \left(\frac{y}{1+y}\right)^m = (1-x)\sum_{m=0}^{\infty} a_m x^m$$

$$= (\Psi a)(x),$$

and

$$\int_0^{\infty} |(\Psi a)(x)|^p (1-x)^{-2} dx = \int_0^{\infty} |(\Phi a)(y)|^p \; dy.$$

Now, let $g$ be in $L^{p'}[0,\infty)$. Then using Hölder's inequality we get:

$$\begin{aligned}
|\,(\Phi a, g)\,| &\le \int_0^\infty |(\Phi a)(y)g(y)|\,dy \\
&= \int_0^\infty \int_0^\infty \left\{ \left|(Ba)(yt)t^{1/pp'}\right| \right\} \left\{ |g(y)|\,t^{-1/pp'} \right\} dy e^{-t} dt \\
&\le \left[ \int_0^\infty t^{-1/p} e^{-t} dt \int_0^\infty |(Ba)(yt)|^p \, t dy \right]^{1/p} \\
&\times \left[ \int_0^\infty t^{-1/p} e^{-t} dt \int_0^\infty |g(y)|^{p'}\, dy^{1/p'} \right]^{1/p'} \\
&= \left( \int_0^\infty t^{-1/p} e^{-t} dt \right) ||Ba||_p \, ||g||_{p'} = \Gamma(\frac{1}{p'}) \, ||Ba||_p \, ||g||_{p'} ,
\end{aligned}$$

Hence,

$$||\Phi a||_p \le .\Gamma(1/p') \, ||Ba||_p \, .$$

Furthermore, exact equality would imply that $|Ba(yt)|^p \, t = C \, |f(y)|^{p'}$ a.e., for $t > 0,\ y > 0$, and for some constant $C > 0$. Then on fixing $y > 0$, and integrating with respect to $t$ over $[0,\infty)$, we get $\|Ba\|_p = \infty$, implying that $\|a\|_p = \infty$ by $(a)$ in Theorem 2.1, which is a contradiction. This completes the proof of $(a)$ and Lemma 2.2.

The theorem below, the main result of this paper, provides an unexpected extension of Widder's inequality from the Hilbert sequence space $l^2$ to the Banach sequence $l^p$, for $1 < p < \infty$, which is a chain of four sharp $l^p$ inequalities in succession that collapse to a shorter chain of three sharp $l^2$ inequalities when setting $p = 2$, namely, Widder's inequality.

**Theorem 2.3.** *(a) For $a$ in $l^p$ and $1 < p < \infty$, we have*

$$\begin{aligned}
\left[ \sum_{n=0}^\infty \left| \sum_{m=0}^\infty \frac{a_m}{m+n+1} \right|^p \right]^{1/p} &\le \Gamma\left(\frac{1}{p}\right) \left[ \int_0^1 \left| \sum_{m=0}^\infty a_m x^m \right|^p (1-x)^{p-2} dx \right]^{1/p} \\
&\le \pi \csc \frac{\pi}{p} \left[ \int_0^\infty \left| \sum_{m=0}^\infty a_m e^{-y} \frac{y^m}{m!} \right|^p dy \right]^{1/p} \\
&\le \pi \csc \frac{\pi}{p} \, \|a\|_p \, , \ i.e.,
\end{aligned}$$

$$\begin{aligned}
||Ha||_p &\le \Gamma\left(\frac{1}{p}\right) \left[ \int_0^1 |\Psi a|^p \, (1-x)^{-2} dx \right]^{1/p} \\
&\le \pi \csc \frac{\pi}{p} \, \|Ba\|_p\} \le \pi \csc \frac{\pi}{p} \, \|a\|_p \, ,
\end{aligned}$$

*(b) For a in $l^2$, we have:*

$$\sum_{n=0}^{\infty}\sum_{m=0}^{\infty}\frac{a_m\overline{a_n}}{m+n+1}\leq\pi\sum_{n=0}^{\infty}\sum_{m=0}^{\infty}\binom{m+n}{n}\frac{a_m\overline{a}_n}{2^{m+n+1}}\leq\pi\,\|a\|_2\,,\ i.e.,$$

$$(Ha,\overline{a})\leq\pi(Wa,\overline{a})\leq\pi\,\|a\|_2\,.$$

*Moreover, in both (a) and (b) the constants are best possible and the inequalities hold in the strict sense unless the sequence a is null.*

*Proof.* (a) Let $a=\{a_n : n\geq 0\}$, $b=\{b_n : n\geq 0\}$ be sequences in $l^p$ and $l^{p'}$, respectively. Then using Hölder's inequality, $(a)$ in Theorem 2.1 applied to the sequence $b$, and $(a)$ in Lemma 2.2 applied to the sequence $a$, we have:

$$(Ha,b)=\sum_{n=0}^{\infty}\sum_{m=0}^{\infty}\frac{a_m b_n}{m+n+1}=\int_0^1(\Psi a)(x)((\Psi b)(x)(1-x)^{-2}dx,$$

$$|(Ha,b)|\leq\left(\int_0^1|\Psi(a)(x)|^p(1-x)^{-2}dx\right)^{1/p}$$

$$\times\left(\int_0^1|\Psi(b)(x)|^{p'}(1-x)^{-2}dx\right)^{1/p'}$$

$$\leq\Gamma(\frac{1}{p})\,\|b\|_{p'}\left(\int_0^1|(\Psi a)(x)|^p(1-x)^{-2}dx\right)^{1/p}$$

$$=\Gamma(\frac{1}{p})\,\|b\|_{p'}\left(\int_0^{\infty}\left|\int_0^{\infty}(Ba)(yt)e^{-t}dt\right|^p dy\right)^{1/p}$$

$$=\Gamma(\frac{1}{p})\,\|b\|_{p'}\left(\int_0^{\infty}|(\Phi a)(y)|^p dy\right)^{1/p}\leq\Gamma(\frac{1}{p})\Gamma(\frac{1}{p'})\,\|Ba\|_p\,\|b\|_{p'}$$

$$=(\pi\csc\frac{\pi}{p})\,\|Ba\|_p\,.\,\|b\|_{p'}$$

Hence, from the preceding argument and $(b)$ in Theorem 2.1, it follows that

$$\|Ha\|_p\leq\Gamma(\frac{1}{p})\left(\int_0^1|(\Psi a)(x)|^p(1-x)^{-2}dx\right)^{1/p}\leq\Gamma(\frac{1}{p})\Gamma(\frac{1}{p'})\,\|Ba\|_p$$

$$\leq\Gamma(\frac{1}{p})\Gamma(\frac{1}{p'})\,\|a\|_p=\pi\csc\frac{\pi}{p}\,\|a\|_p\ ,$$

Now, we show that the constants are best possible. We claim that $\Gamma(1/p)$ is the best constant between the norms of $Ha$ and $\Psi a$ for all sequences $a$ in $l^p$. Suppose there is a smaller constant $C_1<\Gamma(1/p)$ for this inequality. Then the inequality between the norms of $Ha$ and $a$ would hold for all sequences $a$ in $l^p$ with the constant $C_1\Gamma(1/p)$, which is smaller than $\Gamma(1/p')\Gamma(1/p)$, contradicting the fact that $\Gamma(1/p')\Gamma(1/p)$ is the best constant between the norms of $Ha$ and $a$ by (2). Hence, $\Gamma(1/p)$ is the best constant. Similarly, it follows that $\Gamma(1/p')$ is the best constant between the norms of $\Psi a$ and $Ba$, since $\Gamma(1/p')$ is the best constant for between the norms of $\Psi a$ and $a$ by $(a)$ in Theorem 2.1, and 1 is the best constant between the norms of $Ba$ and $a$ by $(b)$ in Theorem 2.1

Finally, we show that each inequality holds in the strict sense unless the sequence $a$ is null. Clearly, the inequality between the norms of $Ba$ and $a$ holds in the strict sense by $(b)$ in Theorem 2.1, and in the inequality between the norms of $\Psi a$ and $Ba$ by $(a)$ in Lemma 2.2. So it remains to show that this is also true for the inequality between the norms of $Ha$ and $\Psi a$. Suppose there is a sequence $a$ in $l^p$ such that

$$\|Ha\|_p = \Gamma(\frac{1}{p})\left(\int_0^1 |(\Psi a)(x)|^p (1-x)^{-2} dx\right)^{1/p},$$

which is equivalent to the existence of a sequence $b$ in $l^{p'}$ for which

$$|(Ha,b)| = \Gamma(\frac{1}{p})\,\|b\|_{p'}\left(\int_0^1 |(\Psi a)(x)|^p (1-x)^{-2} dx\right)^{1/p}.$$

Since

$$|(Ha,b)| \le \left(\int_0^1 |(\Psi a)(x)|^p (1-x)^{-2} dx\right)^{1/p} \times \left(\int_0^1 |(\Psi b)(x)|^{p'} (1-x)^{-2} dx\right)^{1/p'},$$

and

$$\left(\int_0^1 |(\Psi b)(x)|^{p'} (1-x)^{-2} dx\right)^{1/p'} \le \Gamma(\frac{1}{p})\,\|b\|_{p'}$$

by (a) in Theorem 2.3. It follows that

$$\Gamma(\frac{1}{p})\,\|b\|_{p'} \le \left(\int_0^1 |(\Psi b)(x)|^{p'} (1-x)^{-2} dx\right)^{1/p'},$$

implying that

$$\Gamma(\frac{1}{p})\,\|b\|_{p'} = \left(\int_0^1 |(\Psi b)(x)|^{p'} (1-x)^{-2} dx\right)^{1/p'},$$

which contradicts ($a$) in Theorem 2.1 unless the sequence $b$ is null, and proves ($a$).

(b) The assertion that the constant 1 is best possible for the inequality $(Wa, \overline{a}) \leq \|a\|_2^2$ on the far right follows from the corresponding inequality in ($a$) for the Borel transform $Ba$ when $p = 2$. Next, $\pi$ is the best constant for the inequality $(Ha, \overline{a}) \leq \pi \; \|a\|_2^2$. This follows from ($a$) in Theorem 2.1, because

$$(Ha, \overline{a}) = \int_0^1 \left| \sum_{m=0}^{\infty} a_m x^m \right|^2 dx \leq \pi \; \|a\|_2^2$$

by (a) in Theorem 2.1.

Finally, the assertion about the intermediate inequality $(Ha, \overline{a}) \leq \pi(Wa, \overline{a})$ between Hilbert's and Widder's quadratic forms follows from the fact that 1 is the best constant between $(Wa, \overline{a})$ and $\|a\|_2$, $\pi$ is the best constant between $(Ha, \overline{a})$ and $\|a\|_2$, as was just shown, and then arguing as before above for ($a$).

Next, we examine the possibility of exact equality in the inequalities in ($b$). Since $(Wa, a) = ||Ba||_2^2$, the inequality on the far right in ($a$) holds in the strict sense unless the sequence $a$ is null by ($a$) for $p = 2$. Next, suppose $(Ha, a) = \pi \; ||a||_2^2$, for some non-null sequence $a$ in $l^2$. Then

$$0 < (Ha, \overline{a}) = \int_0^1 \left( \sum_{m=0}^{\infty} a_m x^m \right) \left( \sum_{n=0}^{\infty} \overline{a}_n x^m \right) dx$$

$$= \int_0^1 \left| \sum_{m=0}^{\infty} a_m x^m \right|^2 dx < \pi \; ||a||_2^2,$$

by ($a$) in Theorem 2.1 for $p = 2$ unless $a$ is the null sequence, a contradiction. This completes the proof of the theorem.

We now show how Theorem 2.3 remains valid when the Hilbert matrix is replaced by certain generalized Hilbert matrices whose quadratic forms are also dominated by Widder's quadratic form in the case of $l^2$. This wider class of matrices are closely related to the Hausdorff matrices. They are generated by the Beta distribution on the unit interval and were studied by Hardy in [10] who showed them to be bounded transformations on $l^p$ and determined their operator norm.

Consider the generalized Hilbert matrix with parameters $\alpha > 0$, $\beta > 0$, and the transformation of generalized Abel means $\Psi^{(\alpha)}$ of order $\alpha$ of a sequence $a$ that are given by

$$H_{\alpha\beta}(n, m) = (\alpha + \beta - 1) \frac{\binom{m+\alpha-1}{\alpha-1} \binom{n+\beta-1}{\beta-1}}{\binom{m+n+\alpha+\beta-1}{\alpha+\beta-1}}, \quad for \; m \geq 0, \; n \geq 0.$$

$$(\Psi^{(\alpha)})(x) = (1-x)^{\alpha} \sum_{m=0}^{\infty} \binom{m+\alpha-1}{\alpha-1} a_m x^m, \quad for\ x\ in\ [0,1).$$

Then using the author's results in [1] and [4], the following theorem can be established.

**Theorem 2.4.** *(a) For a in $l^p$, and $1 < p < \infty$, $\alpha > 1/p$, $\beta > 1/p'$, we have:*

$$\left[ \sum_{n=0}^{\infty} \left| \sum_{m=0}^{\infty} H_{\alpha\beta}(n,m) a_m \right|^p \right]^{1/p}$$

$$\le \frac{(\alpha+\beta-1)\Gamma((\beta-1/p'))}{\Gamma(\beta)} \left[ \int_0^1 \left| \left(\Psi^{(\alpha)} a\right)(x) \right|^p (1-x)^{-2} dx \right]^{1/p}$$

$$\le \frac{(\alpha+\beta-1)\Gamma(\alpha-1/p)/p)\Gamma((\beta-1/p'))}{\Gamma(\alpha)\Gamma(\beta)} \left[ \int_0^{\infty} \left| \sum_{n=0}^{\infty} a_n \frac{y^n}{n!} e^{-y} \right|^p y^s dy \right]^{1/p}$$

$$\le \frac{(\alpha+\beta-1)\Gamma(\alpha-1/p)/p)\Gamma((\beta-1/p'))}{\Gamma(\alpha)\Gamma(\beta)} ||a||_p,$$

*where the constants are best possible and the inequalities hold in the strict sense unless the sequence a is null.*

*(b) If a in $l^2$, and $\alpha > 1/2$, then $H_{\alpha\alpha}$ is self-adjoint and positive definite and we have:*

$$(H_{\alpha\alpha} a, \overline{a}) = \sum_{n=0}^{\infty} \sum_{m=0}^{\infty} H_{\alpha\alpha}(n,m) a_m \overline{a}_n \le \left[ \frac{(2\alpha-1)\left(\Gamma(\alpha-1/2)\right)^2}{(\Gamma(\alpha))^2} \right]^2 (Wa, a)$$

$$\le \left[ \frac{(2\alpha-1)\left(\Gamma(\alpha-1/2)\right)^2}{(\Gamma(\alpha))^2} \right]^2 ||a||_2^2.$$

*Moreover, the constants are best possible in the inequalities in (a) and (b) and are valid in the strict sense unless the sequence a is null.*

This result is a special case of an extension of Hilbert's and Widder's inequality that is included in [5], and is based on the results obtained by the author in both [1] and [4]. Note that the generalized Hilbert matrix $H_{\alpha\beta}$ reduces to the Hilbert matrix $H$ when the parameters $\alpha$ and $\beta$ are equal to 1, and Theorem 2.4 reduces to Theorem 2.3.

**Acknowledgements** The author would like to thank the organizers of the conference at Roosevelt University in honor of my friend and colleague Calixto C. Calderon for the invitation to participate in the conference and the inclusion of this paper in the volume of contributed papers.

## References

1. Georgakis, C.: Bounded sequence-to-function hausdorff transformations. Proc. Am. Math. Soc. **103**, 531–542 (1988)
2. Georgakis, C.: The hausdorff mean of a fourier-stieltjes transform. Proc. Am. Math. Soc. **111**, 465–471 (1992)
3. Georgakis, C.: $L^p$-Inequalities for Hausdorff Transformations and Exchangeable Random Variables. Mathematical Epitheorisis, vol. 54, pp. 52–75. Greek Mathematical Society, Greece (2000) a paper based on an invited lecture in 1994 for the Anniversary of the Greek Mathematical Society
4. Georgakis, C.: An Extension of Hilbert's Inequality, a manuscript
5. Georgakis, C.: An Extension of Widder's Inequality, a manuscript
6. Hardy, G.H.: Two Inequalities. J. Lond. Math. Soc. **2**, 196–201 (1917)
7. Hardy, G.H.: On Hilbert's double series theorem, and some connected theorems concerning the convergence of infinite series and integrals. Messenger Math. **48**, 107–112 (1919)
8. Hardy, G.H.: Note on a theorem of Hilbert. Math. Zeitsch **6**, 314–317 (1920)
9. Hardy, G.H.: Remarks in addition to Dr. Widder's note on inequalities. J. Lond. Math. Soc. **4**,199–202 (1929)
10. Hardy, G.H.: An inequality for Hausdorff means. J. Lond. Math. Soc. **18**, 46–50 (1943)
11. Hardy, G.H., Littlewood, J.E.: Elementary theorems concerning power series with positive coefficients and moment constants of positive function. J. Math. **157**, 141–158 (1927)
12. Hardy, G.H., Littlewood, J.E., Polya, G.: Inequalities. Cambridge University Press, Cambridge (1967)
13. Widder, D.V.: An inequality related to one of Hilbert's. J. Lond. Math. Soc. **4**, 194–198 (1929)
14. Weyl, H.: Singuläre Integralgleichungen mit besonder Berücaksichtigung des Fourier-schen Integral theorems. Inaugural - Dissertation, Gottingen, pp. 83–86 (1908)

# Solyanik Estimates in Harmonic Analysis

**Paul Hagelstein and Ioannis Parissis**

**Abstract** Let $\mathcal{B}$ denote a collection of open bounded sets in $\mathbb{R}^n$, and define the associated maximal operator $M_{\mathcal{B}}$ by

$$M_{\mathcal{B}} f(x) := \sup_{x \in R \in \mathcal{B}} \frac{1}{|R|} \int_R |f|.$$

The *sharp Tauberian constant of* $M_{\mathcal{B}}$ *associated with* $\alpha$, denoted by $C_{\mathcal{B}}(\alpha)$, is defined as

$$C_{\mathcal{B}}(\alpha) := \sup_{E: 0<|E|<\infty} \frac{1}{|E|} \left| \{x \in \mathbb{R}^n : M_{\mathcal{B}} \chi_E(x) > \alpha\} \right|.$$

Motivated by previous work of A. A. Solyanik, we show that if $M_{\mathcal{B}}$ is the uncentered Hardy–Littlewood maximal operator associated with balls, the estimate

$$\lim_{\alpha \to 1^-} C_{\mathcal{B}}(\alpha) = 1$$

holds. Similar results for iterated maximal functions are obtained, and open problems in the field of Solyanik estimates are also discussed.

**Keywords** Halo function • Maximal functions • Tauberian conditions

2010 *Mathematics Subject Classification.* Primary 42B25

P. Hagelstein (✉)
Department of Mathematics, Baylor University, Waco, TX 76798, USA
e-mail: paul_hagelstein@baylor.edu

I. Parissis
Department of Mathematics, Aalto University, P. O. Box 11100, FI-00076, Aalto, Finland
e-mail: ioannis.parissis@gmail.com



A.M. Stokolos et al. (eds.), *Special Functions, Partial Differential Equations, and Harmonic Analysis*, Springer Proceedings in Mathematics & Statistics 108,
DOI 10.1007/978-3-319-10545-1_9

# 1 Introduction

Let $\mathcal{B}$ denote a collection of open sets in $\mathbb{R}^n$, and define the associated geometric maximal operator $M_{\mathcal{B}}$ by

$$M_{\mathcal{B}}f(x) := \sup_{x\in R\in\mathcal{B}} \frac{1}{|R|}\int_R |f|.$$

For some examples, if $\mathcal{B}$ were the collection of all cubes or balls in $\mathbb{R}^n$, $M_{\mathcal{B}}$ would be the uncentered Hardy–Littlewood maximal operator; while if $\mathcal{B}$ were the collection of all rectangles in $\mathbb{R}^n$ with sides parallel to the axes, $M_{\mathcal{B}}$ would be the strong maximal operator. Due to the importance of these classical operators we adopt the special notation $M_{\mathrm{HL}}$ for the uncentered Hardy–Littlewood maximal operator and $M_{\mathrm{S}}$ for the strong maximal operator. To avoid confusion, we will sometimes let $M_{\mathrm{HL},b}$ denote the uncentered Hardy–Littlewood maximal operator $M_{\mathrm{HL}}$ with respect to balls, and let $M_{\mathrm{HL},c}$ denote the uncentered Hardy–Littlewood maximal operator with respect to cubes.

We will also consider the *centered* Hardy–Littlewood maximal operator defined as

$$M^{\mathrm{c}}_{\mathrm{HL},b}f(x) := \sup_{r>0} \frac{1}{|B(x,r)|}\int_{B(x,r)} |f|,$$

where $B(x,r)$ denotes the Euclidean ball of radius $r > 0$, centered at $x \in \mathbb{R}^n$. A similar definition gives $M^{\mathrm{c}}_{\mathrm{HL},c}$, defined with respect to centered cubes. Observe that, strictly speaking, these centered operators do not fall under the scope of our general definition for $M_{\mathcal{B}}$ as there is no collection $\mathcal{B}$ that will generate $M^{\mathrm{c}}_{\mathrm{HL},b}$ or $M^{\mathrm{c}}_{\mathrm{HL},c}$. This is essentially due to the centered nature of the sets defining $M^{\mathrm{c}}_{\mathrm{HL}}$.

Given a collection $\mathcal{B}$ as above, we are typically interested in determining if the associated maximal operator $M_{\mathcal{B}}$ is bounded on $L^p(\mathbb{R}^n)$ for some $1 < p < \infty$ and also what are the optimal weak type $(p,p)$ estimates that $M_{\mathcal{B}}$ satisfies. For instance, it is well known that the uncentered Hardy–Littlewood maximal operator $M_{\mathrm{HL}}$ is bounded on $L^p(\mathbb{R}^n)$ for all $1 < p \leq \infty$ and that it satisfies the weak type $(1,1)$ estimate:

$$|\{x \in \mathbb{R}^n : M_{\mathrm{HL}}f(x) > \alpha\}| \leq \frac{3^n}{\alpha}\|f\|_1.$$

Even weaker conditions on geometric maximal operators are so-called *Tauberian conditions*. The maximal operator $M_{\mathcal{B}}$ is said to satisfy a *Tauberian condition with respect to* $\alpha \in (0,1)$ if there is some constant $C$ such that

$$|\{x \in \mathbb{R}^n : M_{\mathcal{B}}\chi_E(x) > \alpha\}| \leq C|E|$$

holds for all measurable sets $E$. Note that the previous condition is only supposed to hold for some *fixed* $\alpha \in (0, 1)$. Now, if $M_{\mathcal{B}}$ is known to satisfy a weak type $(1, 1)$ estimate or to be bounded on $L^p$ for some $1 < p < \infty$, then it is easily seen that $M_{\mathcal{B}}$ must satisfy a Tauberian condition with respect to $\alpha$, for *all* $0 < \alpha < 1$. However, a maximal operator $M_{\mathcal{B}}$ can in fact satisfy a Tauberian condition with respect to some $0 < \alpha < 1$ without being $L^p$ bounded for any finite $p$. A quick example of this type of behavior can be exhibited by, say, letting $\mathcal{B}$ be the collection of all sets of the form $[0, 1] \cup (x, x+2)$ and observing that, while $M_{\mathcal{B}}$ satisfies a Tauberian condition with respect to 4/5, it is not bounded on $L^p(\mathbb{R})$ for any $1 < p < \infty$.

A Tauberian condition on a maximal operator, although quite weak, is still very useful, as was shown by A. Córdoba and R. Fefferman in their work [4] relating the $L^p$ bounds of certain multiplier operators to the weak type $\big((\frac{p}{2})', (\frac{p}{2})'\big)$ bounds of associated geometric maximal operators; see [4] for details. Moreover, Hagelstein and Stokolos have shown in [8] that, provided $\mathcal{B}$ is a homothecy invariant basis of convex sets in $\mathbb{R}^n$, if $\mathcal{B}$ satisfies a Tauberian condition with respect to *some* $0 < \alpha < 1$, then $M_{\mathcal{B}}$ must be bounded on $L^p(\mathbb{R}^n)$ for sufficiently large $p$. This work has recently been extended by Hagelstein, Luque, and Parissis in [7] to yield weighted $L^p$ bounds on maximal operators satisfying a Tauberian condition with respect to a weighted basis.

The issue of *sharp Tauberian constants* is one that has received very little attention until recently. For specificity, given a maximal operator $M_{\mathcal{B}}$, we define the Tauberian constant $C_{\mathcal{B}}(\alpha)$ by

$$C_{\mathcal{B}}(\alpha) := \sup_{E \subset \mathbb{R}^n : 0 < |E| < \infty} \frac{1}{|E|} |\{x \in \mathbb{R}^n : M_{\mathcal{B}}\chi_E(x) > \alpha\}|. \tag{1}$$

We note here that, in the relevant literature, the function $\phi_{\mathcal{B}} : [1, \infty) \to \mathbb{R}$ defined as $\phi_{\mathcal{B}}(\lambda) := C_{\mathcal{B}}(1/\lambda)$, $\lambda > 1$, is many times called the *Halo function* of the collection $\mathcal{B}$, as for example, in [6]. Obviously, it is equivalent to study the function $C_{\mathcal{B}}(\alpha)$ for $\alpha < 1$ which is the setup we adopt in this paper.

We will use the special notation $C_{\mathrm{HL},b}$, $C_{\mathrm{HL},c}$, and $C_{\mathrm{S}}$ for the sharp Tauberian constants corresponding to the basis of balls, cubes, and axes parallel rectangles, respectively. For the centered Hardy–Littlewood maximal operator we denote the corresponding sharp Tauberian constants by $C^{\mathrm{c}}_{\mathrm{HL},b}$ and $C^{\mathrm{c}}_{\mathrm{HL},c}$.

Now, if the maximal operator $M_{\mathcal{B}}$ satisfies a weak type $(1, 1)$ estimate

$$|\{x \in \mathbb{R}^n : M_{\mathcal{B}}f(x) > \alpha\}| \leq \frac{C}{\alpha}\|f\|_1,$$

then the associated sharp Tauberian constant $C_{\mathcal{B}}(\alpha)$ must satisfy

$$C_{\mathcal{B}}(\alpha) \leq \frac{C}{\alpha}.$$

However, we might expect in many situations $C_{\mathcal{B}}(\alpha)$ to be significantly smaller than $C/\alpha$. For example, even though the weak type $(1,1)$ bound of the uncentered Hardy–Littlewood maximal operator $M_{\mathrm{HL}}$ acting on functions on $\mathbb{R}$ is 2, we would suspect it is unlikely to find a set $E$ contained in $\mathbb{R}$ such that $|\{x \in \mathbb{R} : M_{\mathrm{HL}}\chi_E(x) > .99\}| = 2|E|$. We will show momentarily that this indeed cannot be the case, and in fact that we must have

$$\lim_{\alpha\to 1^-} C_{\mathrm{HL}}(\alpha) = \lim_{\alpha\to 1^-} \sup_{E\subset\mathbb{R}^n:0<|E|<\infty} \frac{1}{|E|}\left|\{x \in \mathbb{R}^n : M_{\mathrm{HL}}\chi_E(x) > \alpha\}\right| = 1.$$

The first estimates along the lines of the one above were obtained by A.A. Solyanik in [9]. In his honor, we call a result of the form

$$\lim_{\alpha\to 1^-} C_{\mathcal{B}}(\alpha) = 1$$

a *Solyanik estimate.*

**Theorem 1 (Solyanik).** *We have the following Solyanik estimates:*

$$\lim_{\alpha\to 1^-} C_{\mathrm{HL},c}(\alpha) = 1 \quad \textit{and} \quad \lim_{\alpha\to 1^-} C_{\mathrm{S}}(\alpha) = 1.$$

*In particular,*

$$C_{\mathrm{HL},c}(\alpha) - 1 \sim_n \left(\frac{1}{\alpha} - 1\right)^{\frac{1}{n}} \quad \textit{and} \quad C_{\mathrm{S}}(\alpha) - 1 \sim_n \left(\frac{1}{\alpha} - 1\right)^{\frac{1}{n}}.$$

*For the sharp Tauberian constant of the centered Hardy–Littlewood maximal operator (with respect to cubes or balls) we have*

$$\lim_{\alpha\to 1^-} C^{\mathrm{c}}_{\mathrm{HL},b}(\alpha) = 1 \quad ; \quad \lim_{\alpha\to 1^-} C^{\mathrm{c}}_{\mathrm{HL},c}(\alpha) = 1$$

*and in particular*

$$C^{\mathrm{c}}_{\mathrm{HL},b}(\alpha) - 1 \sim_n \frac{1}{\alpha} - 1 \quad ; \quad C^{\mathrm{c}}_{\mathrm{HL},c}(\alpha) - 1 \sim_n \frac{1}{\alpha} - 1$$

*as* $\alpha \to 1^-$.

*Remark.* Note that Solyanik's theorem does not include an estimate for $C_{\mathrm{HL},b}$ associated with the uncentered Hardy–Littlewood maximal operator with respect to balls, $M_{\mathrm{HL},b}$. Indeed, Solyanik concludes the estimate for $C_{\mathrm{HL},c}$ as a corollary of estimate for $C_S$ and thus the methods in his paper do not readily apply to non-centered maximal operators defined with respect to balls. However, the method of Solyanik for centered maximal operators deals equally well with balls or cubes. This is because the basic underlying ingredient for these estimates in the case of centered operators is the Besicovitch covering lemma which works equally well for balls or cubes.

We now introduce the directional maximal operator $M_j$, $j = 1, \ldots, n$, acting on $\mathbb{R}^n$ and defined by

$$M_j f(x_1, \ldots, x_n) := \sup_{s<x_j<t} \frac{1}{t-s} \int_s^t |f(x_1, \ldots, x_{j-1}, u, x_{j+1}, \ldots, x_n)| \, du.$$

In the next section we will prove a Solyanik estimate for the iterated maximal operator $M_1 \cdots M_n$, namely that we have

$$\lim_{\alpha \to 1^-} \sup_{E \subset \mathbb{R}^n : 0 < |E| < \infty} \frac{1}{|E|} \left| \{x \in \mathbb{R}^n : M_1 \cdots M_n \chi_E(x) > \alpha\} \right| = 1.$$

This will be done by proving a Solyanik estimate for $M_{\mathrm{HL}}$ on $\mathbb{R}^1$ by a means different than Solyanik did in [9] but one enabling us to afterwards apply induction to get the desired estimate for $M_1 \cdots M_n$. Subsequently we will provide a Solyanik estimate for the uncentered maximal operator $M_{\mathrm{HL},b}$ by utilizing the circle of ideas developed by A. Córdoba and R. Fefferman in their work [3] relating covering lemmas to weak type bounds of geometric maximal operators. Afterwards we will visit the issue of generalizing Solyanik estimates to encompass maximal operators $M_{\mathcal{B}}$ where $\mathcal{B}$ is a homothecy invariant collection of convex sets. Throughout this paper we will indicate open problems and directions for further research.

**Notation.** We write $A \lesssim B$ whenever there is a numerical constant $c > 0$ such that $A \leq cB$. We also write $A \sim B$ if $A \lesssim B$ and $B \lesssim A$. If the constant $c$ depends, for example, on the dimension $n$, we will write $A \sim_n B$.

## 2 Solyanik Estimates for Iterated Maximal Functions

The key result in this section is the following lemma.

**Lemma 1.** *Let $M_{\mathrm{HL}}$ denote the uncentered Hardy–Littlewood maximal operator acting on functions on $\mathbb{R}$. Let $E \subset \mathbb{R}$, where $|E| < \infty$, and let $0 \leq \gamma < \alpha < 1$. Then*

$$|\{x \in \mathbb{R} : M_{\mathrm{HL}}(\chi_E + \gamma \chi_{E^c})(x) > \alpha\}| \leq \left(1 + 4\frac{1-\alpha}{\alpha - \gamma}\right)|E|. \tag{2}$$

*Proof.* We first prove the lemma for $\gamma > 0$. Let $f_{E,\gamma}$ be the function defined on $\mathbb{R}$ by

$$f_{E,\gamma}(x) = \chi_E(x) + \gamma \chi_{E^c}(x).$$

Let $\{I_j\}$ be a countable collection of intervals such that

$$\{x \in \mathbb{R} : M_{\mathrm{HL}}(\chi_E + \gamma \chi_{E^c})(x) > \alpha\} = \cup_j I_j$$

and such that, for each $j$,

$$\frac{1}{|I_j|}\int_{I_j} f_{E,\gamma} > \alpha.$$

We now fix some $\epsilon > 0$. As $M_{\mathrm{HL}}$ is of weak type $(1,1)$ we must have that

$$\{x \in \mathbb{R} : M_{\mathrm{HL}}(\chi_E + \gamma\chi_{E^c})(x) > \alpha\}$$

is of finite measure. Indeed, if $M_{\mathrm{HL}}(\chi_E + \gamma\chi_{E^c})(x) > \alpha$ then we must have $M_{\mathrm{HL}}\chi_E(x) > \alpha - \gamma$. Accordingly there exists a finite subcollection $\{I_j'\}$ of $\{I_j\}$ such that

$$|\{x \in \mathbb{R} : M_{\mathrm{HL}}(\chi_E + \gamma\chi_{E^c})(x) > \alpha\} \setminus \cup_j I_j'| < \epsilon.$$

Arguing as in [5, p. 24] we see that there exists a collection of intervals $\{\tilde{I}_j\}_j$ contained in $\{I_j'\}_j$ such that $\cup_j \tilde{I}_j = \cup_j I_j'$ and $\sum_j \chi_{\tilde{I}_j} \leq 2$. Since $\frac{1}{|\tilde{I}_j|}\int_{\tilde{I}_j} f_{E,\gamma} > \alpha$, we have

$$|E \cap \tilde{I}_j| + \gamma|\tilde{I}_j \setminus E| > \alpha|\tilde{I}_j|,$$

implying

$$\frac{|E \cap \tilde{I}_j|}{|\tilde{I}_j|} > \frac{\alpha - \gamma}{1 - \gamma}.$$

So

$$\begin{aligned} \big|\{x \in \mathbb{R} : M_{\mathrm{HL}} f_{E,\gamma}(x) > \alpha\}\big| &\leq |E| + \frac{1-\alpha}{1-\gamma}\sum |\tilde{I}_j| + \epsilon \\ &\leq |E| + 2\frac{1-\alpha}{1-\gamma}|\cup \tilde{I}_j| + \epsilon. \end{aligned}$$

As we have shown that $\frac{1}{|\tilde{I}_j|}\int_{\tilde{I}_j} f_{E,\gamma} > \alpha$ implies

$$\frac{|E \cap \tilde{I}_j|}{|\tilde{I}_j|} > \frac{\alpha - \gamma}{1 - \gamma},$$

we have

$$\cup_j \tilde{I}_j \subset \left\{x \in \mathbb{R} : M_{\mathrm{HL}}\chi_E(x) > \frac{\alpha - \gamma}{1 - \gamma}\right\}.$$

So by the weak type $(1,1)$ bound of 2 of $M_{\mathrm{HL}}$ on $\mathbb{R}^1$, we have

$$|\cup \tilde{I}_j| \leq 2\frac{1-\gamma}{\alpha-\gamma}|E|$$

and accordingly

$$\left|\left\{x \in \mathbb{R} : M_{\mathrm{HL}} f_{E,\gamma}(x) > \alpha\right\}\right| \leq \left(1 + 4\frac{1-\alpha}{\alpha-\gamma}\right)|E| + \epsilon.$$

As $\epsilon > 0$ was arbitrary we obtain the desired result in the case $\gamma > 0$.

Now observe that for any $\alpha, \delta > 0$ we have

$$|\{x \in \mathbb{R} : M_{\mathrm{HL}}(\chi_E) > \alpha\}| \leq |\{x \in \mathbb{R} : M_{\mathrm{HL}} f_{E,\delta} > \alpha\}| \leq \left(1 + 4\frac{1-\alpha}{\alpha-\delta}\right)|E|$$

by the case already proved. Since the left-hand side of the estimate above does not depend on $\delta$ we can let $\delta \to 0^+$ to get the lemma for $\gamma = 0$ as well. □

We now iterate the above estimate to yield a Solyanik estimate for the iterated maximal operator $M_1 \cdots M_n$.

**Lemma 2.** *Setting* $\alpha_0 = 0$ *and* $0 < \alpha_1 < 1$, *define* $\alpha_j$, $j = 2, 3, 4, \ldots, n$ *by*

$$\alpha_j = 1 - (1 - \alpha_1)^j.$$

*Then*

$$\left|\left\{x \in \mathbb{R}^n : M_1 \cdots M_n \chi_E(x) > \alpha_n\right\}\right| \leq \left(1 + 4\frac{1-\alpha_1}{\alpha_1}\right)^n |E|$$

*holds for every measurable set* $E$ *in* $\mathbb{R}^n$.

*Proof.* We proceed by proving

$$|\{x \in \mathbb{R}^n : M_1 \cdots M_N \chi_E(x) > \alpha_N\}| \leq \left(1 + 4\frac{1-\alpha_1}{\alpha_1}\right)^N |E|, \quad N = 1, \ldots, n,$$

by induction on $N$. Note

$$|\{x \in \mathbb{R}^n : M_1 \chi_E(x) > \alpha_1\}| \leq \left(1 + 4\frac{1-\alpha_1}{\alpha_1}\right)|E|$$

holds by Lemma 1, seen by setting $\alpha = \alpha_1$, $\gamma = 0$.

Suppose now

$$|\{x \in \mathbb{R}^n : M_1 \cdots M_j \chi_E(x) > \alpha_j\}| \leq \left(1 + 4\frac{1-\alpha_1}{\alpha_1}\right)^j |E|.$$

Let

$$E_j := \{x \in \mathbb{R}^n : M_1 \cdots M_j \chi_E(x) > \alpha_j\}.$$

Observe that the $\alpha_j$ satisfy

$$\frac{1-\alpha_{j+1}}{\alpha_{j+1}-\alpha_j} = \frac{1-\alpha_j}{\alpha_j - \alpha_{j-1}},$$

implying

$$\frac{1-\alpha_{j+1}}{\alpha_{j+1}-\alpha_j} = \frac{1-\alpha_j}{\alpha_j - \alpha_{j-1}} = \cdots = \frac{1-\alpha_1}{\alpha_1}.$$

Also, for any $j$ we have

$$\begin{aligned} M_{j+1}M_1 \cdots M_j \chi_E(x) &= M_{j+1}(\chi_{E_j} M_1 \cdots M_j \chi_E + \chi_{E_j^c} M_1 \cdots M_j \chi_E)(x) \\ &\leq M_{j+1}(\chi_{E_j} + \alpha_j \chi_{E_j^c})(x). \end{aligned}$$

Hence

$$\begin{aligned} &\left|\{x \in \mathbb{R}^n : M_{j+1}M_1 \cdots M_j \chi_E(x) > \alpha_{j+1}\}\right| \\ &\leq \left|\{x \in \mathbb{R}^n : M_{j+1}(\chi_{E_j} + \alpha_j \chi_{E_j^c})(x) > \alpha_{j+1}\}\right| \\ &\leq \left(1 + 4\frac{1-\alpha_{j+1}}{\alpha_{j+1}-\alpha_j}\right)|E_j| \qquad \text{(by Lemma 1)} \\ &\leq \left(1 + 4\frac{1-\alpha_1}{\alpha_1}\right)\left(1 + 4\frac{1-\alpha_1}{\alpha_1}\right)^j |E| \\ &\leq \left(1 + 4\frac{1-\alpha_1}{\alpha_1}\right)^{j+1} |E|. \end{aligned}$$

Since this holds for every measurable set $E$ in $\mathbb{R}^n$, by symmetry we have

$$\left|\{x \in \mathbb{R}^n : M_1 \cdots M_{j+1} \chi_E(x) > \alpha_{j+1}\}\right| \leq \left(1 + 4\frac{1-\alpha_1}{\alpha_1}\right)^{j+1} |E|.$$

Setting $j = n - 1$ yields the desired result. □

**Theorem 2.** *Let* $0 < \alpha < 1$. *Then*

$$\left|\{x \in \mathbb{R}^n : M_1 \cdots M_n \chi_E(x) > \alpha\}\right| \leq \left(1 + 4\frac{(1-\alpha)^{1/n}}{1-(1-\alpha)^{1/n}}\right)^n |E|.$$

*Accordingly, letting* $C_{1\cdots n}(\alpha)$ *denote the sharp Tauberian constant with respect to* $\alpha$ *of* $M_1 \cdots M_n$, *we have*

$$C_{1\cdots n}(\alpha) - 1 \sim_n \left(\frac{1}{\alpha} - 1\right)^{1/n}.$$

*Proof.* Using the notation of the previous lemma, we let $\alpha_n = \alpha$. The corresponding $\alpha_1$ satisfies

$$\alpha = 1 - (1-\alpha_1)^n,$$

implying that

$$\alpha_1 = 1 - (1-\alpha)^{1/n}.$$

The result follows by Lemma 2. □

## 3 Solyanik Estimates for the Uncentered Hardy–Littlewood maximal operator

The primary goal in this section is to provide a Solyanik estimate for the uncentered Hardy–Littlewood maximal operator $M_{\mathrm{HL},b}$.

**Theorem 3.** *Let* $M_{\mathrm{HL},b}$ *denote the non-centered Hardy–Littlewood maximal operator, defined with respect to balls in* $\mathbb{R}^n$. *Then we have the corresponding Solyanik estimate*

$$\lim_{\alpha \to 1^-} C_{\mathrm{HL},b}(\alpha) = 1 .$$

*In particular we have that*

$$C_{\mathrm{HL},b}(\alpha) - 1 \lesssim_n \left(\frac{1}{\alpha} - 1\right)^{\frac{1}{n+1}}$$

*as* $\alpha \to 1^-$.

*Proof.* Let $0 < \alpha < 1$, and let $E$ be a set of finite measure in $\mathbb{R}^n$. Let $\{B_j\}$ be a collection of balls such that

$$\{x \in \mathbb{R}^n : M_{\mathrm{HL},b}\chi_E(x) > \alpha\} = \cup_j B_j,$$

where every $B_j$ satisfies

$$\frac{1}{|B_j|}\int_{B_j}\chi_E > \alpha.$$

Without loss of generality we may assume that $\{B_j\}_j$ is a finite collection $\{B_j\}_{j=1}^N$ as our estimates of $|\cup B_j|$ will be independent of $N$. For later convenience we also assume without loss of generality that no two of these balls have the same size. We reorder the balls $B_j$ so that they are nonincreasing in size, i.e.

$$|B_1| \geq |B_2| \geq \cdots \geq |B_N|.$$

We will now obtain a subcollection $\{\tilde{B}_j\}_j$ using a selection algorithm motivated by ideas of A. Córdoba and R. Fefferman in [3]. Let $1 > \delta > 0$; here we think of $\delta$ as being very close to 0. We choose $\tilde{B}_1 = B_1$. Assume $\tilde{B}_1, \ldots, \tilde{B}_k$ have been selected and suppose that $\tilde{B}_k = B_M$ for some positive integer $M < N$. We let $\tilde{B}_{k+1}$ be the first $B_j$ on the list $B_{M+1}, B_{M+2}, \ldots, B_N$ such that

$$|B_j \cap (\cup_{i=1}^k \tilde{B}_i)| \leq (1-\delta)|B_j|.$$

If such a $B_j$ does not exist, the list of selected balls terminates with $\tilde{B}_k$.

Let now $x \in \{x \in \mathbb{R}^n : M_{\mathrm{HL},b}\chi_E(x) > \alpha\}$ so $x$ necessarily lies in one of the balls $B_j$. Suppose for the moment that $B_j$ is *not* one of the selected balls. Let $B_x$ be a ball of volume $\delta|B_j|$ containing $x$ and contained in $B_j$. Since $B_j$ was not selected, $B_x$ must intersect a $\tilde{B}_k$ of size larger than that of $B_j$. As the radius of $B_x$ is less than $\delta^{1/n}$ times the radius of $\tilde{B}_k$, by the triangle inequality we have $x \in (1+2\delta^{1/n})\tilde{B}_k$, where for a ball $B$ in $\mathbb{R}^n$ we let $cB$ denote the $c$-fold concentric dilate of $B$. So

$$\{x \in \mathbb{R}^n : M_{\mathrm{HL},b}\chi_E(x) > \alpha\} \subset \cup_j (1+2\delta^{1/n})\tilde{B}_j.$$

Let now

$$\tilde{E}_j := \tilde{B}_j \setminus \cup_{i=1}^{j-1} \tilde{B}_i.$$

We have that

$$|\{x \in \mathbb{R}^n : M_{\mathrm{HL},b}\chi_E(x) > \alpha\}| \leq \sum_j (1+2\delta^{1/n})^n |\tilde{E}_j|.$$

Since for each $j$ we have $\frac{1}{|\tilde{B}_j|}\int_{\tilde{B}_j}\chi_E > \alpha$ and moreover $|\tilde{E}_j|/|\tilde{B}_j| > \delta$, we conclude

$$\begin{aligned}\frac{1}{|\tilde{E}_j|}\int_{\tilde{E}_j}\chi_E &\geq [\alpha|\tilde{B}_j| - (|\tilde{B}_j| - |\tilde{E}_j|)]/|\tilde{E}_j| \\ &\geq 1-(1-\alpha)|\tilde{B}_j|/|\tilde{E}_j| \geq 1-(1-\alpha)\delta^{-1} \\ &\geq [\delta-(1-\alpha)]/\delta.\end{aligned}$$

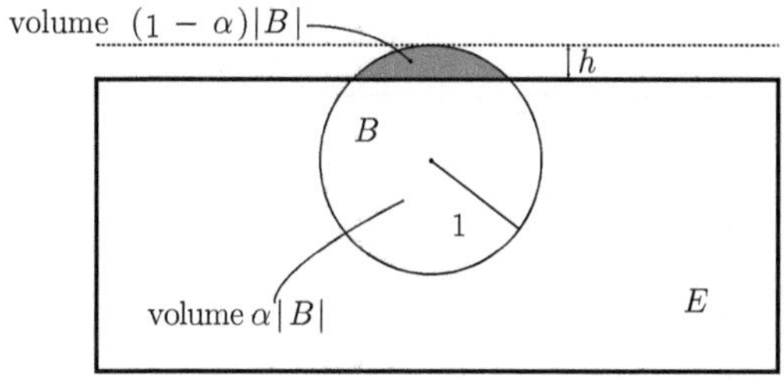


**Fig. 1** A ball $B$ intersecting the slab $E$

Placing an additional restriction on $\delta$ by requiring that $1 > \delta > 1 - \alpha$, we have

$$|\tilde{E}_j| < \frac{\delta}{\delta - (1-\alpha)} |E \cap \tilde{E}_j|.$$

As the $\tilde{E}_j$ are disjoint, we then have

$$|\{x \in \mathbb{R}^n : M_{\mathrm{HL},b}\chi_E(x) > \alpha\}| \leq (1 + 2\delta^{1/n})^n \frac{\delta}{\delta - (1-\alpha)} |E|.$$

Setting $\delta = (1-\alpha)^{\frac{n}{n+1}}$ then yields the desired estimate. □

We strongly suspect the bound $(\frac{1}{\alpha} - 1)^{1/(n+1)}$ is not sharp, as indicated by the following example.

*Example 1.* Let $E$ be the $n$-dimensional rectangle

$$E := [-100, 100] \times \cdots \times [-100, 100] \times [-1, 1].$$

Consider a ball $B$ of radius 1 intersecting the rectangle $E$ on one of its long sides and away from its corners, so that a $(1-\alpha)$ portion of $|B|$ lies outside $E$ (Fig. 1). One can calculate that the union of all such balls constitutes a region of measure approximately $(1+h)|E|$ with $h \simeq_n (\frac{1}{\alpha} - 1)^{\frac{2}{n+1}}$. We conclude

$$C_{\mathrm{HL,b}}(\alpha) - 1 \gtrsim_n \left(\frac{1}{\alpha} - 1\right)^{\frac{2}{n+1}}.$$

In contrast, by doing a similar calculation with a unit cube $Q$ meeting the set $E$ at an angle $\pi/4$ we get $h \simeq_n (\frac{1}{\alpha} - 1)^{\frac{1}{n}}$ (Fig. 2). This proves the lower bound

$$C_{\mathrm{HL},c}(\alpha) - 1 \gtrsim_n \left(\frac{1}{\alpha} - 1\right)^{\frac{1}{n}}.$$

Observe that the latter calculation indicates that the Solyanik estimate for iterated maximal functions provided by Theorem 2 is sharp. Moreover, the fact that the slab example provides a *better* Solyanik estimate for $M_{\mathrm{HL},b}$ inclines us to believe that Theorem 3 is not sharp, and a more refined argument might prove the following:

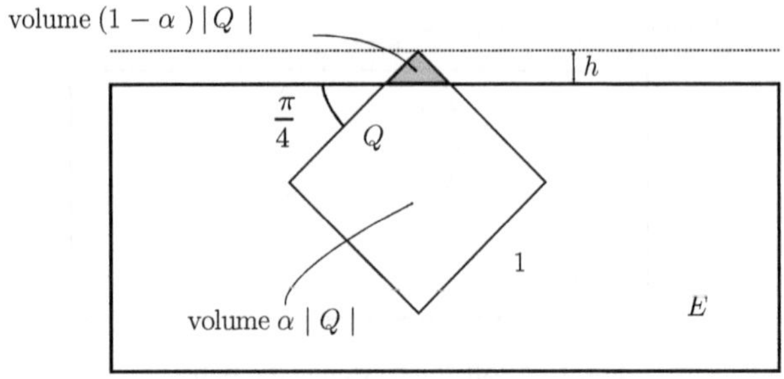


**Fig. 2** A cube $Q$ intersecting the slab $E$

**Conjecture 1.** (a) We have the asymptotic estimate

$$C_{\mathrm{HL},b}(\alpha)-1\sim_n\left(\frac{1}{\alpha}-1\right)^{\frac{1}{n}}$$

as $\alpha\to 1^-$. The exponent here is a natural one to consider, as $\left(\frac{1}{\alpha}-1\right)^{\frac{1}{n}}$ is the sharp Solyanik exponent associated with $M_{\mathrm{HL},c}$ and $M_1\cdots M_n$.

(b) A stronger asymptotic estimate, motivated by Example 1 above, would be that

$$C_{\mathrm{HL},b}(\alpha)-1\sim_n\left(\frac{1}{\alpha}-1\right)^{\frac{2}{n+1}}$$

as $\alpha\to 1^-$.

## 4 Solyanik Estimates for Homothecy Invariant Bases of Convex Sets

With the Solyanik estimates associated with Theorems 1–3 in hand, it is natural to try to extend these types of results to encompass maximal operators such as the maximal operator with respect to rectangles along lacunary directions. Rather than focus our attention on a particular maximal operator, we will here consider the following more general problem:

**Problem.** Let $\mathcal{B}$ denote a collection of open bounded sets in $\mathbb{R}^n$ and $M_{\mathcal{B}}$ the associated geometric maximal operator. Define the associated Tauberian constants $C_{\mathcal{B}}(\alpha)$ by

$$C_{\mathcal{B}}(\alpha):=\sup_{E:0<|E|<\infty}\frac{1}{|E|}|\{x\in\mathbb{R}^n: M_{\mathcal{B}}\chi_E(x)>\alpha\}|.$$

For which $\mathcal{B}$ do we have

$$\lim_{\alpha \to 1^-} C_{\mathcal{B}}(\alpha) = 1?$$

We would expect that the maximal operator $M_{\mathcal{B}}$ should be somewhat well-behaved in order to have $\lim_{\alpha\to 1^-} C_{\mathcal{B}}(\alpha) = 1$, as such an estimate would not hold if $\mathcal{B}$ were, say, the collection of all rectangles in $\mathbb{R}^2$. However, simple $L^p$ boundedness of $M_{\mathcal{B}}$ or even weak type (1,1) bound on $M_{\mathcal{B}}$ is not enough to guarantee that $M_{\mathcal{B}}$ satisfies a Solyanik estimate, as is indicated by the following example of Beznosova and Hagelstein found in [1].

*Example 2.* Let $\mathcal{B}$ consist of all the homothecies of sets in $\mathbb{R}$ in the collection

$$\{((0,1) \cup (x, x+\epsilon)) \cap (0,2) : x \in (0,2),\ \epsilon > 0\}.$$

The operator $M_{\mathcal{B}}$ is dominated by twice the Hardy–Littlewood maximal operator and hence is bounded on $L^p(\mathbb{R})$ for $1 < p \leq \infty$ and is of weak type $(1,1)$. Observe however that $M_{\mathcal{B}}\chi_{(0,1)} = 1$ on $(0,2)$ and hence we have that $\lim_{\alpha\to 1^-} C_{\mathcal{B}}(\alpha) \geq 2$.

Note that the sets in the collection $\mathcal{B}$ above are not all *convex*. We have previously seen convexity plays an important role in problems involving Tauberian conditions, examples including the previously mentioned work of Hagelstein and Stokolos [8] and Hagelstein, Luque, and Parissis [7]. This naturally leads us to the following conjecture involving convex density bases. (Recall that a *density basis* $\mathcal{B}$ in $\mathbb{R}^n$ is a collection of sets for which

$$\lim_{\substack{x \in R \in B \\ \operatorname{diam}(R)\to 0}} \frac{1}{|R|}\int_R \chi_E = \chi_E(x)$$

holds for a.e. $x \in \mathbb{R}^n$, for every set $E \subset \mathbb{R}^n$ of finite measure. An important result of Busemann and Feller is that the maximal operator $M_{\mathcal{B}}$ associated with a homothecy invariant density basis $\mathcal{B}$ satisfies a Tauberian condition with respect to $\alpha$ for every $\alpha > 0$. See [2, 6] for details.)

**Conjecture 2.** Let $\mathcal{B}$ be a homothecy invariant density basis of bounded convex sets in $\mathbb{R}^n$. Then the associated Tauberian constants $C_{\mathcal{B}}(\alpha)$ satisfy

$$\lim_{\alpha \to 1^-} C_{\mathcal{B}}(\alpha) = 1.$$

The following theorem provides some evidence that the above conjecture is on the right track.

**Theorem 4.** *Let $\mathcal{B}$ be a homothecy invariant density basis of convex sets in $\mathbb{R}^n$. Then*

$$\left|\{x \in \mathbb{R}^n : M_{\mathcal{B}}\chi_E(x) = 1\}\right| = |E|$$

*holds for every measurable set $E$ in $\mathbb{R}^n$.*

To appreciate the role that convexity plays in the following argument, observe that the conclusion of this theorem does *not* hold when $\mathcal{B}$ is the homothecy invariant collection of sets indicated in Example 2 above. □

*Proof.* Let us fix some measurable set $E \subset R^n$ with $|E| > 0$. Since $\mathcal{B}$ is a density basis, for a.e. $x \in \mathbb{R}^n$ we have that

$$\lim_{j\to\infty} \frac{1}{|R_{x,j}|}\int_{R_{x,j}} \chi_E = \chi_E(x),$$

where $R_{x,j}$ is any sequence of sets in $\mathcal{B}$ containing $x$ whose diameters tend to 0; for this and other basic properties of density bases, see [6, Ch. III]. So

$$E \subset \{x \in \mathbb{R}^n : M_{\mathcal{B}}\chi_E(x) = 1\} \quad \text{a.e.}$$

and in particular

$$|E| \leq |\{x \in \mathbb{R}^n : M_{\mathcal{B}}\chi_E(x) = 1\}|.$$

Thus it suffices to show that

$$|E| \geq |x \in \mathbb{R}^n : M_{\mathcal{B}}\chi_E(x) = 1|. \tag{3}$$

If $|E| = \infty$ the theorem automatically holds so we may assume without loss of generality that $|E| < \infty$.

The rest of the proof is by way of contradiction and the argument is divided into two basic steps.

*Step 1:* Suppose that (4) fails. Then there exists a set $A \subset E^c$ with $|A| > 0$ such that, for every $x \in A$ there exists a sequence of sets $\{R_{x,j}\}_j \subset \mathcal{B}$ satisfying $x \in R_{x,j}$ for all $j$, $\lim_{j\to+\infty} \operatorname{diam}(R_{x,j}) = +\infty$ and

$$\frac{1}{|R_{x,j}|}\int_{R_{x,j}} \chi_E > 1 - \frac{1}{j}, \quad j = 2, 3, \ldots. \tag{4}$$

We now prove this claim. Assuming that (4) fails and letting

$$\mathcal{H}_E := \{x \in \mathbb{R}^n : M_{\mathcal{B}}\chi_E(x) = 1\} \setminus E$$

we have that $|\mathcal{H}_E| > 0$. Now let $A$ denote the set

$$A = \mathcal{H}_E \cap \left\{x \in E^c : \lim_{\substack{x\in R\in\mathcal{B}\\ \operatorname{diam}(R)\to 0}} \frac{1}{|R|}\int_R \chi_E = 0\right\}.$$

Since $\mathcal{B}$ is a density basis we have that $|A| = |\mathcal{H}_E| > 0$. We fix $x \in A$. Since $x \in \mathcal{H}_E$ we conclude that for every positive integer $j \geq 2$ there exists a sequence

$\{R_{x,j}\}_j \subset \mathcal{B}$, $x \in R_{x,j}$ for each $j$ and (4) holds. It remains to show that $\lim_{j\to+\infty} \operatorname{diam}(R_{x,j}) = +\infty$. By the definition of $A$ there exists $\delta = \delta_x > 0$ such that

$$x \in R \in \mathcal{B},\ \operatorname{diam}(R) < \delta \Rightarrow \frac{1}{|R|}\int_R \chi_E < \frac{1}{2}.$$

Furthermore, it is clear that $\inf_j \operatorname{diam}(R_{x,j}) \geq c > 0$ otherwise the averages in (4) would have a subsequence converging to 0. The previous discussion and the convexity hypothesis for the collection $\mathcal{B}$ imply that there exists a homothetic copy $S_{R_j}$ of $R_{x,j}$ with $\operatorname{diam}(S_{R_j}) = \frac{1}{2}\min(c,\delta)$ that satisfies

$$x \in S_{R_j} \subset R_{x,j} \quad \text{and} \quad \frac{|E \cap S_{R_j}|}{|S_{R_j}|} < \frac{1}{2}.$$

It is essential to notice here that the diameter of $S_{R_j}$ is independent of $j$. We have

$$\begin{aligned}
1 - \frac{1}{j} \leq \frac{|E \cap R_{x,j}|}{|R_{x,j}|} &= \frac{|E \cap S_{R_j}|}{|R_{x,j}|} + \frac{|E \cap R_{x,j} \setminus S_{R_j}|}{|R_{x,j}|} \\
&\leq \frac{|E \cap S_{R_j}|}{|R_{x,j}|} + \frac{|R_{x,j}| - |S_{R_j}|}{|R_{x,j}|} \\
&= \frac{|E \cap S_{R_j}|}{|S_{R_j}|}\left(\frac{\operatorname{diam}(S_{R_j})}{\operatorname{diam}(R_{x,j})}\right)^n + 1 - \left(\frac{\operatorname{diam}(S_{R_j})}{\operatorname{diam}(R_{x,j})}\right)^n \\
&\leq 1 - \frac{1}{2}\left(\frac{\operatorname{diam}(S_{R_j})}{\operatorname{diam}(R_{x,j})}\right)^n.
\end{aligned}$$

Thus we have

$$\operatorname{diam}(R_{x,j}) \geq \frac{\operatorname{diam}(S_{R_j})}{2^{\frac{1}{n}}} j^{\frac{1}{n}} = \frac{\frac{1}{2}\min(c,\delta)}{2^{\frac{1}{n}}} j^{\frac{1}{n}} \to +\infty \quad \text{as} \quad j \to +\infty.$$

This proves the claim of the first step.

*Step 2:* Suppose that $\{R_j\}_j$ is a sequence of convex sets whose diameters satisfy $\operatorname{diam}(R_j) \to +\infty$ and $\sup_j |R_j| < +\infty$. Then for any bounded set $B$ we have that $\lim_{j\to+\infty} |B \cap R_j| = 0$.

To see this note that every convex set in $\mathbb{R}^n$ is contained in a rectangle of comparable volume. Thus we can assume that $\{R_j\}_j$ is a sequence of rectangles in $\mathbb{R}^n$. Since $\sup_j |R_j| < +\infty$ and the diameters of the rectangles $R_j$ tend to infinity we conclude that there is a one-dimensional side $I_j$ of $R_j$ such that $\lim_{j\to+\infty} |I_j| = 0$. The claim now follows since

$$|R_j \cap B| \leq |I_j||\operatorname{diam}(B)|^{n-1} \to 0 \quad \text{as} \quad j \to +\infty.$$

We can now conclude the proof of theorem. Assuming (4) does not hold let us consider the set $A$ provided by the first step above. We fix some ball $B(0,r)$ and $x \in A$ and $R_{x,j} \ni x$ as in the first step. Note that, necessarily, $\sup_j |R_{x,j}| < +\infty$ because of the validity of (4). Thus

$$\frac{|B(0,r)^{\mathrm{c}} \cap E \cap R_{x,j}|}{|R_{x,j}|} \geq \frac{|E \cap R_{x,j}|}{|R_{x,j}|} - \frac{|B(0,r) \cap R_{x,j}|}{|R_{x,j}|} \to 1 \quad \text{as} \quad j \to +\infty$$

by (4) and the statement of the second step. This implies that for any $r > 0$ and $0 < \lambda < 1$ we have

$$A \subset \{x \in \mathbb{R}^n : M(\chi_{E \cap B(0,r)^{\mathrm{c}}}) > \lambda\}.$$

However $\mathcal{B}$ is a homothecy invariant density basis so by the Tauberian condition we should have

$$0 < |A| \leq |\{x \in \mathbb{R}^n : M(\chi_{E \cap B(0,r)^{\mathrm{c}}}) > \lambda\}| \leq c(\lambda)|E \cap B^{\mathrm{c}}(0,r)|$$

which is clearly a contradiction since $|E| < +\infty$ and thus $|E \cap B(0,r)^{\mathrm{c}}| \to 0$ as $r \to +\infty$. □

We are quickly exhausting all that we know at the moment regarding Solyanik estimates in harmonic analysis. As a closing remark, it is worth noting that Theorem 4 provides a viable strategy to proving Conjecture 2. Namely, to prove Conjecture 2 it now suffices to prove the following:

**Conjecture 3.** Let $\mathcal{B}$ be a homothecy invariant density basis of convex sets in $\mathbb{R}^n$. Suppose for some $\gamma > 1$ we have that, for every $0 < \alpha < 1$, there exists a set $E_{\alpha,\gamma}$ such that

$$\left|\left\{x \in \mathbb{R}^n : M_{\mathcal{B}}\chi_{E_{\alpha,\gamma}}(x) > \alpha\right\}\right| \geq \gamma |E_{\alpha,\gamma}|.$$

Then there exists a set $E_\gamma$ and a constant $c(\gamma) > 1$ such that

$$\left|\left\{x \in \mathbb{R}^n : M_{\mathcal{B}}\chi_{E_\gamma}(x) = 1\right\}\right| \geq c(\gamma)|E_\gamma|.$$

**Acknowledgements** P. H. is partially supported by a grant from the Simons Foundation (#208831 to Paul Hagelstein).

I. P. is supported by the Academy of Finland, grant 138738.

## References

1. Beznosova, O.V., Hagelstein, P.A.: Continuity of halo functions associated to homothecy invariant density bases. Colloq. Math. **134**(2), 235–243 (2014)

2. Busemann, H., Feller, W.: Zur Differentiation der Lebesgueschen Integrale. Fundam. Math. **22**(1), 226–256 (1934) (ger)
3. Córdoba, A., Fefferman, R.: A geometric proof of the strong maximal theorem. Ann. Math. **102**(1), 95–100 (1975)
4. Córdoba, A., Fefferman, R.: On the equivalence between the boundedness of certain classes of maximal and multiplier operators in Fourier analysis. Proc. Natl. Acad. Sci. USA **74**(2), 423–425 (1977)
5. Garnett, J.B.: Bounded Analytic Functions, 1st edn. Graduate Texts in Mathematics, vol. 236. Springer, New York (2007)
6. de Guzmán, M. (1975) Differentiation of integrals in $\mathbf{R}^n$. In: Measure Theory: Proc. Conf., Oberwolfach, 1975. Lecture Notes in Mathematics, vol. 541, pp. 181–185. Springer, Berlin (1976)
7. Hagelstein, P.A., Luque, T., Parissis, I.: Tauberian conditions, Muckenhoupt weights, and differentiation properties of weighted bases, to appear in Trans. Am. Math. Soc.
8. Hagelstein, P.A., Stokolos, A.: Tauberian conditions for geometric maximal operators. Trans. Am. Math. Soc. **361**(6), 3031–3040 (2009)
9. Solyanik, A.A.: On halo functions for differentiation bases. Mat. Zametki **54**(6), 82–89, 160 (1993). doi:10.1007/BF01209086 (Russian, with Russian summary); English transl. Math. Notes **54**(5–6), 1241–1245 (1994)

# Some Open Problems Related to Generalized Fourier Series

**Kazaros S. Kazarian**

**Summary.** Let $1 \le n_1 < n_2 < \cdots$ and let $\Omega = \{-n_k\}_{k=1}^{\infty} \cup \{n_k\}_{k=1}^{\infty}$ be an infinite set of natural numbers such that $\operatorname{card} \Omega^c = \infty$, where $\Omega^c = \mathbb{Z} \setminus \Omega$. We study conditions for which $\{M(x)e^{ikx} : k \in \Omega\}$ should be complete and minimal in $L^p(\mathbb{T}), 1 \le p < \infty$. The problem of describing pairs $(\Omega, M)$ for which $\{M(x)e^{ikx} : k \in \Omega\}$ is complete and minimal in $L^p(\mathbb{T}), 1 \le p < \infty$ is open.

## 1 Introduction

The term *generalized Fourier series* has been used in different contexts in the literature. Many authors call the series expansion of a function from $L^2$ with respect to a complete orthonormal system a generalized Fourier series, while in the theory of orthogonal series, it is called a Fourier series with respect to a given orthonormal system. If $w \ge 0$ is a measurable function on a measurable set $E \subseteq \mathbb{R}$, then we say that $\phi \in L^p(E, w), 1 \le p < \infty$, if $\phi : E \to \mathbb{C}$ is measurable on $E$ and the norm is defined by

$$\|\phi\|_{L^p(E,w)} := \left( \int_E |\phi(t)|^p w(t) dt \right)^{\frac{1}{p}} < +\infty.$$

Furthermore, in this chapter we denote $\mathbb{T} = \mathbb{R}/2\pi\mathbb{Z}$. The closed linear span in $L^p(E, w)$ of a system of functions $F = \{f_n\}_{n=1}^{\infty}$ will be denoted by $\overline{\operatorname{span}}_{L^p(E,w)}(F)$ or by $\overline{\operatorname{span}}_{L^p(E,w)}\{f_n : n \in \mathbb{N}\}$. We denote by $L^0(E)$ the space of all measurable finite almost everywhere functions defined on a measurable set $E \subseteq \mathbb{R}, |E| > 0$. If the closure is considered in the sense of convergence in measure, then we will use the following notations: $\overline{\operatorname{span}}_{L^0(E)}(F)$ or $\overline{\operatorname{span}}_{L^0(E)}\{f_n : n \in \mathbb{N}\}$. We say that $F = \{f_n\}_{n=1}^{\infty}$ is complete in $L^p(E, w)$ if $\overline{\operatorname{span}}_{L^p(E,w)}(F) = L^p(E, w)$. The term "generalized Fourier series" was used by A. Zygmund (see [21, p. 84]) for series

K.S. Kazarian (✉)
Departamento de Matemáticas, 01.17, Universidad Autónoma de Madrid, Madrid 28049, Spain
e-mail: kazaros.kazarian@uam.es

 
A.M. Stokolos et al. (eds.), *Special Functions, Partial Differential Equations, and Harmonic Analysis*, Springer Proceedings in Mathematics & Statistics 108,
DOI 10.1007/978-3-319-10545-1_10

$$\sum_{n=1} b_n(f)\sin nx, \qquad \text{where} \quad b_n(f) = \frac{1}{\pi}\int_{\mathbb{T}} f(t)\sin nt\,dt \tag{1}$$

and $f$ is an odd function that may not be Lebesgue integrable. In [2], R. P. Boas studied coefficients defined by equations

$$a_n(f) = \frac{1}{\pi}\int_{\mathbb{T}} f(t)[\cos nt - 1]dt, \quad b_n(f) = \frac{1}{\pi}\int_{\mathbb{T}} f(t)\sin nt\,dt \tag{2}$$

for $f$ odd and even, respectively. A. Y. Petrovich [15]—defining generalized Fourier coefficients—subtracted from $\cos nt$ and $\sin nt$ the first terms of their Taylor series at $x_0 = 0$. The following definition can be considered an extension of the preceding definitions.

Let $X = \{x_j\}_{j=1}^s \subset \mathbb{T}$ be a given collection of distinct points, and let $\Lambda = \{\alpha_j\}_{j=1}^s \subset \mathbb{N}$, $|\Lambda| = \sum_{j=1}^s \alpha_j$. We define a system of fundamental trigonometric interpolation polynomials $T(x_j, \lambda, x)$ $(1 \le j \le s, 0 \le \lambda \le \alpha_j - 1)$ as follows. If $|\Lambda| = 2N + 1 (N = 0, 1, \ldots)$, then choose $T(x_j, \lambda, x)$ to be the trigonometric polynomial of order at most $N$ with real coefficients such that

$$T^{(\nu)}(x_j, \lambda, x_l) = \delta_{ij} \cdot \delta_{\nu\lambda} \quad (1 \le l, j \le s; 0 \le \nu, \lambda \le \alpha_j - 1), \tag{3}$$

where $\delta_{ij}$ is the Kronecker symbol, $T^{(\nu)}$ the $\nu$th derivative of $T$, and $T^{(0)} = T$. For $|\Lambda| = 2N$, we choose $T_{j,\lambda}(X, \Lambda, x)$ to be the trigonometric polynomial of order at most $N$ such that (3) holds. If $T(x_j, \lambda, x)$ is not a constant, then

$$T(x_j, \lambda, x) = a_N^{(j,\lambda)} \cos Nx + b_N^{(j,\lambda)} \sin Nx + \ldots$$

and

$$\frac{b_m^{(j,\lambda)}}{a_m^{(j,\lambda)}} = -\kappa_\omega, \tag{4}$$

where $\kappa_\omega$ can be specified from the trigonometric polynomial

$$\omega(x) = \prod_{1 \le j \le s} \sin^{\alpha_j} \frac{x - x_j}{2} = a_N \cos Nx + b_N \sin Nx + \ldots \tag{5}$$

as the quotient of its leading coefficients

$$\kappa_\omega = \frac{a_N}{b_N}, \quad \kappa_\omega = \infty, \quad \text{if} \quad b_N = 0.$$

If $\kappa_\omega = \infty$, then (4) is to be understood as $a_N^{(j,\lambda)} = 0$. For any $f \in L^1(\mathbb{T}, |\omega|)$ and $n \in \mathbb{N}$, we define

$$a_n(f) = \frac{1}{\pi}\int_{\mathbb{T}} f(t)[\cos nt - \sum_{j=1}^{s}\sum_{\lambda=0}^{\alpha_j-1} \frac{d^\lambda}{dt^\lambda}\cos nt\Big|_{t=x_j} T(x_j,\lambda,t)]dt, \tag{6}$$

$$b_n(f) = \frac{1}{\pi}\int_{\mathbb{T}} f(t)[\sin nt - \sum_{j=1}^{s}\sum_{\lambda=0}^{\alpha_j-1} \frac{d^\lambda}{dt^\lambda}\sin nt\Big|_{t=x_j} T(x_j,\lambda,t)]dt. \tag{7}$$

We bring the following theorem (see [6]) to show that formulas (6) and (7) make sense.

**Theorem 1.1 (A).** *Let $f \in L^1(\mathbb{T}, |\omega|)$. Then the generalized Fourier series*

$$\sum_{n=1}^{\infty} a_n(f)\cos nx + b_n(f)\sin nx,$$

*where the coefficients are defined by (6), (7), is summable to $f$ by the Abel–Poisson method. Moreover, the harmonic function*

$$U_r(f,x) = \sum_{n=1}^{\infty}(a_n(f)\cos nx + b_n(f)\sin nx)r^n \tag{8}$$

*converges almost everywhere to $f$ along nontangential paths.*

Theorem A and its subsequent generalizations may have a connection with the following:

**Problem 1 (Fourier-Luzin).** Given an arbitrary measurable function by its values, determine the coefficients of the trigonometric series that represents it.

In the monograph [13], N. N. Luzin gives considerable attention to this problem, calling it Fourier's problem. However, taking into account that in the general formulation this problem has been investigated for the first time by Luzin, it is rightful to call it the Fourier–Luzin problem. A similar problem can also be considered for other orthonormal systems (see [13], pp. 49–51).

To explain the role of generalized Fourier series in the weighted norm spaces, let's introduce some definitions.

**Definition 1.1.** We say that a weight function $\psi(x) \geq 0$ has a singularity of degree $p(1 \leq p \leq \infty)$ at a point $x_0 \in \mathbb{R}$ if $\psi^{-1} \notin L^{1/(p-1)}(I)$ for any open interval I containing $x_0$ [if $p = 1$, then $1/(p-1) = \infty$].

**Definition 1.2.** We say that a weight function $\psi(x) \geq 0$ has a singularity of degree $p\ (1 \leq p \leq \infty)$ and order $\alpha(\alpha = 1, 2, \ldots)$ at a point $x_0 \in \mathbb{R}$ if for any open interval I containing $x_0$,

$$[\psi(x)]^{-1}|x - x_0|^{p(\alpha-1)} \notin L^{1/(p-1)}(I)$$

and there exists an open interval $I_0 = (x_0 - \delta_0, x_0 + \delta_0)$, $\delta_0 > 0$, such that

$$[\psi(x)]^{-1}|x - x_0|^{p\alpha} \in L^{1/(p-1)}(I_0).$$

Results obtained in [7–11] show that $\sum_{n=1}^{\infty} a_n(f)\cos nx + b_n(f)\sin nx$, where the coefficients defined by (6), (7) play basically the same role in the $L^p$-spaces with a weight having at any point $x_j (1 \leq j \leq s)$ a singularity of degree $p$ and order $\alpha_j$ that the Fourier series play in the $L^p$-spaces with Lebesgue measure.

The concept of generalized Fourier series can be extended for a given complete orthonormal system $\Phi = \{\varphi_n(x)\}_{n=1}^{\infty}$ of functions defined on a set $E, |E| > 0$.

Let $\Phi_1$ be a finite or infinite subsystem of the system $\Phi$, and let $P$ be an operator defined on finite linear combinations of the elements of $\Phi$. Then the series $\sum_{n=1}^{\infty} a_n(f)\varphi_n(x)$ is called a generalized Fourier series of the function $f$ with respect to the system $\Phi$, where $f$ is a measurable function defined on $E$ such that

$$a_n(f) = \int_E f(t)[\varphi_n(t) - P(\varphi_n)(t)]dt \tag{9}$$

exists for all $n \in \mathbb{N}$.

## 2 Infinitely Many Singularities

Let $1 \leq n_1 < n_2 < \cdots < n_k < n_{k+1} < \cdots$, and let $\Omega = \{-n_k\}_{k=1}^{\infty} \cup \{n_k\}_{k=1}^{\infty}$ be an infinite set of natural numbers such that $\Omega^c \neq \emptyset$, where $\Omega^c = \mathbb{Z} \setminus \Omega = \{-m_k\}_{k=1}^{\infty} \cup \{0\} \cup \{m_k\}_{k=1}^{\infty}$. Let $S_\Omega = \overline{\text{span}}_{L^1(\mathbb{T})}\{e^{ikx} : k \in \Omega^c\}$.

**Theorem 2.1.** *Let $1 \leq p < \infty$ and $M \in L^p(\mathbb{T})$. Then the system $\{M(x)e^{ikx} : k \in \Omega\}$ is complete in $L^p(\mathbb{T})$ if and only if the following condition holds:*
*If $\frac{g}{M} \in L^{p'}(\mathbb{T})$ for some $g \in S_\Omega$, where $\frac{1}{p} + \frac{1}{p'} = 1$, then $g(x) = 0$ a.e. on $\mathbb{T}$.*

*Proof.* Suppose that $\{M(x)e^{ikx} : k \in \Omega\}$ is complete in $L^p(\mathbb{T})$, and let $g \in S_\Omega$ be a nontrivial function such that $\frac{g}{M} \in L^{p'}(\mathbb{T})$. Then for any $k \in \Omega$,

$$\int_{\mathbb{T}} M(x)e^{ikx}\overline{\frac{g(x)}{M(x)}}dx = \int_{\mathbb{T}} e^{ikx}\overline{g(x)}dx = 0,$$

which contradicts the completeness of the system $\{M(x)e^{ikx} : k \in \Omega\}$. Hence, (C) holds.

Now suppose that (C) holds, but $\{M(x)e^{ikx} : k \in \Omega\}$ is not complete in $L^p(\mathbb{T})$. Then there exists a nontrivial $\varphi \in L^{p'}(\mathbb{T})$ such that

$$\int_{\mathbb{T}} M(x)e^{ikx}\overline{\varphi(x)}dx = 0 \qquad \text{for all} \quad k \in \Omega,$$

which yields $\overline{M}\varphi \in S_\Omega$.

**Theorem 2.2.** *Let $1 \leq p < \infty$ and $M \in L^p(\mathbb{T})$. The system $\{M(x)e^{ikx} : k \in \Omega\}$ is minimal in $L^p(\mathbb{T})$ if and only if the following condition holds:*
*For any $k \in \Omega$, there exists a function $g_k \in S_\Omega$ such that $\frac{e^{ikx}+g_k(x)}{\overline{M(x)}} \in L^{p'}(\mathbb{T})$.*

*Proof.* Suppose that $\{M(x)e^{ikx} : k \in \Omega\}$ is minimal in $L^p(\mathbb{T})$. Then there exists a system $\{\varphi_j\}_{j\in\Omega} \subset L^{p'}(\mathbb{T})$ such that

$$\int_{\mathbb{T}} M(x)e^{ikx}\overline{\varphi_j(x)}dx = \delta_{kj} \qquad \text{for all} \quad k, j \in \Omega.$$

Hence, for any $j \in \Omega$, we have that

$$\int_{\mathbb{T}} e^{ikx}[M(x)\overline{\varphi_j(x)} - \frac{1}{2\pi}e^{-ijx}]dx = 0 \qquad \text{for all} \quad k \in \Omega,$$

which yields $\overline{M(x)}\varphi_j(x) - \frac{1}{2\pi}e^{ijx} = g_j(x) \in S_\Omega$. The proof of the necessity is complete.

If (M) holds, then if we denote

$$\varphi_k(x) = \frac{\frac{1}{2\pi}e^{ikx} + \frac{1}{2\pi}g_k(x)}{\overline{M(x)}} \qquad \text{for} \quad k \in \Omega,$$

then it is easy to check that the system $\{\varphi_j\}_{j\in\Omega} \subset L^{p'}(\mathbb{T})$ will be biorthogonal to $\{M(x)e^{ikx} : k \in \Omega\}$.

*Remark 1.* The functions $g_k, k \in \Omega$ in the formulation of Theorem 2.2 will be unique if the system $\{M(x)e^{ikx} : k \in \Omega\}$ is complete in $L^p(\mathbb{T})$.

*Remark 2.* If, in the formulation of Theorem 2.2, we suppose that $M \in L^\infty(\mathbb{T})$, then we will have that $g_k \in S_\Omega^{p'} = \text{span}_{L^{p'}(\mathbb{T})}\{e^{ikx} : k \in \Omega^c\}$ and $S_\Omega^\infty \subset C(\mathbb{T})$, the space of continuous functions on $\mathbb{T}$.

The following problem is open.

**Problem 2.** Describe pairs $(\Omega, M)$ with card $\Omega^c = \infty$ such that conditions (C) and (M) hold simultaneously.

## 2.1 *Comments on Problem 2*

We recall some statements on quasi-analytic classes (see [14] and also [1, 12]) that can be useful for studying Problem 2 and are adopted for our case.

**Definition 2.1.** It is said that a function $M \in L^1(\mathbb{T})$ has a right-hand mean-value zero of exponential order $\rho$ at a point $x_0 \in \mathbb{T}$ if

$$\rho = \limsup_{a\to 0+} \frac{\log(-\log \int_{x_0}^{x_0+a} |M(t)|dt)}{-\log a},$$

and $M \in L^1(\mathbb{T})$ has a left-hand mean-value zero of exponential order $\rho$ at a point $x_0 \in \mathbb{T}$ if

$$\rho = \limsup_{a\to 0+} \frac{\log(-\log\int_{x_0-a}^{x_0} |M(t)|dt)}{-\log a}.$$

**Definition 2.2.** It is said that a subsequence $1 \le m_1 < m_2 < \cdots < m_k < m_{k+1} < \cdots$ of natural numbers has an index of convergence $\sigma$ if $\sigma = \inf b$, where $b > 0$ is such that

$$\sum_{k=1}^{\infty} \frac{1}{m_k^b} < +\infty.$$

**Theorem 2.3 (Mandelbrojt).** *Let $f : \mathbb{T} \to \mathbb{R}$ be such that $f \in L^1(\mathbb{T})$, and suppose that at a point $x_0 \in \mathbb{T}$, the function $f$ has a right-hand (or a left-hand) mean-value zero of exponential order $\rho > O$. Moreover,*

$$\int_{\mathbb{T}} f(x)e^{in_jx}dx = 0, \quad j \in \mathbb{N}, \textit{ where } \quad 1 \le n_1 < n_2 < \cdots,$$

*$\{m_k\}_{k=1}^{\infty} = \mathbb{N} \setminus \{n_j\}_{j=1}^{\infty}$, $1 \le m_1 < m_2 < \cdots$, and the index of convergence of $\{m_k\}_{k=1}^{\infty}$ is $\sigma, 0 \le \sigma < 1$. Then $f(x) = 0$ almost everywhere on $\mathbb{T}$ if $\rho > \frac{\sigma}{1-\sigma}$.*

By the above theorem and Theorem 2.1, we obtain

**Theorem 2.4.** *Let $\{m_k\}_{k=1}^{\infty} = \mathbb{N}\setminus\{n_j\}_{j=1}^{\infty}$, $1 \le m_1 < m_2 < \cdots$, be a subsequence of natural numbers with index of convergence $\sigma, 0 \le \sigma < 1$. Let $x_0 \in \mathbb{T}$ and*

$$M(x) = e^{-\frac{1}{|x-x_0|^\rho}} \qquad x \in \mathbb{T}, \quad \textit{where} \quad \rho > \frac{\sigma}{1-\sigma}.$$

*Then the system $\{M(x)e^{ikx} : k \in \Omega\}$, where $\Omega = \{-n_j\}_{j=1}^{\infty} \cup \{n_j\}_{j=1}^{\infty}$, is complete in $L^p(\mathbb{T})$ for all $p$ $(1 \le p < \infty)$.*

*Proof.* We have to show that condition (C) holds. Suppose that $\frac{g}{M} \in L^{p'}(\mathbb{T})$ for some $g \in S_\Omega$. We have that for some $0 < a_0 < 1$,

$$\int_{x_0}^{x_0+a_0} \frac{|g(t)|}{|M(t)|}dt \le 1,$$

which yields

$$e^{\frac{1}{a^\rho}} \int_{x_0}^{x_0+a} |g(t)|dt \le 1 \qquad \text{for all} \quad 0 < a \le a_0.$$

Hence,

$$\frac{\log(-\log\int_{x_0}^{x_0+a} |g(t)|dt)}{-\log a} \ge \rho,$$

and we obtain that $g \in S_\Omega$ has a right-hand mean-value zero of exponential order greater than or equal to $\rho$ at the point $x_0$. Thus, by Mandelbrojt's theorem we finish the proof.

**Corollary 2.1.** *Let $\{m_k\}_{k=1}^{\infty} = \mathbb{N} \setminus \{n_j\}_{j=1}^{\infty}$, $1 \le m_1 < m_2 < \cdots$, be a subsequence of natural numbers with index of convergence $\sigma, 0 \le \sigma < 1$. Then the subsystem of the trigonometric system $\{e^{ikx} : k \in \Omega\}$, where $\Omega = \{-n_j\}_{j=1}^{\infty} \cup \{n_j\}_{j=1}^{\infty}$, is complete in $L^0(\mathbb{T})$.*

The last corollary can be improved using known results on the completeness of sets of complex exponentials [20]. We say that a subsequence $1 \le m_1 < m_2 < \cdots < m_k < m_{k+1} < \cdots$ of natural numbers has density $D(\mathcal{M})$ if

$$\lim_{k\to\infty} \frac{k}{m_k} = D(\mathcal{M})$$

exists. Let $\mathcal{M} = \{\mu_n\}_{n=1}^{\infty}$ be a sequence of real or complex numbers. The completeness radius of $\mathcal{M}$ is defined to be the number

$$R(\mathcal{M}) = \sup\{a : \{e^{i\mu_n x}\}_{n=1}^{\infty} \text{ is complete in } C[-a,a]\}.$$

We consider that $R(\mathcal{M}) = 0$ if $\{e^{i\mu_n x}\}_{n=1}^{\infty}$ is not complete in $C[-a,a]$ for any $a > 0$. By Carleman's theorem [4], it follows that

$$R(\mathcal{M}) \ge \pi D(\mathcal{M}). \tag{10}$$

**Lemma 2.1.** *Let $\mathcal{M} = \{m_k\}_{k=1}^{\infty}$, $1 \le m_1 < m_2 < \cdots$, be a subsequence of natural numbers with $D(\mathcal{M}) = 0$. Then $D(\mathcal{N}) = 1$, where $\mathcal{N} = \{n_j\}_{j=1}^{\infty} = \mathbb{N} \setminus \mathcal{M}$, $1 \le n_1 < n_2 < \cdots$.*

*Proof.* Let $\nu = m_k$ for some $k \in \mathbb{N}$. Then evidently $n_\nu = m_k + k$ if

$$m_k + j \notin \mathcal{M} \quad (1 \le j \le k). \tag{11}$$

For any $\varepsilon > 0$, there exists $N = N(\varepsilon)$ such that $k < \varepsilon m_k$ if $k > N$. Hence, if $k > N$, it follows that $k < \varepsilon m_k < \varepsilon m_{k+1}$ and (11) holds if $0 < \varepsilon < \frac{1}{2}$. Thus, for $\nu = m_k, k > N$,

$$\frac{\nu}{n_\nu} = \frac{m_k}{m_k + k} \to 1 \quad \text{if} \quad k \to +\infty. \tag{12}$$

Let's consider the case when $m_k < \nu m_{k+1}, k > N$. It is easy to check that

$$\frac{m_k}{m_k + k} < \frac{m_k + 1}{m_k + k + 1} < \cdots < \frac{m_{k+1} - 1}{m_{k+1} + k - 1}.$$

Hence, by (12), it follows that

$$\lim_{\nu\to\infty} \frac{\nu}{n_\nu} = 1.$$

Lemma 2.1 and (10) yield

**Proposition 2.1.** *Let $\mathcal{M} = \{m_k\}_{k=1}^{\infty}$, $1 \le m_1 < m_2 < \cdots$, be a subsequence of natural numbers with $D(\mathcal{M}) = 0$. Then the subsystem of the trigonometric system $\{e^{ikx} : k \in \mathcal{N}\}$, $\mathcal{N} = \mathbb{N} \setminus \mathcal{M}$, $1 \le n_1 < n_2 < \cdots$, is complete in $L^0(\mathbb{T})$.*

The above proposition is stronger than Corollary 7. It follows from the following:

**Theorem 2.5 (Schwartz [18]).** *Let $\mathcal{M} = \{m_k\}_{k=1}^{\infty}$ be a sequence of real numbers such that $\sum_{k=1}^{\infty} \frac{1}{|m_k|} < +\infty$; then $R(\mathcal{M}) = 0$.*

The following open problem may be classified as a known problem but not as a well-known problem.

**Problem 3.** Describe subsequences $1 \le n_1 < n_2 < \cdots < n_k < n_{k+1} < \cdots$ of natural numbers such that subsystems of the trigonometric system $\{e^{ikx} : k \in \Omega\}$, where $\Omega = \{-n_j\}_{j=1}^{\infty} \cup \{n_j\}_{j=1}^{\infty}$, are complete in $L^0(\mathbb{T})$.

The first result on multiplicative completion was obtained by Boas and Pollard [3]. Later Price and Zink [16] gave a complete characterization of sets of functions that can be multiplicatively completed. Talalyan [19] studied general orthogonal systems that are complete in measure. In particular, he proved that if a system of functions $F$ is complete in measure, then any family obtained by deleting a finite number of elements of $F$ again will be complete in measure. Later, Goffman and Waterman [5] gave an elegant short proof of this result and pointed out that it is always possible to delete certain infinite subsets. Subsequently, Price [17] proved that there are subsystems of Walsh, Haar, and trigonometric systems that have density zero and yet are complete in the sense of measure.

## References

1. Belov, A.S.: Quasi-analyticity of the sum of a lacunary series. Mat. Sbornik **99**(141), 3, 433–467 (1976) (in Russian); English translation in Math. USSR-Sb. 28(3), 389–419 (1976)
2. Boas, R.P.: Integrability Theorems for Trigonometric Transforms. Springer, New York (1967)
3. Boas Jr., R.P., Pollard, H.: The multiplicative completion of sets of functions. Bull. Am. Math. Soc. **54**, 518–522 (1948)
4. Carleman, T.: "Uber die Approximation analytischer Funktionen durch lineare Aggregate von vorgegebene Potenzen. Ark. Mat. Astron. Fys. **17**(9), 1–30 (1922)
5. Goffman, C., Waterman, D.: Basic sequences in the space of measurable functions. Proc. Am. Math. Soc. **11**, 211–213 (1960)
6. Kazarian, K.S.: Summability and convergence almost everywhere of generalized Fourier and Fourier–Haar series (in Russian). Izv. Akad. Nauk Arm. SSR Ser. Mat. **20**(2), 145–162 (1985) (in Russian); English translation in Sov. J. Contemp. Math. Anal. **18**, 63–82 (1985)

7. Kazarian, K.S.: Uniform continuity in weight spaces $L^p, 1 \leq p < \infty$ of families of operators generated by truncated kernels. Doklady AN USSR **272**(5), 1048–1052 (1983); English transl. Soviet Math. Doklady **28**(2), 482–486 (1983)
8. Kazarian, K.S.: Summability of generalized Fourier series in a weighted metric and almost everywhere. Doklady AN USSR **287**(3), 543–546 (1986); English transl. Sov Math. Doklady **33**(2), 416–419 (1986)
9. Kazarian, K.S.: Summability of generalized Fourier series and Dirichlet's problem in $L^p(d\mu)$ and weighted $H^p$-spaces ($p > 1$). Anal. Math. **13**, 173–197 (1987)
10. Kazarian, K.S.: Weighted norm inequalities for some classes of singular integrals. Stud. Math. **86**, 97–130 (1987)
11. Kazarian, K.S.: Weighted inequalities for families of operators generated by truncated Cesàro kernels (Russian). Izv. Akad. Nauk SSSR Ser. Mat. **52**(2), 287–309 (1988); transl. Math. USSR-Izv. **32**(2), 289–311 (1989)
12. Lévine, B., Lifschetz, M.: Quasi-analytic functions represented by Fourier series. Rec. Math. [Mat. Sbornik] N.S. **9**(51), 3, 693–711 (1941)
13. N.N. Luzin, Integral and Trigonometric Series (in Russian). Gostekhizdat, Moscow (1951)
14. Mandelbrojt, S.: Series de Fourier et classes quasi-analytiques de foncions. Gauthier-Villars, Paris (1935) (in French)
15. Petrovich, A.Y.: On the summability of generalized Fourier series. Anal. Math. **4**(4), 303–311 (1978) (in Russian)
16. Price, J.J., Zink, R.E.: On sets of functions that can be multiplicatively completed. Ann. Math. **82**(1) 139–145 (1965)
17. Price, J.J.: Sparse subsets of orthonormal systems. Proc. Am. Math. Soc. **35**(1), 161–164 (1972)
18. Schwartz, L.: Etude des Sommes d'Exponentielles. Publicacions de l'Institute de Matématique de l'Univesité de Strasbourg, 2nd edn., vol. 5. Hermann, Paris (1959) (in French)
19. Talalyan, A.A.: On the convergence almost everywhere of subsequences of partial sums of general orthogonal series. Isv. Akad. Nauk Armyan SSR **10**, 17–34 (1957)
20. Young, R.M.: An Introduction to Nonharmonic Fourier Series. Academic Press, San Diego (2001)
21. Zygmund, A.: Trigonometric Series, vol. 1. Cambridge University Press, Cambridge (1959)

# Computational Analysis for a Mathematical Model of the Mechanics of Aneurysm Development and Rupture

**Tor A. Kwembe**

*In honor of Professor Calixto P. Calderón from whom I studied and learned mathematical biology*

**Summary.** This paper in honor of Professor Calixto P. Calderón focuses on the computational analysis of a mathematical model describing the mechanics of aneurysm development and rupture potentials of aneurysms contained within a cylindrical region of blood vessels. We described the transformation of the membrane from pseudoelastic isotropic material behavior to anisotropic material behavior as an explanation for an aneurysm potential to rupture. The results presented can be used to improve noninvasive means for predicting aneurysm rupture, treatment and management decisions after rupture.

## 1 Introduction

In this paper, we have provided a computational analysis of a mathematical model of the mechanics of aneurysm development and material characteristics that can be used to understand the rupture potentials of aneurysms. In essence, the model is a description of a slow flow quasi-static, non-convectional acceleration, axi-symmetric Navier–Stokes equations in cylindrical coordinates coupled with the Camenschi–Fung type elastrodynamic system of equations with filtration. The characteristics of the solutions of the system of equations described may provide insights into the development of noninvasive means for detecting when an aneurysm may rupture and in deciding the best treatment and management strategies of ruptured aneurysms. The significance of the work is in the use of computational techniques to solve and analyze the dimensionless solutions of the set of differential equations for the fluid velocity and membrane displacement components and pressure. In this approach, the Camenschi–Fung system of equations are first

T.A. Kwembe (✉)
Department of Mathematics and Statistical Sciences, Jackson State University, Jackson, MS 39217, USA
e-mail: tor.a.kwembe@jsums.edu



A.M. Stokolos et al. (eds.), *Special Functions, Partial Differential Equations, and Harmonic Analysis*, Springer Proceedings in Mathematics & Statistics 108,
DOI 10.1007/978-3-319-10545-1_11

nondimensionalized to a set of dimensionless system of differential equations involving the fluid velocity and membrane displacement components and pressure. The components of the fluid velocity are then expressed in terms of pressure, the membrane components, and their partial derivatives. The Membrane displacement components are expressed in terms of pressure. On further eliminating the fluid velocity and the membrane displacement components from these expressions, a nonlinear partial differential equation in pressure is obtained. The perturbation method is used to approximate the solution and then MATLAB® codes are developed to solve and study the responses of blood flow and the aneurysm wall stresses.

How aneurysm evolves depends on any condition that causes the weakening of the walls of the blood vessel [7, 13, 14, 32, 33, 35, 36]. The forms of aneurysms that have been discovered and studied are the aortic, cerebral artery and intracranial aneurysm[7, 12, 14, 32–35, 37]. The development of the cerebral aneurysm, for example, depends on various physical factors associated with blood flow [7, 32, 34]. Research suggests that the inertial forces of the bloodstream result in the local elevation of intravascular pressure and the flow impact [12]. This means that the impacting forces and the local pressure elevation at the aneurysm have a large contribution to the development, for example, of cerebral aneurysms. The other contingent factor is the wall shear stress, which is the viscous friction of the bloodstream that acts parallel to the vessel wall [7, 12, 21]. We shall discuss the overall impact of these forces on the thinning of the aneurysm wall. We shall see from the numerical results that when the lateral contraction of the membrane wall is in balance with the longitudinal extension, an aneurysm may rupture and that at the rupture point about 80 % of the membrane wall thins out. We shall present numerical results which affirm what is known in the literature [7, 12, 34] that the intra-aneurysmal pressure and the bloodstream flow factors into the evolution and development of aneurysms.

First, let's discuss the existing efforts in the literature regarding how aneurysm evolves, developed, and leads to rupture and the available treatment options. In this era of high efficient technology, modern neuroimaging techniques are often used to detect unruptured cerebral artery aneurysm, which is estimated to be present in $3\,\% - 6\,\%$ of the population [12, 34]. Aggressive rupture preventing treatment is often an option, but may lead to morbidity. The specific risk for rupture of an aneurysm is unknown and risk assessments are based on the general knowledge of factors leading to subarachnoid hemorrhage deduced from epidemiological studies [12, 33, 34]. Aneurysms of large size, proximal location, and small neck, or fundus ratio are associated with increased risk for rupture [12, 32, 34]. Therefore, there is the need to seek more reliable parameters for predicting the risk of aneurysmal rupture. Parameters such as intra-aneurysmal pressure gradients, bloodstream flow profiles, membrane displacement profiles, membrane thickness, and Poisson's ratio could provide additional information regarding the risk of rupture. It is hypothesized in [23] that Phase Contrast with vastly undersampled isotropic projection reconstruction could accurately assess intra-aneurysmal pressure gradients in a canine aneurysmal model when compared with invasive measurements. Computational model for simulation of fluid–structure interaction in cerebral aneurysms based on patient specific lesion geometry, with emphasis on wall tension is developed in [12].

In this model, we have introduced the Poisson's ratio and membrane wall thickness, as additional parameters which can be used for predicting the potential for an aneurysm to rupture. We have utilized the Camenschi dimensionless variable transforms and quasi-static conditions to reduce the problem to parameters that can be measured by noninvasive means. We used the resulting differential and integral equations and MATLAB® to develop numerical analysis and computational simulations of the solutions based on experimental data of parameters derived from the cited literature. The analysis confirms that the profile of the deformed radius and the displacement components of the membrane become discontinuous as the Poisson's ratio approaches anisotropic material values. Computational analysis of intra-aneurysmal pressure, membrane displacement and thickness affirm that their profile before and after the rupture of an aneurysm is consistent with in vitro and in vivo observations. In particular, it shows that when 80 % of the membrane wall thins out then the aneurysm may rupture.

Many contingent hypotheses regarding aneurysm enlargement and rupture [7, 12–14, 32–36] do not address the roles of the Poisson's ratio and the thinning of the membrane wall. Others have suggested [33] that studies on the mechanics of saccular aneurysms, for example, should be focused on quasi-static analyses that investigate the roles of lesion geometry and material properties including growth and remodeling. In this model, material properties, growth, and remodeling are taken into account. Remodeling of the membrane wall in responses to tears is incorporated in the filtration coefficient and retains the normal balance in the microvascular mechanism in supplying tissues or the surrounding fluids with nutrients and clearing waste products. However, the model does not take into consideration the healing process of the membrane layers. Computational simulations also find that when the Poisson's ratio lies within the anisotropic material values, the aneurysm may stretch or shrink but not rupture. In fact, most experimental studies choose values of Poisson's ratio in the isotropic material range $0 < \nu_0 < 0.5$ and do not factor the filtration process and anisotropic material composition of the membrane's inner and outer layers, which remain in place after the media has deflated (aneurysm has ruptured), into consideration. Poisson's ratio in the anisotropic material ranges is considered and they indicate that the membrane extensional rigidity is weakened by generating a substantial anisotropy in stiffness. The inner and outer layers of the membrane surrounding the media which hold the aneurysm sac are predominantly made of polyurethane material which is anisotropic [17].

In Fig. 1, we depicted a normal artery wall which consists of three layers. The inner most endothelial layer is called the intima, the middle layer consisting of smooth muscle is called the media, and the outer layer consisting of connecting tissues is called the adventitia [9]. In Fig. 2, we depicted the deformed curvature of the blood vessel or aneurysmal sac which is composed of only the intima and adventitia. The material composition of the intima and adventitia is dominantly of anisotropic polyurethane. The intima, based on in vitro and in vivo observations [3, 28] remains normal but subintimal cellular proliferation also occurs when

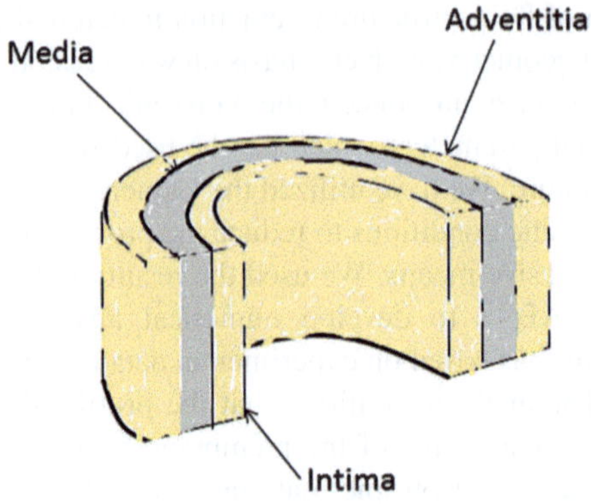


**Fig. 1** Normal membrane wall

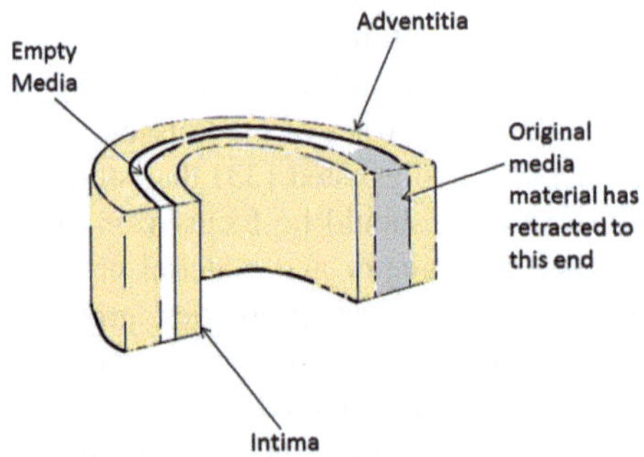


**Fig. 2** Ruptured aneurysm membrane wall

aneurysm developed. The internal elastic membrane responsible for the thinning of the artery wall is either reduced in size or absent causing the media to retract to the junction of the aneurysm neck with the parent blood vessel as shown in the figure. This development transforms the membrane within the aneurysm region from pseudoelastic isotropic to anisotropic media. Thus, the deformation transforms the curvature of the weakened portion of the blood vessel into an aneurysmal sac and hence tolerating the extension of the membrane and transverse shear, curving and pressure difference. We shall consider the blood vessels as a permeating deforming shell filled with a pulsating incompressible fluid and constrained by the surrounding tissue and fluid. The extensional rigidity D of the membrane depends on its stiffness and the deformation grows in the direction of the space created by the surrounding tissue due to impacting forces on the vessel wall. The deformation of the vessel is not volume preserving but the biological material properties of the membrane are more or less the same everywhere and transition from isotropic to anisotropic media as aneurysm evolved. It is also biologically known that the interior of a cell is anisotropic due to intracellular organelles. This characteristic may be responsible for the development of two or more aneurysms on the same artery. Multiple aneurysms are due to defects in the arterial wall and may be congenital [4]. Nevertheless, we shall consider the membrane as a pseudoelastic isotropic material in the biological sense [9] that the material properties of the membrane remain the same throughout the deformation process becoming anisotropic only during the development of aneurysm sac. This feature holds only for nontraumatic aneurysms. In general though, the theory of isotropic elasticity allows the Poisson's ratio in the range $-1 \leq \nu_0 \leq 0.5$ for an object with surface with no constraint. Physically, this means that for the material to be stable, the stiffness must be positive. That is, it is required that both the bulk and shear modulus be greater than or equal to zero [11, 16]. On the other hand, isotropic objects that are constrained at the surface can have Poisson's ratios outside the above range and be stable [11, 16]. Since the blood vessels are constrained by the surrounding tissue and fluid, and living biological tissues and lifeless mechanical materials have different tensional responses, we shall consider values of Poisson's ratio to include values outside the isotropic range. This consideration also includes the regime of anisotropic deformation, when the constituted material of the membrane within the aneurysm region is made up of the intima and adventitia, since the concept of Poisson's ratio can be extended to anisotropic materials with values outside the isotropic range [11, 16, 17, 27, 38]. Finally, we have provided the post rupture analysis and demonstrated how the membrane extensional rigidity deduced from the membrane wall thickness and Poisson's ratio can be used to design a grading scale for measuring the severity of aneurysm rupture.

## 2 The Mathematical Model

### *2.1 Derivation*

It is known that inertial forces of the bloodstream result in local elevation of intravascular pressure and the flow impacting force together with local pressure elevation at the aneurysm contributes to the development of aneurysms [21, 34]. It is also hypothesized that aneurysms rupture when the wall tension exceeds the strength of the wall tissue. We shall consider an axi-symmetric motion of equations in two media; the Newtonian fluid, coupled with the elastic membrane. Due to the nature of the origin of the blood flow into the blood vessels, which is imposed by the heart pumping blood into the circulatory system, we shall consider non-convection acceleration, axi-symmetric, viscous, incompressible pressure driven Navier–Stokes equations in cylindrical coordinates (a slow flow problem) coupled with the Camenschi–Fung type elastic membrane equations [5, 9, 18]. We shall assume that the aneurysm occurs within a cylindrical portion of the vessel of length $L$. We let $a_0$ be the undeformed radius of the cylinder and $a(z,t)$ be the deformed radius describing the curvature of the aneurysm. We let $P(z,t)$ be the intra-aneurysmal pressure with $P_1(t)$ and $P_2(t)$ the end pressures at $z = 0$ and $z = L$, respectively.

If we let $\overrightarrow{v} = (v, u)$ be the velocity vector of the fluid flowing through the portion of the vessel containing the aneurysms where $v(z, r, t)$ and $u(z, r, t)$ are the transversal and longitudinal components, and let $\overrightarrow{\eta} = (\xi, \zeta)$ be the displacement vector of the membrane, where $\xi(z, t)$ and $\zeta(z, t)$ are the transversal and longitudinal components of the membrane displacement respectively and furthermore, if $t_{rz}$ and $t_{rr}$ are the stress tensor components in the fluid then, we have the Camenschi–Fung type system of equations:

$$\frac{\partial v}{\partial t} = -\frac{1}{\rho}\frac{\partial P}{\partial r} + \frac{\mu}{\rho}\left[\frac{\partial^2 v}{\partial r^2} + \frac{1}{r}\frac{\partial v}{\partial r} + \frac{\partial^2 v}{\partial z^2} - \frac{v}{r^2}\right]; \tag{1}$$

$$\frac{\partial u}{\partial t} = -\frac{1}{\rho}\frac{\partial P}{\partial z} + \frac{\mu}{\rho}\left[\frac{\partial^2 u}{\partial r^2} + \frac{1}{r}\frac{\partial u}{\partial r} + \frac{\partial^2 u}{\partial z^2}\right]; \tag{2}$$

$$\frac{\partial v}{\partial r} + \frac{v}{r} + \frac{\partial u}{\partial z} = 0 \tag{3}$$

$$\rho_m h \frac{\partial^2 \zeta}{\partial t^2} = D\left(\frac{\nu_0}{a_0}\frac{\partial \xi}{\partial z} + \frac{\partial^2 \zeta}{\partial z^2}\right) + t_{rz}\mid_{r=a_0}; \tag{4}$$

$$\rho_m h \frac{\partial^2 \xi}{\partial t^2} = -D\left(\frac{\nu_0}{a_0}\frac{\partial \zeta}{\partial z} + \frac{\xi}{a_0^2}\right) - t_{rr}\mid_{r=a_0}; \tag{5}$$

$$t_{rz} = -\mu\left(\frac{\partial u}{\partial r} + \frac{\partial v}{\partial z}\right)\mid_{r=a_0} \tag{6}$$

$$t_{rr} = \left(P - 2\mu\frac{\partial u}{\partial r}\right)\mid_{r=a_0} \tag{7}$$

defined in the region $0 < r \le a_0 \le a(z, t)$; $0 < z < L$; $t \ge 0$ where $\rho$ is the mass density of the fluid, $\mu$ is the dynamic viscosity, $\rho_m$ is the mass density of the membrane, $h$ is the thickness of the membrane wall, $D = \frac{Eh}{1-\nu_0^2}$ is the extensional rigidity of the membrane, $E$ is the Young's modulus, and $\nu_0$ is the Poisson's ratio.

We shall also assume that the natural vascular process of exchange of nutrients and waste from the evolution and development of aneurysms are by filtration. Thus, the boundary conditions expressing the adherence of the fluid to the membrane wall and the fluid filtration through the membrane are:

$$v(a(z, t)) = \frac{\partial \xi}{\partial t} + \frac{v_{n,f}}{\sqrt{1 + \left(\frac{\partial a}{\partial z}\right)^2}} \tag{8}$$

$$u(a(z, t)) = \frac{\partial \zeta}{\partial t} - \frac{v_{n,f}\left(\frac{\partial a}{\partial z}\right)}{\sqrt{1 + \left(\frac{\partial a}{\partial z}\right)^2}}. \tag{9}$$

where $v_{nf}$ is the filtration velocity normal to the membrane wall, and we supposed that it is governed by the Darcy–Starling Law given as:

$$v_{n,f} = \frac{k^*}{\mu h}(P(z,t) - P_e) \tag{10}$$

where $k^*$ is the permeability coefficient and $P_e$ is the constant exterior pressure.

Now Eq. (10) is derived from the fact that in a normal vascular system of functions, there is a free exchange of nutrients, water, electrolytes, and microphage between the intravascular and extra-vascular components of the blood vessels. Several mechanisms are responsible for this critical function of the vascular system. Physiologists including Pappenheimer, Landis, Starling, Renkin, Michel, Curry, Rippe, and Bates [19–21] investigated the mechanism by which plasma and its solutes cross the vascular barrier. They discovered that capillaries are the vascular segment responsible for molecular exchange in normal tissues and that gases, water, and microphages cross the capillary endothelial cell barrier freely, but the passage of larger molecules such as plasma proteins is tightly restricted. They also discovered that several mechanisms are involved in this exchange. The most important, though, are the bulk flow and diffusion. The rate of change in either direction is determined by physical factors such as hydrostatic pressure, osmotic pressure, and the physical nature of the barrier separating the blood and the interstitium of the tissue. That is, the permeability of the membrane wall. While the diffusion process is deemed the most important mechanism in this exchange, the diffusion coefficient in the Fick equation depends on molecular size [26]. It is important for the exchange of small molecules and is driven by molecular concentration gradient across vascular endothelium defined by the Fick equation $J_0 = k(C_v - C_i)$. $C_v - C_i$ is the concentration difference. Sample and Golovin employed this condition in [29,30] to study the dynamics of a double-lipid bilayer membrane by coupling intermembrane separation and the lipid chemical composition of a two-component membrane and dependence on the membrane curvatures. They focused on the thermodynamical equilibrium in [29] and non-equilibrium in [30] of fluxes across the membrane. Here, we are considering the impacting forces on the membrane, rather than the concentration of the fluid content of the blood vessels and hence admit the exchange of large molecular fluxes, such as plasma proteins. Consequently, filtration is much more important than diffusion for flux of large molecules such as plasma proteins and is governed by the Starling equation [19–21,23]:

$$v_{nf} = \frac{k^*}{\mu h}\left[(P(z,t) - P_e) - \sigma\,(\pi - \pi_e)\right],$$

where $k^*$, the filtration coefficient, is a property of the membrane wall and a measure of the permeability of the membrane to water, $P - P_e$ and $\pi - \pi_e$ are hydrostatic and osmotic pressure differences, respectively, between the plasma and the interstitium, $\sigma$ is the osmotic reflection coefficient and varies from zero to one. High values of $\sigma$ indicate little plasma-protein escape [26]. When $(P(z,t) - P_e) - \sigma(\pi - \pi_e)$ is positive, filtration takes place and when it is negative reabsorption takes place. The amount of fluid filtered or reabsorbed per unit time, the filtration velocity or flux is determined by the permeability of the membrane and by the surface area available for the exchange.

Vascular permeability is essential for the health of normal tissues and it is also an important characteristic of many disease state in which it is greatly increased [19–21, 26]. Since aneurysm is caused by the weakening of the membrane which results in high levels of plasma-protein escape activity, we shall take the osmotic reflection coefficient to be zero and, therefore, arrive at the Darcy–Starling filtration velocity given in Eq. (10). That is, when an aneurysm evolved, the reabsorption process is stopped.

The values of the pressure at the ends of the aneurysm region are given by:

$$P(0,0,t) = P_1(t); \qquad P(L,0,t) = P_2(t). \tag{11}$$

We further assume that, the end at $z = 0$, the beginning of the aneurysm region is fixed and the end of the aneurysm region $z = L$ is free of stress, we have:

$$\zeta \mid_{z=0} = 0; \qquad D\left(\frac{\partial \zeta}{\partial z} + \nu_0 \frac{\xi}{a_0}\right) \mid_{z=L} = 0; t \geq 0 \tag{12}$$

The deformed radius is given by

$$a(z,t) = a_0 + \xi(z,t); \quad 0 < z < L; \quad t \geq 0. \tag{13}$$

We further make the following assumptions: (1) the length $L$ of the aneurysm region must be much larger than the undeformed radius of the blood vessel and (2) the following slow flow conditions holds:

$$\left(\frac{a_0}{L}\right)^2 \ll 1; \qquad \frac{\rho u_* a_0}{\mu} \ll 1; \qquad \frac{\rho_m h u_*^2}{D} \ll 1. \tag{14}$$

where $u_*$ is the characteristic longitudinal speed.

Hence, Eqs. (1)–(14) constitute the mathematical model describing the mechanics of the evolution and development of aneurysms contained within a cylindrical portion of the blood vessel. If any form of aneurysm occurs within the cylindrical portion of the arteries or blood vessels, in general, then the deformed radius $a(z,t)$ defines the geometry of the aneurysm. Since, the extension, or stretch, of $a(z,t)$ is measured through the neck of the aneurysm, or the opening into the aneurysm (fundus aspect ratio), from the bloodstream vessel. Consequently, the model given here can be used to study the characteristics of all forms of aneurysms within a cylindrical portion of the vessel.

### 2.1.1 Computational Simulations Method

Human aneurysms are formed in different sizes and shapes, and they exhibit a variety of material behavior [23, 33, 34, 36]. As indicated earlier much of the work geared towards understanding the mechanics of the evolution and development of an aneurysm has focused mostly on the fluid content dynamics, shape, and size. The current interests are to determine more reliable parameters to predict the potential for aneurysmal rupture. We employ here, computational simulations methods to help understand the characteristics of the fluid flow and material responses to the impacting forces. Towards that end, we will nondimensionalized Eqs. (1–13) by using (14) and the following Camenschi [5] dimensionless variable transformations:

$$t = \frac{L}{u_*}t^0; \qquad z = Lz^0; \qquad r = a_0 r^0; v = \frac{u_* a_0}{L}v^0; u = u_* u^0;$$

$$P = \frac{\mu u_* L}{a_0^2}P^0; \zeta = \epsilon L\zeta^0; \qquad \xi = \epsilon a_0 \xi^0;\ a(z,t) = a_0 a^0(z^0, t^0)$$

where $\epsilon < 1$ is a dimensionless parameter given as

$$\epsilon = \frac{P_* a_0}{D} = \frac{\mu u_* L}{a_0 D}.$$

The expression for $\epsilon$ and condition (14) are natural scaling obtained from the nondimensional analysis process and agreed with those of fluid flow in permeable media and Camenschi [5]. The superscript index "0" is used for dimensionless quantities. Upon dropping the index after transformation, Eqs. (1–13) reduce to the following dimensionless equivalent equations:

$$\frac{\partial P}{\partial r} = 0; \qquad 0 < r \le 1 \le a(z,t); \qquad t \ge 0 \tag{15}$$

$$\frac{\partial P}{\partial z} = \frac{1}{r}\frac{\partial}{\partial r}\left(r\frac{\partial u}{\partial r}\right); \qquad 0 < r \le 1 \le a(z,t) \qquad 0 < z < 1; t \ge 0 \tag{16}$$

$$\frac{1}{r}\frac{\partial}{\partial r}(rv) + \frac{\partial u}{\partial z} = 0; \qquad 0 < r \le a_0 \le a(z,t); \qquad 0 < z < 1 \tag{17}$$

$$v_0\frac{\partial \xi}{\partial z} + \frac{\partial^2 \zeta}{\partial z^2} = \frac{\partial u}{\partial r}\,|_{r=1}; \qquad 0 \le z \le 1 \tag{18}$$

$$v_0\frac{\partial \zeta}{\partial z} + \xi = P(z,t); \qquad 0 \le z \le 1; \qquad t \ge 0 \tag{19}$$

along with the boundary conditions:

$$v(z,r,t)\mid_{r=a(z,t)} = \epsilon\frac{\partial \xi}{\partial t} + \delta(P - P_e); \quad 0 \le z \le 1; \quad t \ge 0 \tag{20}$$

where $\delta = \frac{k^* L^2}{h a_0^3}$.

$$u(z,r,t)\mid_{r=a(z,t)} = \epsilon\frac{\partial \zeta}{\partial t}; \quad t \ge 0; \quad 0 \le z \le 1 \tag{21}$$

$$\zeta\mid_{z=0} = 0; t \ge 0 \tag{22}$$

$$\epsilon\left(\frac{\partial \zeta}{\partial z} + \nu_0 \xi\right)\mid_{z=1} = 0; t \ge 0 \tag{23}$$

$$P(0,t) = P_1(t); \qquad P(1,t) = P_2(t); \qquad t \ge 0 \tag{24}$$

$$a(z,t) = 1 + \epsilon\xi(z,t); \quad 0 \le z \le 1; \quad t \ge 0. \tag{25}$$

We note here that the dimensionless system of equations constitute only the intravascular pressure, the components of the fluid velocity and membrane displacement, the Poisson ratio $\nu_0$ and the filtration coefficient $\delta$ as the inertial forces impacting the membrane wall. Since in simulation, $\epsilon$ will always be taken to be less than one, the mass densities and fluid viscosity do not play a major role in the evolution and development of an aneurysm and in its potential to rupture. In dimensionless analysis, the evolution of an aneurysm will always depend on the size of the transversal displacement of the membrane and wall thickness, $h$.

The solutions to the system of Eqs. (15)–(24) as with Camenschi in [5] are obtained by direct integration under the conditions that $v(z,r,t)$ and $u(z,r,t)$ remain bounded at $r = 0$ and noting from Eq. (15) that P = P(z,t) only. Hence, we have the following expression for the fluid velocity and membrane displacement components. The argument (z, t) is omitted in P, $\xi$, and $\zeta$ in some cases for brevity.

$$v(z,r,t) = \left[\frac{-r^3}{16} + \frac{r}{8}(1+\epsilon\xi)^2\right]\frac{\partial^2 P}{\partial z^2} + \frac{\epsilon r}{4}(1+\epsilon\xi)\frac{\partial \xi}{\partial z}\frac{\partial P}{\partial z} - \frac{\epsilon r}{2}\frac{\partial^2 \zeta}{\partial z \partial t} \tag{26}$$

$$u(z,r,t) = \frac{1}{4}[r^2 - (1+\epsilon\xi(z,t))^2]\frac{\partial P(z,t)}{\partial z} + \epsilon\frac{\partial \zeta(z,t)}{\partial t} \tag{27}$$

$$\xi(z,t) = \frac{2-\nu_0}{2(1-\nu_0^2)}P(z,t) + \frac{\nu_0}{2(1-\nu_0^2)}P_2(t) \tag{28}$$

$$\zeta(z,t) = \frac{1-2\nu_0}{2(1-\nu_0^2)}\int_0^z P(\tau,t)d\tau - \frac{P_2(t)}{2(1-\nu_0^2)}z \tag{29}$$

We know that the influence of the stress tensor components $t_{rz}$ and $t_{rr}$ in the fluid has been incorporated in solving for the components of the membrane displacement. Hence, the genesis of the aneurysm is in the weakening of the membrane at $r = a_0$. Therefore, (28) and (29) constitute the components that measure the dynamics of the aneurysm wall and the deformed radius given in (25) radially and the change with respect to P(z,t) may not be linear.

$P = P(z,t)$ is the intra-aneurysmal pressure and from (15), (16), (19), (20), and (26)–(28) it satisfies the nonlinear partial differential equation

$$\begin{aligned}
&\frac{1}{16}\left\{1+\epsilon\left[\frac{2-\nu_0}{2(1-\nu_0^2)}P(z,t)+\frac{\nu_0}{2(1-\nu_0^2)}P_2(t)\right]\right\}^3\frac{\partial^2 P(z,t)}{\partial z^2}\\
&+\frac{\epsilon(2-\nu_0)}{8(1-\nu_0^2)}\left\{1+\epsilon\left[\frac{2-\nu_0}{2(1-\nu_0^2)}P(z,t)+\frac{\nu_0}{2(1-\nu_0^2)}P_2(t)\right]\right\}^2\left(\frac{\partial P(z,t)}{\partial z}\right)^2\\
&-\frac{\epsilon}{2}\left\{1+\epsilon\left[\frac{2-\nu_0}{2(1-\nu_0^2)}P(z,t)+\frac{\nu_0}{2(1-\nu_0^2)}P_2(t)\right]\frac{1-2\nu_0}{2(1-\nu_0^2)}+\frac{2-\nu_0}{(1-\nu_0^2)}\right\}\frac{\partial P(z,t)}{\partial t}\\
&=-\frac{\epsilon}{4(1-\nu_0^2)}\left\{1+\epsilon\left[\frac{2-\nu_0}{2(1-\nu_0^2)}P(z,t)+\frac{\nu_0}{2(1-\nu_0^2)}P_2(t)\right]-2\nu_0\right\}P_2'(t)\\
&+\delta(P(z,t)-P_e)
\end{aligned} \tag{30}$$

We shall employ the perturbation method of [2, 5] by developing the pressure function $P(z,t)$ in a power series with respect to the small parameter $\epsilon$ and neglecting the $O(\epsilon^2)$ terms. That is, we assume that

$$P(z,t)=\sum_{n=0}^{\infty}\frac{\epsilon^n P_n(z,t)}{n!} \tag{31}$$

Consequently, we have the following system of partial differential equations in $P_0(z,t)$ and $P_1(z,t)$:

$$\frac{1}{16}\frac{\partial^2 P_0(z,t)}{\partial z^2}=\delta[P_0(z,t)-P_e] \tag{32}$$

$$\begin{aligned}
\frac{1}{16}\frac{\partial^2 P_1(z,t)}{\partial z^2}-\delta P_1(z,t)=&-\frac{1}{16}\left[3\frac{2-\nu_0}{2(1-\nu_0^2)}P_0(z,t)+\frac{3\nu_0 P_2(t)}{2(1-\nu_0^2)}\right]\frac{\partial^2 P_0(z,t)}{\partial z^2}\\
&-\frac{(2-\nu_0)}{8(1-\nu_0^2)}\left(\frac{\partial P_0(z,t)}{\partial z}\right)^2\\
&+\frac{1}{4}\left[\frac{1-2\nu_0}{1-\nu_0^2}\frac{\partial P_0(z,t)}{\partial t}-\frac{P_2'(t)}{1-\nu_0^2}\right]\\
&+\frac{2-\nu_0}{2(1-\nu_0^2)}\frac{\partial P_0(z,t)}{\partial t}+\frac{\nu_0 P_2'(t)}{2(1-\nu_0^2)}
\end{aligned} \tag{33}$$

The solution of Eqs. (32)–(33) are obtained by direct integration, using the boundary conditions in (24) for $P_0$ and $P_1$ to get the perturbation solution $P(z,t) = P_0(z,t) + \epsilon P_1(z,t)$. The approximate aneurysm wall development using (25) and (28) is given by the equation

$$a(z,t) = 1 + \epsilon \left[ \frac{2 - \nu_0}{1 - \nu_0^2} P(z,t) + \frac{\nu_0 P_2(t)}{2(1 - \nu_0^2)} \right] \tag{34}$$

We will now use the data collected in the literature [12, 15, 23, 31, 33, 34, 36] to fit the model. We will consider an aneurysm within a cylindrical portion of a blood vessel of length $L = 2cm$. The membrane thickness $h = 0.25\,\text{cm}$, the undeformed radius $a_0 = 0.5\,\text{cm}$, the permeability coefficient $K^* = 2.5 \times 10^{-7}$. The Darcy–Starling coefficient $\delta = \frac{k^* L^2}{h a_0^3}$. For the end pressure $P_1(t)$ and $P_2(t)$, we shall consider the Ferguson [7, 15, 33] pressure formula described by the Fourier series

$$P_i(t) = P_m + \sum_{n=1}^{N} [A_n \cos(n\omega t) + B_n \sin(n\omega t)] \quad i = 1, 2.$$

Milnor [22] suggested that $N = 10$ is sufficient to describe systemic blood pressure in general with $N < 10$ in the diastole vasculature. Here, we shall take $P_1(t)$ to represent the diastolic pressures and $P_2(t)$ the systolic pressures. All pressure measurements considered here are in mmHg. We shall take $P_m = 65.7\,\text{mmHg}$, see [33], the coefficients of the Fourier series given in row vector form are:

$$A_n = [-7.13 \quad -3.08 \quad -0.130 \quad -0.346 \; 0.0662 \; \ldots]$$

$$B_n = [4.64 \quad -1.18 \quad -0.564 \quad -0.346 \quad -0.120 \; \ldots]$$

and $P_e = 70\,\text{mmHg}$ where we have used the Shah and Humphrey data in [33].

Since the heart beats about 72 times per minute, we shall take $\frac{2\pi}{\omega} = \frac{60}{72}$, or $\omega = 8\,\text{rad/s}$ for a regular heart beat. In dimensionless parameters, the fluid characteristics such as density and viscosity do not figure in the numerical solutions. The surviving parameters are size of the aneurysm region, $L$, permeability and filtration coefficient $\delta$, undeformed radius $a_0$, membrane thickness, $h$, the Poisson's ratio $\nu_0$, and the time scale. All of which can be measured by noninvasive means. Intra-aneurysmal pressure $P(z,t)$, fluid flow patterns, membrane displacement, and the dynamics of the deformed radius are expressed in terms of the surviving parameters. As can be seen, the computational graphics indicate that the Poisson's ratio, intra-aneurysmal pressure gradients, flow profiles, and membrane displacement could provide additional information regarding the rupture potential of aneurysms.

We have used the MATLAB® platform version 7.10, 207B, 208B to numerically and graphically understand the roles the bloodstream impact forces, intra-aneurysmal pressure elevations and Poisson's ratio play in the rupture potential of aneurysms contained within cylindrical portions of blood vessels. The method begins with the solutions of the fluid velocity components $v(z,r,t)$, $u(z,r,t)$; the membrane displacement components $\xi(z,t)$, $\zeta(z,t)$; the deformed radius $a(z,t)$; the intra-aneurysmal pressure, $P(z,t)$; the filtration velocity, $v_{nf}$; the rate of filtration $Q_{nf}$, and the cross-sectional fluid flux $Q(z,t)$. MATLAB® codes of the solutions are used to produce numerical data and graphical representation at the circular frequencies $\omega = 2\pi f$ where $f = 0.2, 0.5, 1, 1.2, 5, 8, 10$ to simulate in vitro tests. It is known that most actual laboratory tests are performed at values of $f$ at 0.1 or 0.2 [15, 33]. The impact of the Poisson ratio on the comparative relationship of the deformed radius, $a(z,t)$, with intra-aneurysmal pressure, $P(z,t)$, and with the local elevations of the intra-aneurysmal pressure is expressed for the Poisson's ratio ranges of $0 < \nu < 1$, $1 < \nu \leq 2$, and $\nu > 2$ covering both isotropic and anisotropic material values.

### 2.1.2 Simulation and Analysis

Computations revealed discontinuity in the profiles of the impact fluid velocity and its components, the membrane displacement and its components, the intra-aneurysmal pressure, the deformed radius, the rate of filtration, and the flux on any cross-section of the membrane within the aneurysm region when the Poisson's ratio approaches anisotropic values. We shall interpret the points of discontinuity in the solution profiles as indicators for aneurysm rupture. The dynamic profiles of the bloodstream impacting forces, intra-aneurysmal pressure elevations, and the deformed radius exhibit similar characteristics for $f = 0.1, 0.2, 0.5, 1, 1.2, 5$. However, we will give the results for the cases $f = 1.2$ corresponding to the regular heart beat of 72 beats per minute, and $f = 0.2$, the experimental value at irregular heart beat of 12 beats per minute. In each case, we considered two scenarios, the case where a fixed point $(z_0, t_0)$ is considered within the aneurysm and one where all points $(z,t)$ are considered within the aneurysm region. In both cases, we see that for $f = 0.2$ and $f = 1.2$, time changes of the simulations of the intra-aneurysmal pressure, $P(z,t)$ are in synchrony with the membrane displacement components, its magnitude and the deformed radius (see Figs. 3 and 4). For these figures, we maintained $P_1(t)$ and $P_2(t)$ at the same oscillating values. Even when one end is maintained at higher pulsating pressure values there were no significant differences, in regard to rupture potential indicators, in the characteristics of the profiles of the intra-aneurysmal pressure, membrane displacement, deformed radius, filtration velocity, filtration rate, and fluid flux.

In Figs. 3 and 4, the profile of the deformed radius a(z,t) is flat before and after the discontinuity. However, in Fig. 5 which is associated with the case of a regular heart beat, it shows local elevations before the discontinuity at the time $\nu_0 = 1$ and then oscillate with time afterwards to vanish. In Fig. 6 corresponding to the case of irregular heart beat, the profile of the deformed radius increases before the discontinuity and then drops gradually after the discontinuity and then oscillate and die out. In Figs. 5 and 6, the intra-aneurysmal pressure, $P(z,t)$

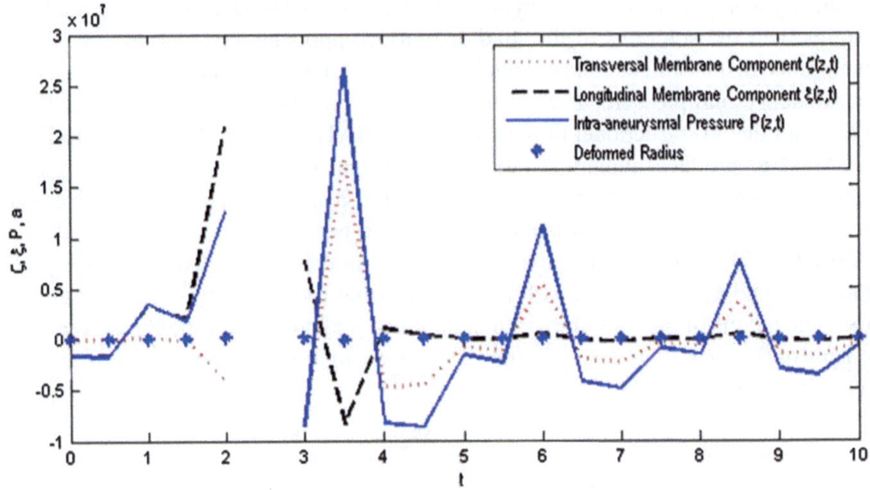


**Fig. 3** Membrane displacement, deformed radius, pressure vs time where f = 1.2 with z and $\nu_0$ varying

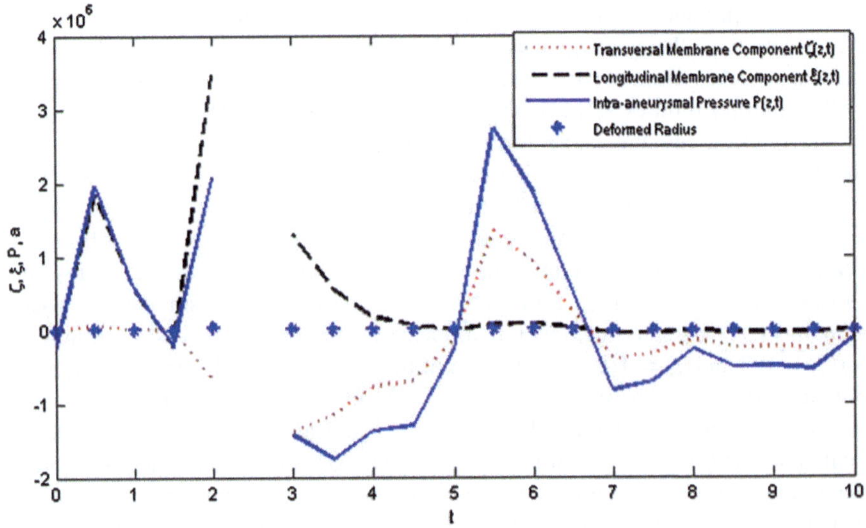


**Fig. 4** Membrane displacement, deformed radius, pressure vs time where f = 0.2 with z and $\nu_0$ varying

shows local pressure elevations before the discontinuity. Following the point of discontinuity, the pressure profile dropped and then elevates and then oscillates and die out. The transversal component of the membrane displacement shows the same characteristics with the deformed radius, $a(z,t)$. But this is not surprising, since $a(z,t) = 1 + \epsilon\xi(z,t)$. The longitudinal component of the membrane displacement

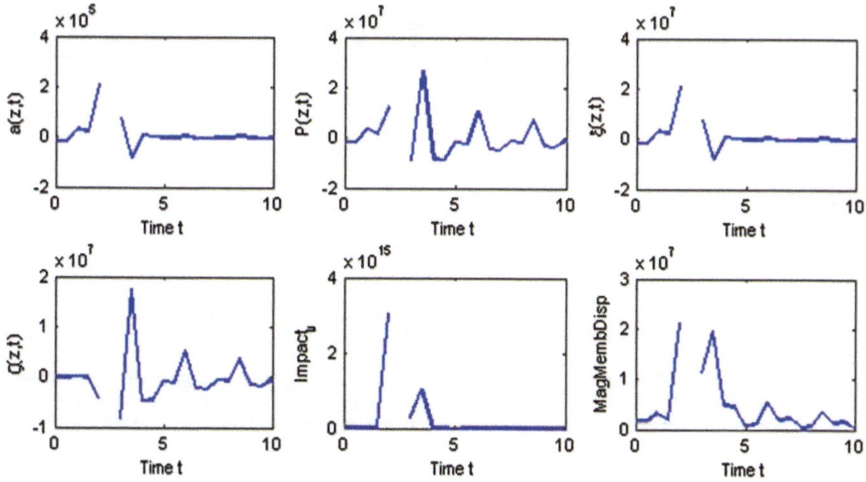


**Fig. 5** Profile of solutions when $\nu_0$ is varying and f = 1.2

elevated slightly followed by a drop leading into the point of discontinuity, then delayed afterwards slightly and elevates, then oscillate to vanish. The profiles of the impact fluid velocity in Figs. 5 and 6 show elevated values leading into the point of discontinuity and oscillate to zero afterward. Similarly, the profile of the magnitude of the membrane displacement elevates rapidly leading to the point of discontinuity and then oscillate to zero afterward. There are more oscillations following the point of discontinuity for the case of a regular heart beat than the case of irregular heart beat.

Figures 7 and 8 are the time varying profiles of the filtration velocity, filtration rate, and the fluid flux, respectively, for the regular and irregular heart beat. The graphs have a discontinuity at the time of aneurysm rupture. We observe that the graphs exhibit a discontinuity at the membrane thickness of about 0.05 cm from 0.25 cm. That is, the aneurysm may rupture when 80 % of the membrane thins out.

## *2.2 Post Rupture Discussions*

Aneurysm rupture leads to the development of vasospasm and vasospasm is the leading cause of disability and death from aneurysm rupture [3, 10, 28]. Ruptured aneurysms are most likely to rebleed within the first day with the risk remaining very high for the first two weeks if left untreated [3, 4, 10, 28]. Open surgery and clipping, endovascular coiling and catheter or stent insertion, and aggressive treatment of vasospasm are the most common methods of treatment that have been correlated with improved outcome [3, 25, 28]. The problem is most aneurysms do not cause symptoms until they rupture. When they rupture, they are associated with

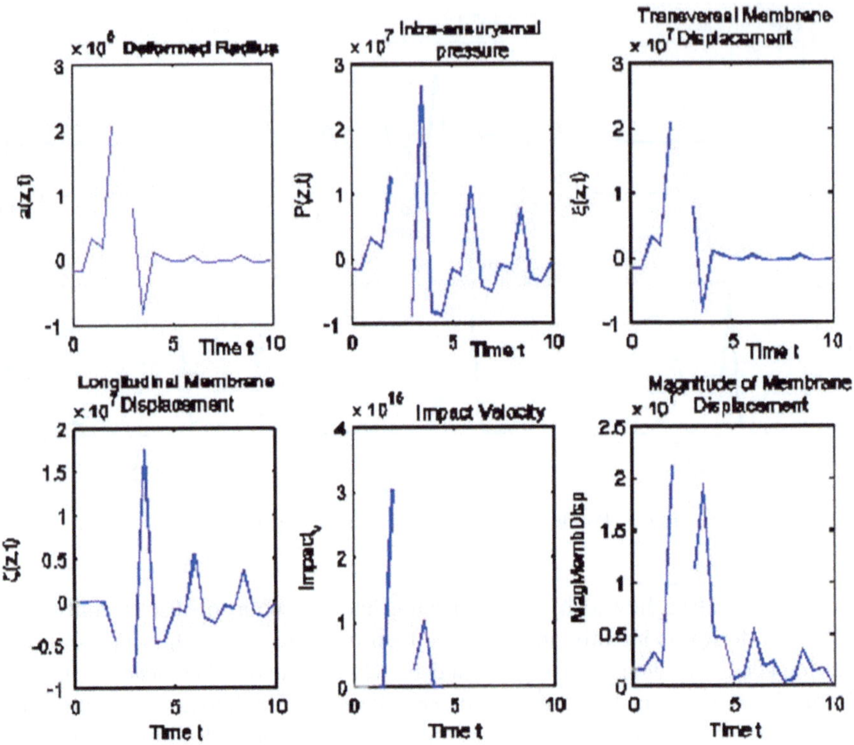

**Fig. 6** Profile of solutions when $\nu_0$ is varying and f = 0.2

significant morbidity and mortality if treatment is not sought on time. Therefore, there is need for a clinical method that measures the levels of severity of aneurysmal subarachnoid hemorrhage and which enables the decision on the best clinical approach for stabilizing the affected portion of the vessel and the method of treatment. Currently, the two widely used methods for grading the clinical severity of ruptured aneurysms are the Hunt and Hess six-point scale and the Fisher four-point grades [3, 8, 14]. The grade levels for these scales are determined by the level of blood on a CT scan. However, an aneurysm rupture can be severe and the CT scan may have no blood images. For example, the vasospasm is most common in Fisher grade level three which indicates the presence of localized clot or vertical layers less than 1mm thick, but it is rarely found in patients with no blood on CT scan [1, 3, 4, 25].

The profiles of the intra-aneurysmal pressure P(z,t), the membrane thickness $h$, and the Poisson's ratio after an aneurysm rupture given in Figs. 3, 4, 5, 6, 7, 8 can provide a means for an alternative grading scale for measuring the severity of an aneurysm rupture that is not predicated by the presence of blood on a CT scan. The parameters can be used to noninvasively measure the membrane extensional

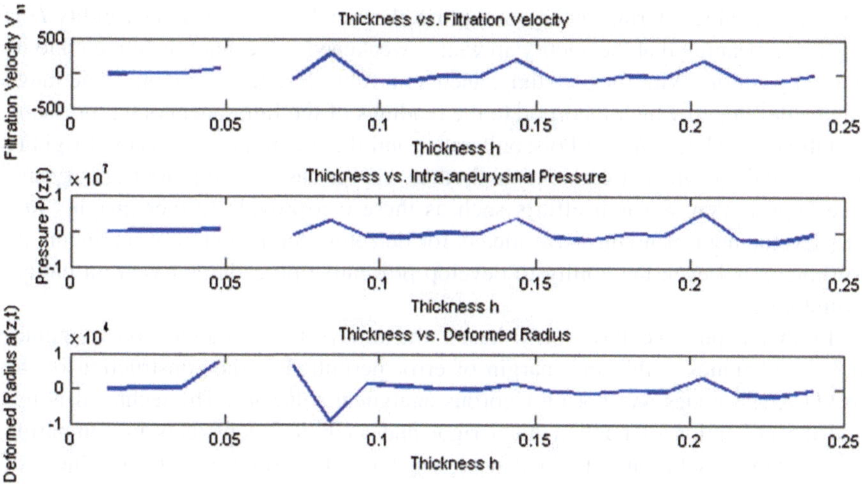


**Fig. 7** Plots of Intra-aneurysmal pressure, the deformed radius, and filtration velocity with the thickness of the membrane wall with varying $\nu_0$ where f = 1.2

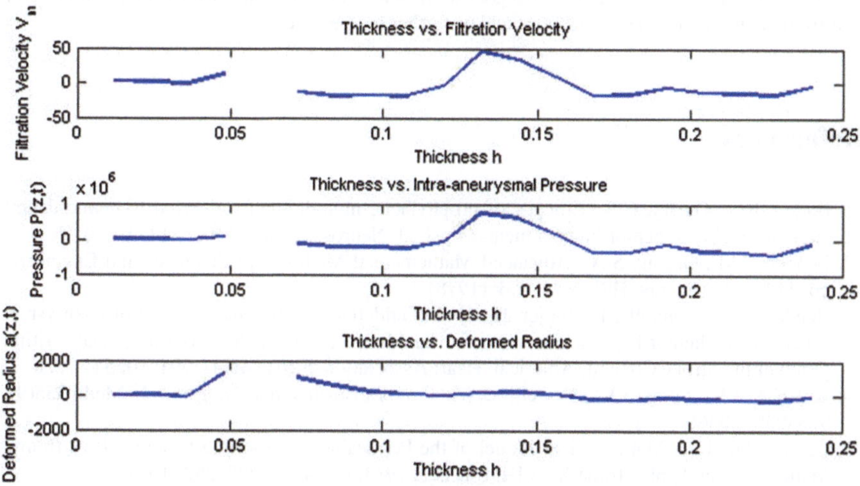


**Fig. 8** Plots of Intra-aneurysmal pressure, the deformed radius, and filtration velocity with the thickness of the membrane wall with varying $\nu_0$ where f = 0.2

rigidity $D = \frac{Eh}{1-\nu_0^2}$ after an aneurysm has ruptured and therefore, determine the severity of the rupture according to the values of the Poisson's ratio extrapolated from the readings of the intra-aneurysmal pressure and membrane thickness. In the range $0 < \nu_0 < 1$, the extensional rigidity $D$ is positive. It is positive and large for small values of $\nu_0$, meaning that the membrane wall can be repaired or retracted to its original form [6, 24]. For values of the Poisson's ratio in the range $1 < \nu_0 < 2$, anisotropic values, during aneurysm rupture, the membrane extensional rigidity $D$ is negative, meaning that the aneurysm wall is weakened and cannot be retracted to its normal position. With further study, aneurysm rupture severity grading scale range can be developed which incorporate the readings of the Intra-aneurysmal pressure, membrane thickness $h$, the Poisson's ratio, and the membrane extensional rigidity $D$ in deciding which treatment method to use in cases of ruptured aneurysms. The hope is that research efforts such as these can provide further insights into the development of noninvasive means for detecting the potential of an aneurysm to rupture and even the ability to develop preventive procedures based on patient symptoms.

In this paper, we have only based the analysis of solutions on computational techniques within the margin of error permitted in the construction of the MATLAB® codes and not on rigorous analytical solutions. The technical issues and the utilization of mathematical rigor and analytical constructs in comparing the numerical solutions obtained here and the actual solutions are a subject of another work.

**Acknowledgements** This paper is from the work supported by an NSF Award DMS: 1214359. My sincere thanks and appreciation also goes to the anonymous reviewer for pointing out areas of improvement to the original manuscript and of further investigation.

## References

1. Barker, F.G., Ogilvy, C.S.: Efficacy of prophylactic nimodipine for delayed ischemic deficit after subarachnoid hemorrhage: a metaanalysis. J. Neurosurg. **84**(3), 405–414 (1996)
2. Bender, C.M., Orszag, S.A.: Advanced Mathematical Methods for Scientists and Engineers, pp. 319–361. McGraw Hill, New York (1978)
3. Benderson, J., Connolly, E., Batjer, H., et al.: Guidelines for the management of aneurysmal subarachnoid hemorrhage: a statement for healthcare professionals from a special writing group of the Stroke Council, American Heart Association. Stroke **40**(3), 994–1025 (2009)
4. Brisman, J.L., Song, J.K., Newell, D.W.: Cerebral aneurysms. N. Engl. J. Med. **355**(9), 928–939 (2006)
5. Camenschi, G.: A Mathematical Model of the Permeable Transporting Systems, Rev. Roum. Math. PUres et Appli., Tome XXVIII, Bucarest 1983, vol. 4, pp. 275–282 (1981)
6. Evans, E., Mohandas, N., Leung, A.: Static and dynamic rigidities of normal and sickle erythrocytes: major influence of cell hemoglobin concentration. J. Clin. Invest. **73**, 477–488 (1984)
7. Ferguson, G.G.: Physical factors in the initiation, growth, and rupture of human intracranial saccular aneurysms. J. Neurosurg. **37** 666–677 (1972)
8. Fisher, C.M., Kristler, J.P., Davis, J.M.: Relation of cerebral vasospasm to subarachnoid hemorrhage visualized by computerized tomographic scanning. Neurosurgery. **6**(1), 1–9 (1980)
9. Fung, Y.C., Biomechanics Mechanical Properties of Living Tissues. Springer, New York (1993)
10. Greenberg, M.S.: Subarachnoid hemorrhage and aneurysms. In: Greenberg, M.S. (ed.) Handbook of Neurosurgery, 5th edn, pp. 754–803. Thieme Medical Publishers, Newy York (2001)
11. Greaves, G.N., Greer, A.L., Lakes, R.S., Rouxel, T.: Poisson's ratio and modern materials. Nat. Mater. **10**, 823–836 (2011)

12. Isaken, J.G., Bazileus, Y., Kuamsdal, T., Zhang, Y., Kaspersen, J.H., Waterloo, K., Romner, B., Ingebrigtsen, T.: Determination of wall tension in cerebral artery aneurysm by numerical simulation. Stroke **39**, 3172–3178 (2008)
13. Kwembe, T.A., Jones, S.N.: A mathematical model of cylindrical shaped aneurysms. In: BIOMAT 2005 Proceedings of the International Symposium on Mathematical and Computational Biology, pp. 35–48. World Scientific Publishers, Rio de Janeiro (2006)
14. Kwembe, T.A.: Mathematical and computational modeling of physiological disorders: a case study of the IUPS human physiome project and aneurysm models. In: BIOMAT 2006, Proceedings of the International Symposium on Mathematical and Computational Biology, pp. 181–192, Manaus, 27–30 November 2006, World Scientific Publishers, Singapore (2007)
15. Kyriacou, S.K., Humphrey, J.D.: Influence of size, shape and properties on the mechanics of axisymmetric saccular aneurysms. J. Biomech. **29**(8), 1015–1022 (1996)
16. Lakes, R.S., Wineman, A.: On Poisson's Ration in Linearly Viscoelastic Solids. J. Elasticity **85**, 45–63 (2006)
17. Lee, T., Lakes, R.S.: Anisotropic polyurethane foam with Poisson's ratio greater than 1. J. Mater. Sci. **32**, 2397–2401 (1997)
18. Mazumdar, J.: An Introduction to Mathematical Physiology and Biology, 2nd edn, pp. 136–161. Cambridge University Press, New York (1999)
19. Michel, C.C.: Fluid exchange in the microcirculation. J. Physiol. **557**(3), 701–702 (2004)
20. Michel, C.C.: Microvsular permeability, ultrafiltration, and restricted diffusion. Am. J. Physiol. Heart Circ. Physiol. **287** H1887–H1888 (2004)
21. Michel, C.C., Curry, F.E.: Microvascular permeability. Physiol. Rev. **79**, 703–761 (1999)
22. Milnor, W.R.: Hemodynamics. Williams & Wilkens, Baltimore (1982)
23. Moftaker, R., Aagaard-Kienitz, B., Johnson, K., Turki, P.A., Turk, A.S., Niemann, D.B., Consigny, D., Grinde, J., Wiebeno, O., Mistretto, C.A.: Noninvasive Measurement of Intraaneurysmal Pressure and Flow Pattern using Phase Contrast With Vastly Undersampled Isotropic Projection Imaging. AJNR Am. J. Neuroradiol. **28**, 1710–1714 (2007)
24. Mohandas, N., Evans, E.: Mechanical Properties of the red cell membrane in relation to Molecular Structure and Genetic defects. Annu. Rev. Biophys. Biomol. Struct. **23**, 787–818 (1994)
25. Morey, S.S.: AHA recommendations for the management of intracranial aneurysms. Agency for Health Care Policy and Research. Am. Fam. Physician. **63**(12), 2465–2466 (2001)
26. Nagy, J.A., Benjamin, L., Zeng, H., Dvorak, A.M., Dvorak, H.F.: Vascular permeability, vascular hyperpermeability, and angiogenesis. Angiogenesis **11** 109–119 (2008)
27. Norris, A.N.: Extreme values of Poisson's ratio and other engineering moduli in anisotropic materials. J. Mech. Mater. Struct. **1**(4), 793–812 (2006)
28. Olafsson, E., Hauser, W.A., Gudmundsson, G.A.: population-based study of prognosis of ruptured cerebral aneurysm: mortality and recurrence of subarachnoid hemorrhage. Neurology. **48**(5), 1191–1195 (1997)
29. Sample, C., Golovin, A.A.: Morphological and chemical oscillations in a couple-membrane system. SIAM J. Appl. Math. **71**, 622–634 (2011)
30. Sample, C., Golovin, A.A.: Nonlinear dynamics of a double bilipid membrane. Phys. Rev. E **76**, Article 031925 (2007)
31. Segers, P., Carlier, S., Pasquet, A., Rabben, S.I., Hellevik, I.R., Remme, E., De Backer, T., De Sutter J., Thomas, J.D., Verdonck, P.: Individualizing the aorta-radial pressure transfer function: feasibility of a model based approach. Am. J. Physiol. Heart Circ. Physiol. **279**, H542–H549 (2000)
32. Sekhar, L.N., Heros, R.C.: Origgin, growth, and rupture of saccular aneurysms: a review. Neurosurgery **8** 248–260 (1981)
33. Shah, A.D., Humphrey, J.D.: Finite strain elastodynamics of intracranial saccular aneurysms. J. Biomech. **32**, 593–500 (1999)
34. Shojima, M., Oshima, M., Takagi, K., Torii, R., Nagata, K., Ichiro, S., Morita, A., Kirino, T.: Role of the bloodstream impacting force and the local pressure elevation in the rupture of cerebral aneurysms. Stroke **36**, 1933–1938 (2005)

35. Stehbans, W.E.: Etiology of intracranial berry aneurysms. J. Neurysurg. **70**, 823–831 (1989)
36. Steiger, H.J.: Pathophysiology of development and rupture of cerebral aneurysms. Acta Neurochir. Suppl. (Wien) **48**, 1–57 (1990)
37. Taylor, A.J., Bobik, A., Berndt, M.C., Ramsay, D., Jennings, G.: Experimental rupture of artherosclerotic lesions increases distal vascular resistance: a limiting factor to the success of infarct angioplasty. Arterioscler. Thromb. Vasc. Biol. **22**, 153–160 (2002)
38. Ting, T.C.T., Chen, T.: Poisson's ratio for anisotropic elastic materials can have no bounds. Q. J. Mech. Appl. Math. **58**(1), 73–82 (2005)

# Singular Integral Operators on $C^1$ Manifolds and $C^1$ Curvilinear Polygons

**Jeff E. Lewis**

**Summary.** The Multilinear Commutator Theorem of Alberto Calderón, Calixto Calderón, Eugene Fabes, Max Jodeit, and Nestor Rivière (Indiana Univ. Math. J. **49**(1), 1–5, 2000; Bull. Am. Math. Soc. **84**(2):287–290, 1978) can be used to show that an *Analytic Double Layer Kernel* restricted to the boundary of a $C^1$ domain fits into the symbolic calculus for singular integral operators (S.I.O.) with $C^0$ coefficients on $C^1$ manifolds and $C^1$ Curvilinear Polygons. In particular a system of S.I.O. is Fredholm on $L^p$ if and only the symbol is a nonsingular homomorphism. For $C^1$ curvilinear polygons, the symbol is constructed using the Mellin transform and can be extended to handle power weights near the vertices. The talk is a review of joint work with Renata Selvaggi and Irene Sisto (Trans. Am. Math. Soc. **340**(1), 293–308 (1993), MR MR1124170 (94a:58194)) and joint work with Lewis and Cesare Parenti (Commun. Partial Differ. Equ **8**(5), 477–544, 1983, MR MR695401, 86f:35185), Lewis (Proc. Am. Math. Soc. **112**(2), 419–427, 1991, MR MR1043413, 91i:47071).

## 1 CCFJR Multilinear Commutator Theorem

### *1.1 Multilinear Commutator Theorem*

Recall the Multilinear Commutator Theorem of Alberto Calderón, Calixto Calderón, Eugene Fabes, Max Jodeit, and Nestor Rivière [CCF+78, CC00], as presented in [CCF+78]:

J.E. Lewis (✉)
Department of Mathematics, Statistics, and Computer Science, University of Illinois at Chicago, Chicago, IL 60607–7045, USA
e-mail: jlewis@uic.edu

 
A.M. Stokolos et al. (eds.), *Special Functions, Partial Differential Equations, and Harmonic Analysis*, Springer Proceedings in Mathematics & Statistics 108,
DOI 10.1007/978-3-319-10545-1_12

THEOREM 4. *Let $G(W_1, \ldots, W_k)$ be a function of the complex variables $W_1, \ldots, W_k$, holomorphic in $|W_i| < R_i$. Let $\Psi_i(x)$ be functions in $\mathbf{R}^n$ with distribution derivatives in $L^{p_i}$, $1 < p_i \leqslant \infty$, and $\varphi_i(x)$, $i = 1, 2, \ldots, k$ functions with bounded derivatives. Let $k(x, z)$ be homogeneous of degree $-n$, with $\int_{|z|=1} |k(x, z)|^s d\sigma_z < \infty$, where $d\sigma_z$ is the surface area element of $|z| = 1$. Then if $f \in L^q$, $1 < q \leqslant \infty$*

$$\overset{*}{T} = \sup_{\epsilon} \left| \int_{|x-y|>\epsilon} \prod_1^m \frac{\Psi_i(x) - \Psi_i(y)}{|x-y|} \right.$$

$$\left. G\left( \frac{\varphi_1(x) - \varphi_1(y)}{|x-y|}, \ldots, \frac{\varphi_k(x) - \varphi_k(y)}{|x-y|} \right) k(x, x-y) f(y)\, dy \right|$$

*is finite a.e. and belongs to $L^\gamma$, $1/q + \Sigma_1^m 1/p_i = 1/\gamma$ provided that:* (i) $k(x, -z) = (-1)^{m+1} k(x, z)$, (ii) $G(W_1, \ldots, W_k) = G(-W_1, \ldots, -W_k)$, (iii) *If $G(W_1, \ldots, W_k)$ is analytic in $|W_j| < R_j$ then $\|\nabla \varphi_j\| < cR_j/k$ where $c$ is a positive universal constant.*

(iv) $$0 < \frac{1}{q} + \Sigma \frac{1}{p_i} < \frac{s-1}{s} + \frac{m}{n}, \quad q > \frac{s}{s-1}, p_i > 1.$$

*If $k(x, z) = k(z)$ we may take $s = 1$ and condition* (iv) *is replaced by:*

(v) $$0 < \frac{1}{q} + \Sigma \frac{1}{p_i} < 1 + \frac{m}{n}, \quad q > 1, p_i > 1.$$

We shall use a simplified version of **CCFJR Theorem 4**.

**Definition 1.** For $\Psi(x), \phi(x)$ in $C^1$, define

$$\delta\Psi(x, y) \equiv \frac{\Psi(x) - \Psi(y)}{|x-y|}$$

$$\delta\phi(x, y) \equiv \frac{\phi(x) - \phi(y)}{|x-y|}$$

$$\sigma_{x,y} \equiv \frac{x-y}{|x-y|}$$

**Theorem 1.1 (CCFJR Theorem 4).** *(Simplified): Let $Tf(x) =$*

$$\text{p. v.} \int \left[ \prod_{i=1}^m \delta\Psi_i(x, y) \right] G\left(\delta\phi_1(x, y), \ldots, \delta\phi_k(x, y)\right) k(x, x-y) f(y)\, dy$$

*where*

- $\Psi_i(x)$ *is* $C^1$
- $G(W_1, \dots, W_k)$ *is holomorphic and even in* $|W_i| < R_i$, $G(W_1, \dots, W_k) = G(-W_1, \dots, -W_k)$
- $|\delta\phi_i| \leq R_i/10k$
- *For the kernel* $k$, *assume*
  - $k(x, z)$ *homogeneous of degree* $-n$ *in* $z$, $\sup_x |k(x, \sigma)|$ *in* $L^\infty(d\sigma)$
  - $k(x, -z) = (-1)^{m+1}k(x, z)$ *– compensates the parity of* $\delta\Psi_i$
- $f$ *is in* $L^p$ $(\mathbf{R}^n)$, $1 < p < \infty$.

  *Then* $(Tf)^* \equiv \sup_{\epsilon>0} |T_\epsilon f|$ *is in* $L^p$ $(\mathbf{R}^n)$.

$$\|(Tf)^*\|_p \leq c_{G,\phi,k,p} \left[ \prod_{i=1}^{m} \|\Psi_i'\|_\infty \right] \|f\|_p$$

## 1.2 Analytic Double Layer Kernels (ADLK)

**CCFJR Theorem 4,** 1.1, may be used to show that the restriction of an *Analytic Double Layer Kernel in* $\mathbf{R}^{n+1}$ to the boundary of a bounded $C^1$ domain $\Omega$ defines a singular integral operator on the $C^1$ manifold $\partial\Omega$.

**Definition 2.** $k(X)$ is an *Analytic Double Layer Kernel (ADLK) in* $\mathbf{R}^{n+1}$ if

- $k(X)$ is real analytic in $\mathbf{R}^{n+1} \setminus \{0\}$,
- $k(X)$ is homogeneous of degree $-n$,
- $k(X)$ is *odd.*

Examples include the first derivative of the fundamental solution of a second order elliptic operator [system] or the Stokes and elastostatics systems in $\mathbf{R}^{n+1}$.

Let $S$ be a $C^1$ surface,

$$K_0 f(P) = \text{p. v.} \int_S k(P - Q) f(Q)\, dS_Q$$

In local coordinates: let $S$ be the graph $(x, \phi(x))$, $x$ in $\mathbf{R}^n$, $\phi$ in $C^1$,

$$K_0 f(x) = \text{p. v.} \int k(x - y, \phi(x) - \phi(y)) f(y) b(y)\, dy,$$

$$b(y) = \sqrt{1 + |\nabla\phi|^2} \in C^0.$$

With a rotation, assume that $|\nabla\phi|$ is small.

Then $K_0$ is an S.I.O. with the same symbol as

$$\overline{K}_0 f(x) = \text{p. v.} \int k(x-y, \nabla\phi(x)\cdot(x-y)) f(y) b(y)\, dy$$

The main result is Theorem 1.2.

**Theorem 1.2.** *If $k(X)$ is an* Analytic Double Layer Kernel in $\mathbf{R}^{n+1}$, *$b(x)$ is $C^0$, $\phi$ in $C^1$, $|\nabla\phi|$ is sufficiently small,*

$$K_0 f(x) = \text{p. v.} \int k(x-y, \phi(x)-\phi(y)) f(y) b(y)\, dy,$$

$$\overline{K}_0 f(x) = \text{p. v.} \int k(x-y, \nabla\phi(x)\cdot(x-y)) f(y) b(y)\, dy.$$

*Then the operator $K_0 - \overline{K}_0$ is a compact operator on $L^p$.*

*Proof.* The heavy lifting is estimating the remainder term.

For $x, y \in \mathbf{R}^n$, denote $\sigma = \sigma_{x,y} = \dfrac{x-y}{|x-y|}$.

In local coordinates, the operator $K_0$ and the remainder $R$ are expressed as

$$K_0 f(x) = \text{p. v.} \int \frac{k\left(\sigma_{x,y}, \delta\phi(x,y)\right)}{|x-y|^n} f(y) b(y)\, dy,$$

$$R f(x) = \text{p. v.} \int \frac{k\left(\sigma_{x,y}, \delta\phi(x,y)\right) - k\left(\sigma_{x,y}, \nabla\phi\cdot\sigma\right)}{|x-y|^n} f(y) b(y)\, dy.$$

With a rotation, assume that $|\nabla\phi|$ is small and represent

$$k(\sigma,\tau) = \frac{1}{2\pi i}\int_{|\zeta|=R} \frac{k(\sigma,\zeta)}{\zeta-\tau}\, d\zeta.$$

Ignoring $b$, the kernel for the remainder $Rf$ is

$$k_{\phi,\phi} = \frac{1}{2\pi i}\int_{|\zeta|=R} \frac{\delta\phi(x,y) - \nabla\phi\cdot\sigma}{(\zeta-\delta\phi(x,y))\,(\zeta-\nabla\phi\cdot\sigma)} \frac{k(\sigma,\zeta)}{|x-y|^n}\, d\zeta.$$

For $\phi$ fixed, $\psi$ in $C^1$, consider the kernel

$$k_{\psi,\phi} = \frac{1}{2\pi i}\int_{|\zeta|=R} \frac{\delta\psi(x,y) - \nabla\psi\cdot\sigma}{(\zeta-\delta\phi(x,y))\,(\zeta-\nabla\phi\cdot\sigma)} \frac{k(\sigma,\zeta)}{|x-y|^n}\, d\zeta$$

**N.B.** For fixed $\zeta$, the expression in the integrand is not even in $W = \delta\phi(x, y)$.

The plan of the proof is:

- Use the **Integration Trick**

$$\int_{|\zeta|=R} g(\zeta)\, d\zeta = \frac{1}{2} \int_{|\zeta|=R} g(\zeta) - g(-\zeta)\, d\zeta. \tag{1}$$

- Use the oddness of the original kernel: $k(\sigma, -\zeta) = -k(-\sigma, \zeta)$ to transfer the parity in $\zeta$ to a parity in $\sigma$.
- **Goal:** Look for kernels within the expression for $k_{\psi,\phi}$ which are even or odd in $\sigma$.

  Fix $\zeta$ and let

$$G(\sigma, \zeta) = \frac{k(\sigma, \zeta)}{(\zeta - \delta\phi(x, y))\,(\zeta - \nabla\phi \cdot \sigma)}.$$

Using the **Integration Trick** (1),

$$k_{\psi,\phi}(\sigma) = \frac{1}{2\pi i} \frac{1}{2} \int_{|\zeta|=R} (\delta\psi(x, y) - \nabla\psi \cdot \sigma)\, \frac{G(\sigma, \zeta) - G(\sigma, -\zeta)}{|x - y|^n}\, d\zeta.$$

Write

$$\begin{aligned} G(\sigma, \zeta) - G(\sigma, -\zeta) &= \frac{k(\sigma, \zeta)}{(\zeta - \delta\phi(x, y))\,(\zeta - \nabla\phi \cdot \sigma)} \\ &\quad + \frac{-k(\sigma, -\zeta)}{(-\zeta - \delta\phi(x, y))\,(-\zeta - \nabla\phi \cdot \sigma)} \\ &= I + II. \end{aligned}$$

Using $k$ is odd

$$\begin{aligned} II &= \frac{k(-\sigma, \zeta)}{(\zeta + \delta\phi(x, y))\,(\zeta + \nabla\phi \cdot \sigma)} \\ &= \frac{k(-\sigma, \zeta)}{(\zeta + \delta\phi(x, y))\,(\zeta - \nabla\phi \cdot (-\sigma))}. \end{aligned}$$

Let

$$\begin{aligned} h(\sigma, \zeta) &= \frac{k(\sigma, \zeta)}{\zeta - \nabla\phi \cdot \sigma}, \\ W &= \delta\phi(x, y), \\ G_\zeta(W) &= \frac{1}{\zeta^2 - W^2}. \end{aligned}$$

Then

$$\begin{aligned}
G(\sigma,\zeta) - G(\sigma,-\zeta) &= I + II \\
&= \frac{1}{\zeta - W} h(\sigma,\zeta) + \frac{1}{\zeta + W} h(-\sigma,\zeta) \\
&= \frac{(\zeta + W)\, h(\sigma,\zeta) + (\zeta - W)\, h(-\sigma,\zeta)}{\zeta^2 - W^2} \\
&= G_\zeta(W) \cdot (\zeta\, (h(\sigma,\zeta) + h(-\sigma,\zeta) + W\, (h(\sigma,\zeta) - h(-\sigma,\zeta)))) \\
&= G_\zeta(W)\, (\zeta \cdot (\text{even in } \sigma) + W \cdot (\text{odd in } \sigma))\,, \\
&= G_\zeta(W) \cdot \beta(\sigma,\zeta,W), \\
\beta(\sigma,\zeta,W) &= \zeta\, (h(\sigma,\zeta) + h(-\sigma,\zeta) + W\, (h(\sigma,\zeta) - h(-\sigma,\zeta)))
\end{aligned}$$

For each $\zeta$, $|\zeta| = R$, apply the Theorem of **CCFJR** 1.1 four times to the operator with the kernel

$$\frac{k_{\phi,\psi}(\sigma,\zeta)}{|x-y|^n} = (\delta\psi(x,y) - \nabla\psi \cdot \sigma)\, G_\zeta(\nabla\phi \cdot \sigma) \frac{\beta(\sigma,\zeta,W)}{|x-y|^n}$$

$$\begin{aligned}
G_\zeta(W) &= \frac{1}{\zeta^2 - W^2}, \\
\beta(\sigma,\zeta,W) &= \zeta \cdot (\text{even in } \sigma) + \delta\phi(x,y) \cdot (\text{odd in } \sigma)\,.
\end{aligned}$$

Then *add* or *integrate* $d\zeta$ the result. If $\psi = \psi^j$ is smooth, the resulting operators are compact. If $\{\psi^j\}$ is a sequence of smooth functions, $\psi^j \to \phi$, in $C^1$, the resulting operators with $\psi = \psi^j - \phi$ converge to 0 in $L^p$ operator norm.

*Remark 1.* Theorem 1.2 shows that Systems of Analytic Double Layer Kernels on the boundary of a $C^1$ domain fit into the symbolic calculus of singular integral operators on $C^1$ manifolds developed by Lewis, Renata Selvaggi, and Irene Sisto in [LSS93].

## 2 S.I.O. on $C^1$ Manifolds

### *2.1 S.I.O. on $C^1$ Manifolds and Surfaces*

Motivated by the work of Renata Selvaggi and Irene Sisto [SS81], we developed a Symbolic Calculus for operators on compact $C^1$ manifolds of dimension $n$:

- Operators in $\mathbf{R}^n$: $Tf(x) = a(x)If(x) + \text{p.v.} \int k(x, x-y) f(y)\, dy + Cf(x)$
- $k(x,z)$ is a classical s.i.o. kernel:

1. $k(x,z)$ is homogeneous of degree $-n$ in z
2. $\int_{|z'|=1} k(x,z')\,d\sigma = 0$
3. $k(x,z)$ is *analytic* or $C^\infty$ in $z$

- $a(x)$ and $k(x,z')$ are *continuous* in $x$
- $C$ is a compact operator on $L^p$

In local coordinates, the Symbol Is Well Defined:

- Symbol $\sigma(T)(x,\xi) \equiv a(x)I + \mathscr{F}_z(k(x,z))(\xi)$
- Intertwining with *rapid oscillations*

$$\lim_{\lambda\to\infty} e^{i\lambda\xi_0\cdot} T\left(e^{-i\lambda\xi_0\cdot}\phi\right) = \sigma(T)(x,\xi_0)\phi(x)$$

- $$|\sigma(T)(x,\xi)| \leq \inf_{C \text{ compact}} \|T + C\|$$

### 2.1.1 Coordinate Changes

- The symbol is invariant under $C^1$ coordinate changes. *Nontrivial*—uses full power of Multilinear Commutator results of A.P. Calderón or Ronald Coifman, Alan McIntosh, and Yves Meyer [CMM82].
- If $\chi(x)$ is a $C^1$ coordinate change,

$$\begin{aligned} K_\chi f(x) &= \chi^*(K(\chi_* f)(x)) \\ &= \text{p. v.} \int k(\chi(x), \chi(x) - \chi(y))) f(y) J(y)\, dy, \end{aligned}$$

$J$ is the [continuous] Jacobean, is an S.I.O. and

$$\sigma(K_\chi)(x,\xi) = \sigma(K)\left(\chi(x), [(d\chi(x))^{-1}]^t \cdot \xi\right). \tag{2}$$

- The crucial step is to show:

  If $k(x)$ is an odd kernel and $\chi(x)$ is a $C^1$ coordinate change, the operator with kernel

$$k(x, \chi(x) - \chi(y)) - k(x, d\chi(x)\cdot(x-y))$$

is a *compact operator* on $L^p$, $1 < p < \infty$.

#### 2.1.2 S.I.O. on a $C^1$ Manifold

- Using 2, transport the operators to a compact $C^1$ manifold
- Symbol Exact Sequence

$$0 \to \text{Compact} \to \text{Operators T} \xrightarrow{\sigma} \text{Symbols} \to 0$$

- For details, see Lewis, Selvaggi, and Sisto [LSS93].

#### 2.1.3 Elliptic S.I.O. on a $C^1$ Manifold

**Definition 3.** An S.I.O. $T$ is *elliptic* if

$$\sigma(T)(P,\xi) \neq 0, \text{ all } (P,\xi), \xi \neq 0.$$

Systems, Vector Bundles: $\sigma(T)(P,\xi)$ is invertible.

#### 2.1.4 Elliptic S.I.O. are Fredholm

- If $T$ is *elliptic*, there is an S.I.O. (parametrix) $R$ such that $TR - I$ and $RT - I$ are compact operators.
- If $T$ is *elliptic*, $T$ is a Fredholm operator on $L^p$, $1 < p < \infty$.

### 2.2 *Index of Elliptic S.I.O.*

- Elliptic S.I.O. have an *analytic* $\text{Index}_a(T))$ defined as

$$\text{Index}_a(T) = \dim(\ker(T)) - \dim\big(\ker(T^*)\big)$$

- *Needed:* An *Atiyah–Singer Index Theorem* which for smooth manifolds characterizes the *analytic* $\text{Index}_a(T)$ in terms of the topological properties of $\sigma(T)$
- The *Seminar on the Atiyah–Singer Index Theorem* [Pal65, p. 313] states the Atiyah–Singer Index Theorem for *smooth coefficient* S.I.O. as *the analytic index $i_a(T)$ and the topological index $i_t(T)$ are equal.*

*Remark 2.* Discussion of Atiyah–Singer

We do *NOT* have a proof of an extension of an *Atiyah–Singer Index Theorem* to the $C^1$ case.

We have the information:

- The elliptic S.I.O. $T$ on a $C^1$ manifold may be approximated by elliptic operators $T_j$ with regular coefficients in operator norm and in $C^0$ in $P$, $C^\infty$ in $\xi$ in the symbol space. Each $T_j$ has the same analytical index as $T$, and the *Atiyah–Singer Index Theorem* may be applied to each $T_j$.
- In [LSS93] the approximation uses a spherical harmonic representation of an analytic kernel and the symbol so that the kernel and symbol are approximated simultaneously.

Examples—Calculate the Index

- $n = 1$: If $M$ is the $C^1$ image of the circle $S^1$ in the plane, the *analytical index* $i_t(A)$ is related to the winding numbers of the two branches of the symbol $\sigma(A)(P, \xi), \xi = \pm\infty$. See *Seminar on the Atiyah–Singer Index Theorem* [Pal65]. The result can be adapted to $N \times N$ systems. This is done by considering $M$ as a special case of $C^1$ curvilinear polygon $\mathbb{P}$ with two *flat vertices* as discussed in Sect. 3. See Lewis and Cesare Parenti [LP83] and Lewis [Lew91].
- Renata Selvaggi and Irene Sisto [SS97] considered a boundary value problem for harmonic forms. The s.i.o part $K$ of the boundary potential is compact so that $\sigma(I + K) = I$, the identity cross section over the underlying $C^1$ boundary manifold. The analytic index, $\text{Index}_a\,(I + K)$, is expressed in terms of topological invariants (Betti numbers) of the $C^1$ boundary manifold.

### *2.3 Possible Extensions*

VMO Manifolds

*Remark 3 ("Are there better results?").*

- Can one consider a manifold of class $C^{0+\text{VMO}}$with coordinate changes $\chi(X)$ which are continuous with $d\chi(X)$ in VMO? Such a class would have the *coefficients* in VMO. There are results that the commutator of a classical S.I.O. with a VMO function is a compact operator. See Coifman, Rochberg, and Weiss [CRW76] and Ukiyama [Uch78].
- Is there an *Index Theorem* for such operators with invertible symbols?

## 3 S.I.O. on $C^1$ Curvilinear Polygons

Consider a Curvilinear Polygon $\mathbb{P}$ with $N$ vertices $P_0, P_2, \ldots, P_{2N} = P_0$. Assume

- Each edge of $\mathbb{P}$ is a line segment or a $C^1$ curvilinear segment.
- Each edge $\overline{P_{2k} P_{2k+2}}$ will be parameterized as a $C^1$ curve $\gamma_{2k}(t), t \in J \stackrel{\text{def}}{\equiv} [0, 1]$.

The *Goal:* Develop a Symbolic Calculus which would include an *Analytic Double Layer Kernel* restricted to $\partial\mathbb{P}$. This was accomplished for polygons in Lewis and Parenti [LP83] and extended to $C^1$ curvilinear polygons in Lewis [Lew91]

A function $f$ on $\partial\mathbb{P}$ is considered as a [column] vector $[f_1, \ldots, f_N]$ on $J = [0, 1]$.

Consider an operator $cI$ + ADLK on $\partial\mathbb{P}$ as a system of operators on $L^p(J)$, $J = [0, 1]$.

## 3.1 S.I.O. on a Closed Interval $J = [0, 1]$

On $J = [0, 1]$, we construct a Symbolic Calculus which includes operators

- (from an edge to itself)

$$K^{0,0} f(t) = \text{p. v.} \int_0^1 k(\gamma(t) - \gamma(s)) b(s) f(s)\, ds,$$

  $\gamma(s)$ in $C^1(J)$, $b$ in $C^0(J)$, including the finite Hilbert Transform

$$Hf(t) = \text{p. v.} \frac{1}{\pi} \int_0^1 \frac{f(s)}{t - s}\, ds.$$

- (from an edge to an adjacent edge) Hardy Kernel Operator

$$K^{1,2} f(t) = \text{p. v.} \int_0^1 k(\gamma_1(t) - \gamma_2(s)) b(s) f(s)\, ds,$$

  $\gamma_1, \gamma_2$ in $C^1(J)$, $b$ in $C^0(J)$, $\gamma_1'(0) \neq \gamma_2'(0)$, which has the behavior of a Hardy Kernel Operator, including the finite Stieltjes Transform

$$Hf(t) = \frac{1}{\pi} \int_0^1 \frac{f(s)}{t + s}\, ds,$$

  and
- (from an edge to a nonadjacent edge) Compact Operators on $L^p(J)$.

## 3.2 Hilbert Transform on $\mathbf{R}$ and $\mathbf{R}^+$

A typical vertex of $\partial\mathbb{P}$:

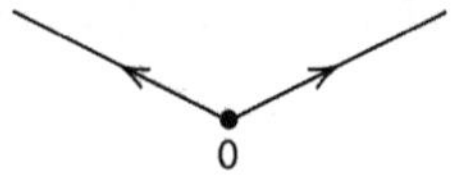

Parameterize the edges on each side by $t \geq 0$, with $t = 0$ at the vertex 0.

To motivate the use of the *Mellin Transform* and *Mellin Multipliers*, consider a special case—a flat vertex:

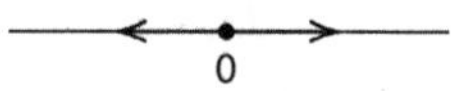

In this case, near the vertex, the restriction of an *Analytic Double Layer Kernel* will result in a multiple of the Hilbert transform.

### 3.2.1 Hilbert Transform on R and $\mathbf{R}^+$

Identify $\mathbf{R}$ with "two copies of $\mathbf{R}^+$ glued together at $t = 0$" and identify a function $f$ on $\mathbf{R}$ with the [column] vector $\mathbf{f} = \left(f^+, f^-\right)$ defined on $\mathbf{R}^+$.

If $f(t) = f^+$ has support in $\mathbf{R}^+$, the Hilbert transform of $f^+$ is

$$Hf(t) = \text{p. v. } \frac{1}{\pi} \int_0^\infty \frac{f(s)}{t-s}\, ds.$$

For $t > 0$,

$$\begin{aligned} (Hf)^-(t) &= (Hf)(-t) \\ &= \frac{1}{\pi} \int_0^\infty \frac{f(s)}{-t-s}\, ds \\ &= -Sf^+(t), \\ Sf(t) &= \frac{1}{\pi} \int_0^\infty \frac{f(s)}{t+s}\, ds. \end{aligned}$$

$Sf$ is the *Stieltjes Transform* of $f$, a Hardy Kernel operator.

### 3.2.2 Hilbert Transform on R as a System on $\mathbf{R}^+$

The full Hilbert Transform on $\mathbf{R}$ can be interpreted as a system on $\mathbf{R}^+$.

The operator is identified with the system

$$\mathbf{H} = \begin{pmatrix} H & S \\ -S & -H \end{pmatrix}$$

## 3.3 *Hardy Kernels and Mellin Multipliers*

### 3.3.1 Hardy Kernels

A typical Hardy Kernel

- A classical Hardy Kernel operator:

$$\begin{aligned} K_\hbar f(t) &= \frac{1}{t}\int_0^t f(s)\,ds \\ &= \int_0^\infty k_\hbar\left(\frac{t}{s}\right) f(s)\,\frac{ds}{s}, \\ k_\hbar(t) &= \begin{cases} t^{-1} & ,t \geq 1 \\ 0 & \text{otherwise.} \end{cases} \end{aligned}$$

- Hilbert Transform on $\mathbf{R}^+$—a singular Hardy Kernel

$$\begin{aligned} Hf(t) &= \text{p.v.}\,\frac{1}{\pi}\int_0^\infty \frac{f(s)}{t-s}\,ds \\ &= \int_0^\infty k_H\left(\frac{t}{s}\right) f(s)\,\frac{ds}{s}, \\ k_H(t) &= \text{p.v.}\,\frac{1}{\pi}\frac{1}{t-1}. \end{aligned}$$

- Stieltjes Transform

$$\begin{aligned} Sf(t) &= \frac{1}{\pi}\int_0^\infty \frac{f(s)}{t+s}\,ds \\ &= \int_0^\infty k_S\left(\frac{t}{s}\right) f(s)\,\frac{ds}{s}, \\ k_S(t) &= \frac{1}{\pi}\frac{1}{t+1}. \end{aligned}$$

### 3.3.2 Mellin Transform

The tool to treat the examples in Sect. 3.3.1 is the *Mellin Transform.*

- For $f \in C_0^\infty(\mathbf{R}^+)$, the *Mellin Transform* of $f$ is

$$\tilde{f}(z) = \int_0^\infty t^{z-1} f(t)\,dt$$

- The *Inverse Mellin Transform*:

$$f(t) = \frac{1}{2\pi i}\int_{\Gamma_c} z^{-t}\tilde{f}(z)\,dz$$
$$= \frac{1}{2\pi}\int_{-\infty}^{\infty} t^{-z}\tilde{f}(c+i\xi)\,d\xi$$
$$\Gamma_c \equiv \{\mathrm{Re}\, z = c\}$$

- For $L^p$ results, the correct choice of $c$ is $c = \frac{1}{p}$. See Lewis Parenti [LP80].

*Remark 4.* The choice of $c = \frac{1}{p}$ is motivated by the observation:
If $f \in L^p(\mathbf{R}^+)$ with norm

$$\|f\|_p \overset{\text{def}}{\equiv} \left(\int_0^\infty |f(t)|^p\,dt\right)^{1/p},$$

let

$$F_{1/p}(x) = e^{-x/p} f(e^{-x}).$$

Then

$$\|f\|_p = \|F_{1/p}\|_{L^p(\mathbf{R})}.$$

The Fourier Transform of $F_{1/p}$ is

$$\mathscr{F}(F_{1/p})(\xi) = \int_{\mathbf{R}} e^{-ix\xi} F_{1/p}(x)\,dx$$
$$= \int_0^\infty t^{i\xi+1/p} f(t)\,\frac{dt}{t}$$
$$= \tilde{f}\left(\tfrac{1}{p} + i\xi\right).$$

The Fourier inversion formula yields

$$e^{-x/p} f(e^{-x}) = \frac{1}{2\pi}\int_{\mathbf{R}} e^{ix\xi}\tilde{f}(1/p + i\xi)\,d\xi$$
$$= \frac{1}{2\pi i}\int_{-\infty}^{\infty} e^{-x/p}[e^{-x}]^{-(1/p+i\xi)}\tilde{f}(1/p+i\xi)\,i d\xi$$
$$= e^{-x/p}\frac{1}{2\pi i}\int_{\Gamma_{1/p}} [e^{-x}]^{-z}\tilde{f}(z)\,dz.$$

### 3.3.3 Mellin Multipliers

Each operator can be represented as a *Mellin Multiplier*

$$\begin{aligned} Kf(t) &= \int_0^\infty k\left(\frac{t}{s}\right) f(s)\,\frac{ds}{s} \\ &= \frac{1}{2\pi i}\int_{\Gamma_{\frac{1}{p}}} z^{-t} m(z)\tilde{f}(z)\,dz \\ m(z) &= \int_0^\infty t^{z-1} k(t)\,dt \end{aligned}$$

The *Mellin Multiplier* is the *Mellin Transform* of the kernel.

*Remark 5.* Note that $m(z)$ is a *Mellin Multiplier* on $L^p(\mathbf{R}^+)$ iff $m\left(\frac{1}{p}+i\xi\right)$ is a *Fourier Multiplier* on $L^p(\mathbf{R})$.

### 3.3.4 Mellin Multiplier Examples

The corresponding Mellin Multipliers for $K_\hbar$, Hilbert Transform, and Stieltjes Transform are

$$\begin{aligned} \tilde{K_\hbar}(z) &= \frac{1}{z-1}, \\ \tilde{H}(z) &= -\frac{\cos \pi z}{\sin \pi z} = -\cot \pi z, \\ \tilde{S}(z) &= \frac{1}{\sin \pi z} = -\csc \pi z. \end{aligned}$$

### 3.3.5 Hilbert Transform on R

The full Hilbert Transform on **R** can be interpreted as a system letting $f$ be a 2–dimensional [column] vector $f = (f^+, f^-)$ on $\mathbf{R}^+$. The operator is identified with

$$\mathbf{H} = \begin{pmatrix} H & S \\ -S & -H \end{pmatrix}$$

The matrix of multipliers is

$$\tilde{\mathbf{H}}(z) = \begin{pmatrix} -\cot \pi z & \csc \pi z \\ -\csc \pi z & \cot \pi z \end{pmatrix}$$

which has determinant $\det \tilde{\mathbf{H}}(z) = -\cot^2 \pi z + \csc^2 \pi z \equiv 1 = -i \cdot i.$

*Remark 6.* For $a$ and $b$ complex numbers

$$\begin{aligned}\det\left(a\mathbf{I} + b\tilde{\mathbf{H}}(z)\right) &= a^2 + b^2\left(-\cot^2 \pi z + \csc^2 \pi z\right)\\ &= a^2 + b^2\\ &= (a+ib)\cdot(a-ib).\end{aligned}$$

The classical symbol of the S.I.O. $aI + bH$ as an operator on $\mathbf{R}$ is $\sigma(aI + bH)(\xi) = a - ib \operatorname{sgn} \xi, \xi \neq 0.$

### 3.3.6 Special Muckenhoupt Weights: Power Weights

We consider the Hardy Kernels and Mellin Multipliers on some weighted $L^p$ spaces.

**Definition 4.**

$$L^p_\alpha = \{f \mid t^\alpha f \in L^p\}.$$

with norm

$$\|f\|_{L^p_\alpha} = \|t^\alpha f(t))\|_{L^p}$$

We write

$$t^\alpha Kf(t) = [\text{p. v.}]\int_0^\infty \left(\frac{t}{s}\right)^\alpha k\left(\frac{t}{s}\right) s^\alpha f(s)\,\frac{ds}{s}.$$

The Mellin Transform of $t^\alpha k(t)$ is $\tilde{k}(\alpha + z)$.

Let $f_\alpha(t) = t^\alpha f(t)$, $(Kf)_\alpha(t) = t^\alpha (Kf)(t)$, $k_\alpha(t) = t^\alpha k(t)$. Then

$$\begin{aligned}(Kf)_\alpha(t) &= [\text{p. v.}]\int_0^\infty k_\alpha\left(\frac{t}{s}\right) f_\alpha(t)\,dt\\ &= \frac{1}{2\pi i}\int_{\Gamma_{1/p}} t^{-z}\tilde{k}(z+\alpha)\tilde{f}_\alpha(z)\,dz.\end{aligned}$$

Thus $K$ is a bounded operator on $L^p_\alpha$ iff $\tilde{k}(z+\alpha)$ is a *Mellin Multiplier* on $L^p$ iff $\tilde{k}\left(\frac{1}{p} + \alpha + i\xi\right)$ is a *Fourier Multiplier* on $L^p(\mathbf{R})$.

### 3.3.7 A Few Muckenhoupt Weights

*Remark 7.* In 1972 Benjamin Muckenhoupt [Muc72] showed that the Hilbert transform is bounded $L^p_\omega$ iff $\omega(t)$ is an $A_p$ weight.

**Definition 5.** $\omega(t)$ is an $A_p$ weight if for all intervals $I$

$$\left(\frac{1}{|I|}\int_I \omega(t)\,dt\right)\left(\frac{1}{|I|}\int_I \omega(t)^{-p'/p}\,dt\right)^{p/p'} \leq C < \infty.$$

If $\omega(t) = t^{\alpha/p}$, $\omega(t)$ is an $A_p$ weight iff $0 < \frac{1}{p} + \alpha < 1$.

The Hilbert Transform on $\mathbf{R}^+$ on $L^p_\alpha$ corresponds to the Mellin Multiplier $\tilde{H}(z) = -\cot \pi z$, $\mathbb{R}z = \frac{1}{p} + \alpha$.

*Mellin Multipliers* can be used to handle special cases of Muckenhoupt's results: $n = 1$, and the weight is a power $\omega(t) = t^{\alpha/p}$.

**N.B.** $z = 0$ and $z = 1$ are poles of $\cot \pi z$.

One of the *reasons* that the Hilbert Transform is not bounded on $L^p_\alpha$, $\frac{1}{p} + \alpha = 0, 1$, is that the corresponding Mellin Multipliers $-\cot \pi z$ and $\csc \pi z$ have poles at $z = 0$ and $z = 1$.

### 3.3.8 Structure of the Operators on $L^p(\mathbf{R}^+)$

To build the operators on $L^p(J)$, consider the operators on $L^p(\mathbf{R}^+)$ are defined as

$$A = c(t)I + d(t)H + e(t)\chi_0 K \chi_0 + C$$

- $c, d, e$ are $C^0$, $\chi_0 = \chi_0(t)$ is a smooth cutoff function, identically 1 near $t = 0$.
- $H$ is the Hilbert Transform $Hf(t) = \text{p. v.}\ \frac{1}{\pi}\int_0^\infty \frac{f(s)}{t-s}\,ds$, with Mellin kernel $\text{p. v.}\ \frac{1}{\pi}\frac{1}{t-1}$ and $\tilde{H}(z) = -\cot(\pi z)$
- $K$ is a Hardy Kernel Operator $Kf(t) = \int_0^\infty k\left(\frac{t}{s}\right) f(s)\,\frac{ds}{s}$

  We assume that on a strip containing $\Gamma_{\frac{1}{p}}$, $\tilde{k}(z)$ is holomorphic and $(1 + |z|)\left|\tilde{k}(z)\right|$ is bounded.
- $C$ is a compact operator on $L^p(\mathbf{R}^+)$.

### 3.3.9 Mellin Multipliers $m(t, z)$ for Operators 3.3.8 on $L^p(\mathbf{R}^+)$

Modulo a compact operator the operator $A$ can be represented as a Mellin Multiplier

$$Af(t) = \frac{1}{2\pi i}\int_{\Gamma_{1/p}} t^{-z} m(t, z)\tilde{f}(z)\,dz,$$

$$m(t, z) = c(t)I + d(t)(-\cot \pi z) + e(t)\tilde{k}(z)$$

### 3.3.10 Description of Symbol

Construct the Symbol: $\text{Symb}^{\frac{1}{p}}(A)$

- Apply Rapid Oscillations: For $\lambda$ real, let $T_\lambda f(t) = t^{i\lambda} f(t)$ so that $\widetilde{T_\lambda f}(z) = \tilde{f}(z + i\lambda)$. If $A_m = c \cdot I + d \cdot H + e \cdot K$ is a Mellin Multiplier operator with multiplier $m(t, z)$,

$$\begin{aligned} T_{-\lambda} A_m T_\lambda f(t) &= \frac{1}{2\pi i} \int_{\Gamma_{1/p}} t^{-z} m(t, z - i\lambda) \tilde{f}(z)\, dz \\ &\to m\left(t, \frac{1}{p} \mp i\infty\right) f(t) \\ &= (c(t) \pm i d(t))\, f(t) \;\text{ as } \lambda \to \pm\infty. \end{aligned}$$

- Apply Extreme Dilations: For $\delta > 0$, let $T_{p,\delta} f(t) = \delta^{-1/p} f(t/\delta)$.
  Since Mellin Multiplier Operators commute with dilations,

$$\begin{aligned} T_{p,\delta^{-1}} A_m T_{p,\delta} f &= c(\delta t) I f + d(\delta t) H f + e(\delta t) K f \\ &\to c(0) f(t) + d(0) H f(t) + e(0) K f(t) \end{aligned}$$

  as $\delta \to 0^+$.

### 3.3.11 Description of Symbol

If $A_m$ were compact, then

- $m\left(t, \frac{1}{p} \pm \infty\right) = 0$, since for fixed $f$, $T_\lambda \rightharpoonup 0$ in $L^p$ as $\lambda \to \pm\infty$.
- $m\left(0, \frac{1}{p} + i\xi\right) = c(0)I + d(0) \cdot \left(-\cot\left(\frac{1}{p} + i\xi\right)\right) + e(0)\tilde{k}\left(\frac{1}{p} + i\xi\right) = 0$, since for fixed $f$, $T_{p,\delta} f \rightharpoonup 0$ in $L^p$ as $\delta \to 0^+$.

The candidates to be part of the symbol: $\text{Symb}^{\frac{1}{p}}(A)(t, z)$:

$$a_0(z) = c(0) + d(0)(-\cot \pi z) + e(0)\tilde{k}(z),\; t = 0, z \in \Gamma_{1/p}$$

$$a_{0+}(t) = c(t) + i\, d(t), t \geq 0, z = \frac{1}{p} + i\,\infty$$

$$a_{0-}(t) = c(t) - i\, d(t), t \geq 0, z = \frac{1}{p} - i\,\infty$$

**N.B.**

- We are assuming that $\tilde{k}\left(\frac{1}{p} \pm i\infty\right) = 0.$

  More precisely, as in Sect. 3.3.8 we assume that on a strip containing $\Gamma_{\frac{1}{p}}$, $\tilde{k}(z)$ is holomorphic and $(1+|z|)\left|\tilde{k}(z)\right|$ is bounded.
-

$$\lim_{\xi\to\pm\infty} -\cot\left(\frac{1}{p} + i\xi\right) = \pm i$$

The above works for $t$ near 0.

### 3.3.12 Operators Near $t = 1$

At the right end of the interval $J = [0, 1]$, we take into account a "Hardy Kernel at $t = 1$"—which we define as an operator $K_1$:

- Let $R$ be the reflection $Ra(t) = a(1-t)$
- $RK_1R$ is a usual "Hardy Kernel Operator at $t = 0$."

### 3.3.13 Structure of the S.I.O. on $J = [0, 1]$

The general operator $A$ on $L^p(J)$ will be

$$Af(t) = c(t)I + d(t)H + \chi K_0 \chi + e(t)R\chi K_1 \chi R + C,$$

with

- $c(t), d(t), e(t)$ are in $C^1(J)$,
- $\chi$ is a smooth cutoff function, $\chi(t) \equiv 1$ near $t = 0$,
- $K_0$ and $K_1$ are Hardy Kernel operators with $(1+|z|)\left|\tilde{K}\right|$ bounded in a strip containing $\Gamma_{1/p}$.

**N.B.**

- $H$ is the finite Hilbert Transform

$$Hf(t) = \text{p. v. } \frac{1}{\pi}\int_0^1 \frac{f(s)}{t-s}\,ds.$$

- $RHR = -H$
- If $K$ is a Hardy Kernel, $\psi \in C_0^\infty((0,1))$, $\psi K \psi$ is compact
- In Lewis [Lew91] the operators are called Op $\Sigma_{\alpha,\beta} \bigotimes C(J)$

### 3.3.14 Construction of Symb$^{\frac{1}{p}}$

The candidates to be part of the symbol

$$a_0\,(z) = c(0) + d(0)\,(-\cot \pi z) + e(0)\tilde{k}(z),\; t = 0, z \in \Gamma_{1/p}$$

$$a_{0+}(t) = c(t) + i\, d(t), t \geq 0, z = \frac{1}{p} + i\,\infty$$

$$a_{0-}(t) = c(t) - i\, d(t), t \geq 0, z = \frac{1}{p} - i\,\infty$$

From the point of view of $t = 1$, obtain

$$a_1\,(z) = c(1) - d(1)\,(-\cot \pi z) + e(1)\tilde{k}_1(z),\; t = 1, z \in \Gamma_{1/p}$$

$$a_{1+}(t) = c(1-t) - i\, d(1-t), t \geq 0, z = \frac{1}{p} + i\,\infty$$

$$a_{0-}(t) = c(1-t) + i\, d(1-t), t \geq 0, z = \frac{1}{p} - i\,\infty$$

The symbol is defined on the boundary of a rectangle, which we label as $\mathscr{R}_J^{\frac{1}{p}}$:

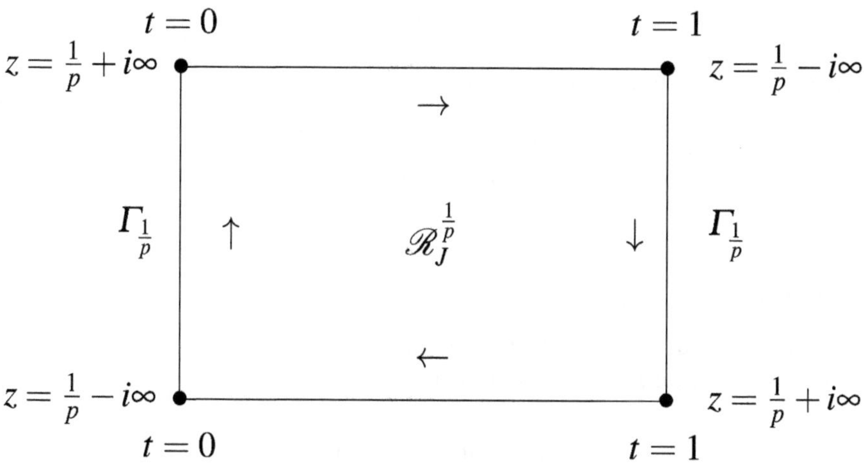


The full Symbol, Symb$^{\frac{1}{p}}(A)$, is:

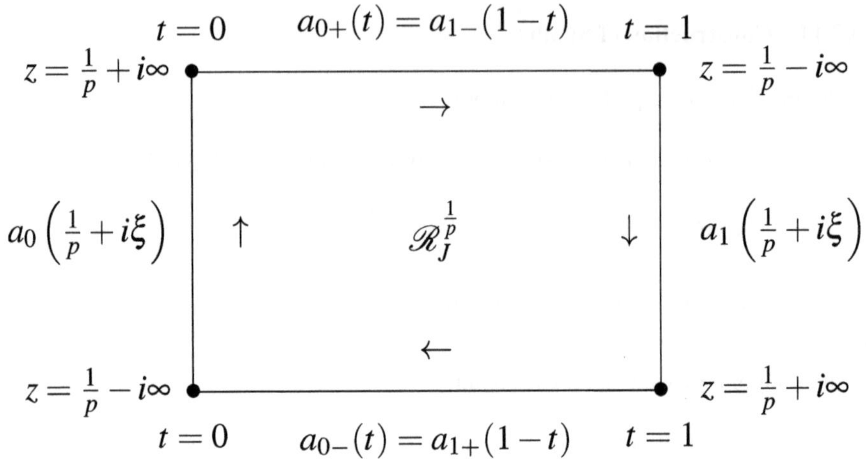


## 3.4 *Elliptic S.I.O. on* $L^p(J)$ *and the* Index$_p$ *Theorem*

**Definition 6.** An operator $A$ is *elliptic on* $L^p(J)$ if

$$\operatorname{Symb}^{\frac{1}{p}}(A)\left(t,\frac{1}{p}+i\xi\right)\neq 0$$

for $(t,\frac{1}{p}+i\xi)$ on the Symbol Space $\mathscr{R}_J^{\frac{1}{p}}$.

**Systems:** A System of operators **A** is *elliptic on* $L^p(J)$ if

$$\det\operatorname{Symb}^{\frac{1}{p}}(\mathbf{A})(t,\frac{1}{p}+i\xi)\neq 0.$$

on the Symbol Space $\mathscr{R}_J^{\frac{1}{p}}$.

*Remark 8.* Ellipticity on $L^p(J)$ depends on $p$.

If $A$ is *elliptic on* $L^p(J)$, we know that

$$\operatorname{Symb}^{\frac{1}{p}}(A)(t,\frac{1}{p}\pm i\infty)\neq 0,$$

but there could be a $p_1\neq p$ such that there is a finite $z_1=\left(\frac{1}{p_1}+i\xi_1\right)$, such that

$$\operatorname{Symb}^{\frac{1}{p_1}}(A)(0,z_1)=0$$

So *$A$ is not elliptic on* $L^{p_1}(J)$

### 3.4.1 Elliptic on $L^p(J)$ iff Fredholm on $L^p(J)$

If $A$ is elliptic on $L^p(J)$, there is an S.I.O. (parametrix) $B_p$ on $L^p(J)$ such that $B_pA - I$ and $AB_p - I$ are compact operators on $L^p(J)$.

**N.B.** The Hardy Kernel parts of $B_p$ may depend on $p$

### 3.4.2 Index Theorem for Elliptic Operators on $L^p(J)$

If $A$ is a Fredholm Operator on $L^p(J)$, the *Analytic Index of A on* $L^p(J)$ is

$$\text{Index}_p(A) = \dim((\ker A) \cap L^p(J)) - \dim((\ker A^*) \cap L^{p/(p-1)}(J))$$

**Theorem 3.3.** *If A is an elliptic S.I.O. on* $L^p(J)$*, then*

$$\text{Index}_p(A) = \frac{1}{2\pi}\Delta_{\mathscr{R}_J^{\frac{1}{p}}}\left(\arg\left(\det \text{Symb}^{\frac{1}{p}}(A)\right)\right).$$

**N.B.** The change in argument, $\Delta(\arg\ldots)$, is calculated as $\mathscr{R}_J^{\frac{1}{p}}$ is traversed in the *clockwise* direction. Theorem 3.3 was shown by Lewis and Parenti in [LP83] and [Lew91].

### 3.4.3 Change in $p$ May Change Index

If the operator $A$ is elliptic on $L^p$ and $L^{\bar{p}}$, $\bar{p} < p$, and $a_0(z)$ has a zero or pole in the strip $\Gamma_{\frac{1}{p},\frac{1}{\bar{p}}} = \left\{\frac{1}{p} < \mathbb{R}z < \frac{1}{\bar{p}}\right\}$,

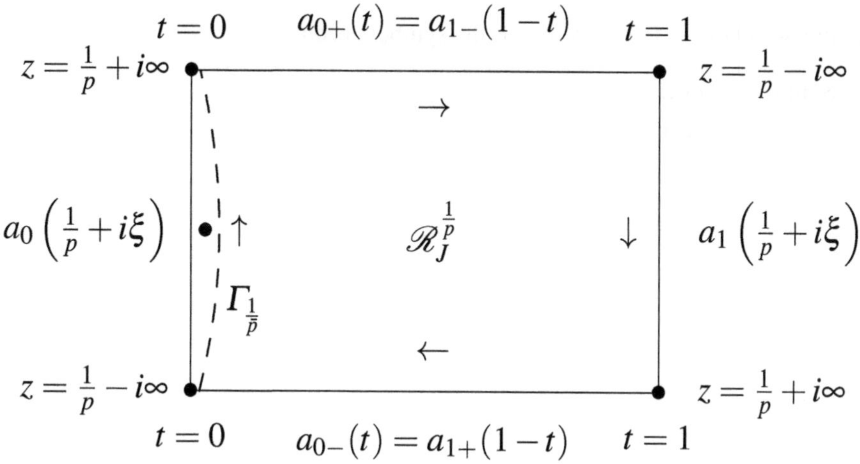

Using the *Argument Principle*:

$$\Delta\,\text{Index} = \#\,\text{Zeroes} - \#\,\text{Poles}$$

A similar adjustment is made if $a_1(z)$ has a pole or zero in the strip $\Gamma_{\frac{1}{p},\frac{1}{p}}$.

## 3.5 Power Weights

The above discussion works with power weights.

Consider the operators on spaces

$$L^p_\alpha = \{f \mid t^\alpha f \in L^p\}.$$

The Hilbert Transform and the Hardy Kernel parts of the operator yield

$$\begin{aligned} t^\alpha A_m f(t) &= \frac{1}{2\pi i}\int_{\Gamma_{1/p}} t^{-z} m(t, z+\alpha)\widetilde{[s^\alpha f(s)]}(z)\,dz, \\ &= \frac{1}{2\pi i}\int_{\Gamma_{1/p}} t^{-z} m(t, z+\alpha)\tilde{f}(z+\alpha)\,dz, \\ &= \frac{1}{2\pi i}\int_{\Gamma_{\alpha+1/p}} t^{-z} m(t, z)\tilde{f}(z)\,dz. \end{aligned}$$

### 3.5.1 Symbol Space $\mathscr{R}_J^{\alpha+\frac{1}{p},\frac{1}{p}}$ for $L^p_\alpha$

With a weight $t^{\alpha/p}$, $0 < \frac{1}{p} + \alpha < 1$, the symbol becomes

$\text{Symb}^{\alpha+\frac{1}{p},\frac{1}{p}}(A)$:

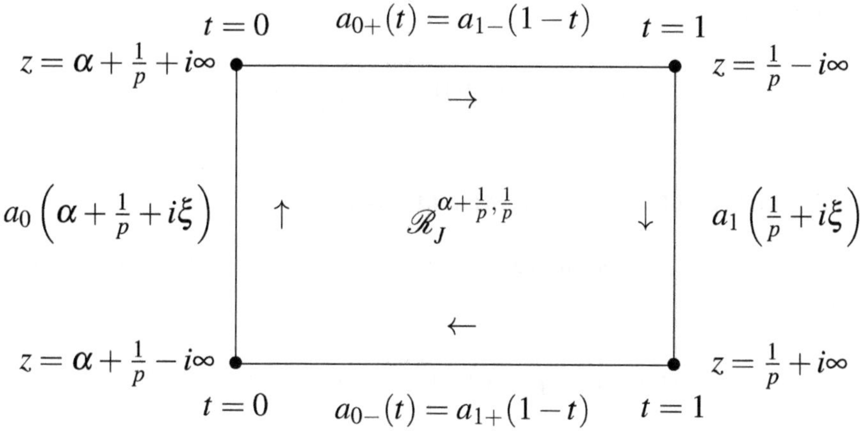


### 3.5.2 Different Weights

*Remark 9.* Martin Costabel [Cos80] had the happy idea of introducing different weights $t^{\alpha_{2k}} \approx (\mathrm{dist}(X, P_{2k}))^{\alpha_{2k}}$ near each vertex $P_{2k}$.

## 3.6 *Analytic Double Layer Kernels on $C^1$ Curvilinear Polygons*

Consider

$$Kf(P) = \mathrm{p.\,v.} \int_{\partial \mathbb{P}} k(P-Q) f(Q) b(Q) ds_Q,$$

$b(Q)$ continuous on each edge $\overline{P_{2k} P_{2k+2}}$.

### 3.6.1 ADLK on $C^1$ Polygons – Same Edge

Let $\gamma_0 = \gamma_0(t)$, $t \in J$, be a $C^1$ curve

$$K_b^{0,0} f(t) = \mathrm{p.\,v.} \int_0^1 k\left(\gamma_0(t) - \gamma_0(s)\right) b(s) f(s) \left|\frac{d\gamma_0}{ds}\right| ds.$$

Then $K_b^{0,0}$ is an S.I.O. with the same symbol as

$$\overline{K_b}^{0,0} f(t) = \mathrm{p.\,v.} \int_0^1 k\left(\gamma_0'(t)(t-s)\right) b(s) f(s) \left|\frac{d\gamma_0}{ds}\right| ds,$$

which has Symbol

$$\mathrm{Symb}^{\frac{1}{p}}(K_b^{0,0}) = k\left(\gamma_0'(t)\right) b(t) \left|\frac{d\gamma_0}{dt}\right| \pi\, \mathrm{Symb}^{\frac{1}{p}}(H).$$

### 3.6.2 ADLK on $C^1$ Polygons: Adjacent Edges

Let $\gamma_1(t), \gamma_2(t), t \in J$, be simple $C^1$ curves, $\gamma_j(0) = (0,0)$, $\gamma_1'(0) = \mathbf{u}_1$, $\gamma_2'(0) = \mathbf{u}_2$, $\mathbf{u_1} \neq \mathbf{u_2}$,

$$K_b^{1,2} f(t) = \text{p. v.} \int_0^1 k\left(\gamma_1(t) - \gamma_2(s)\right) b(s) f(s) \left|\frac{d\gamma_2}{ds}\right| ds.$$

Then $K_b^{1,2}$ is an S.I.O. with the same symbol as the Hardy Kernel

$$\overline{K}_b^{1,2} f(t) = \text{p. v.} \int_0^1 k\left((t/s)\,(\mathbf{u_1} - \mathbf{u_2})\right) b(0) f(s) \left|\gamma_2'(0)\right| \frac{ds}{s}$$

which has kernel

$$k_b^{1,2}(t) = k\left(t\mathbf{u_1} - \mathbf{u_2}\right) b(0) \left|\mathbf{u_2}\right|.$$

### 3.6.3 ADLK on $C^1$ Polygons: Non Adjacent Edges

The operators are compact.

*Remark 10.* The results of Sects. 3.6.1, 3.6.2, and 3.6.3 were shown by the author in [Lew91].

*Remark 11.* ADLK on $C^1$ Polygons—*False Vertex Trick.* To avoid the complications of considering symbols at $t = 1$ there is a *False Vertex Trick* of introducing a flat vertex $P_{2k+1}$ between $P_{2k}$ and $P_{2k+2}$. Then parameterize $\overline{P_{2k} P_{2k+1}}$ with $t = 0$ at $P_{2k}$. Parameterize $\overline{P_{2k+2} P_{2k+1}}$ with $t = 0$ at $P_{2k+2}$. In many cases, Remark 6 shows that the determinant of the $\mathrm{Symb}^{\frac{1}{p}}$ corresponding to each flat vertex $P_{2k+1}$ will be constant.

The symbol and change in argument of the resulting system at $t = 0$ is reduced to the problem of considering blocks associated with each vertex $P_{2k}$.

## 3.7 Related Matters and Applications

### 3.7.1 Index on Sobolev Spaces

With suitable regularity of the coefficients, Lorenza Diomeda and Benedetta Lisena [DL85b] showed that the Index on $\mathbf{W}^{p,k}$ for an elliptic operator [system] depends on

$$\#\text{Zeroes} - \#\text{Poles}$$

in the strip $\Gamma_{-k+\frac{1}{p},\frac{1}{p}} = \left\{-k + \frac{1}{p} < \operatorname{Re} z < \frac{1}{p}\right\}$.

Lorenza Diomeda and Benedetta Lisena [DL85a] treat an oblique derivative problem in a polygonal domain in $W^{p,k}$-spaces.

### 3.7.2 Elliptic Systems in the Plane

Lorenza Diomeda and Benedetta Lisena [DL92] reduce the Dirichlet problem for a large class of strongly elliptic systems to double layer potentials of index 0 for $2 \leq p < \infty$. The two *bad values* of $p$ associated with each corner are calculated and lie in $1 < p < 2$.

### 3.7.3 Inverse Norm Estimates

Irina Mitrea [Mit05] used the representation to find inverse norm estimates for various boundary value problems for the Laplace equation.

### 3.7.4 Spectral Radius

Irina Mitrea [Mit02] treats elastostatic and hydrostatic layer potentials on curvilinear polygons and proves and shows that the spectral radius of the compact $K$ in $I + K$ is less than one, $2 \leq p < \infty$, so that a Neumann series converges.

### 3.7.5 Transmission Problems

Lorenza Diomeda and Benedetta Lisena [DL95] treat the transmission problem for elastostatics—differing Lamé constants in an interior curvilinear polygon.

### 3.7.6 Transmission Problems

Irina Mitrea and Katherine Ott [MO07] give counterexamples to the well-posedness of $L^p$ transmission boundary value problems for the Laplacian.

## 3.8 Possible Extensions

$C^{0+\mathrm{VMO}}$ Curvilinear Polygons

*Remark 12. "Are there better results?"*

- Can the Symbolic Calculus be constructed if each edge of $\partial\mathbb{P}$ is a $C^{0+\mathrm{VMO}}$ curve? It is likely that (one sided) continuity of the tangent vector at each vertex would be required.
- What Multilinear Commutator Theorems are known for Hardy Kernel operators and BMO or VMO?

## References


[CC00] Calderón, A.P., Calderón, C.P.: A representation formula and its applications to singular integrals. Indiana Univ. Math. J. **49**(1), 1–5 (2000). MR1777040 (2001h:32009)

[CCF+78] Calderón, A.P., Calderón, C.P., Fabes, E., Jodeit, M., Rivière, N.M.: Applications of the Cauchy integral on Lipschitz curves. Bull. Am. Math. Soc. **84**(2), 287–290 (1978). MR0460656 (57 #649)

[CMM82] Coifman, R.R., McIntosh, A., Meyer, Y.: L'intégrale de Cauchy définit un opérateur borné sur $L^2$ pour les courbes Lipschitziennes. Ann. Math. **116**(2), 361–387 (1982). MR672839 (84m:42027)

[Cos80] Costabel, M.: Singular integral operators on curves with corners. Integ. Equ. Oper. Theory **3**(3), 323–349 (1980). MR 580713 (81j:45007)

[CRW76] Coifman, R.R., Rochberg, R., Weiss, G.: Factorization theorems for Hardy spaces in several variables. Ann. Math. (2) **103**(3), 611–635 (1976). MR 0412721 (54 #843)

[DL85a] Diomeda, L., Lisena, B.: Boundary integral system for an oblique derivative problem in $W^{p,k}$ -spaces. Boll. Un. Mat. Ital. B (6) **4**(2), 639–654 (1985). MR MR805433 (87h:35065)

[DL85b] Diomeda, L., Lisena, B.: Pseudodifferential operators of Mellin type on Sobolev spaces. Ann. Mat. Pura Appl. **140**(4) 255–284 (1985). MR MR807640 (87b:47052)

[DL92] Diomeda, L., Lisena, B.: The Dirichlet problem for elliptic systems in piecewise $C^1$ plane domains. Indiana Univ. Math. J. **41**(3), 649–670 (1992). MR MR1189905 (94a:35035)

[DL95] Diomeda, L., Lisena, B.: On a transmission problem for two systems of elastostatics with polygonal interface. Integ. Equ. Oper. Theory **23**(4), 416–434 (1995). MR MR1361052 (97b:73018)

[Lew91] Lewis, J.E.: A symbolic calculus for layer potentials on $C^1$ curves and $C^1$ curvilinear polygons. Proc. Am. Math. Soc. **112**(2), 419–427 (1991). MR MR1043413 (91i:47071)

[LP80] Lewis, J.E., Parenti, C.: Pseudodifferential operators and Hardy kernels on $L^p(R^+)$. Ann. Scuola Norm. Sup. Pisa Cl. Sci. (4) **7**(3), 481–503 (1980). MR MR597548 (83d:47052)

[LP83] Lewis, J.E., Parenti, C.: Pseudodifferential operators of Mellin type. Commun. Partial Differ. Equ **8**(5), 477–544 (1983). MR MR695401 (86f:35185)

[LSS93] Lewis, J.E., Selvaggi, R., Sisto, I.: Singular integral operators on $C^1$ manifolds. Trans. Am. Math. Soc. **340**(1), 293–308 (1993). MR MR1124170 (94a:58194)

[Mit02] Mitrea, I.: On the spectra of elastostatic and hydrostatic layer potentials on curvilinear polygons. J. Fourier Anal. Appl. **8**(5), 443–487 (2002). MR MR1921983 (2003g:35163)

[Mit05] Mitrea, I.: Boundary problems for harmonic functions and norm estimates for inverses of singular integrals in two dimensions. Numer. Funct. Anal. Optim. **26**(7–8), 851–878 (2005). MR MR2192025 (2006j:45016)

[MO07] Mitrea, I., Ott, K.: Counterexamples to the well-posedness of Lp transmission boundary value problems for the Laplacian. Proc. Am. Math. Soc. **135**(7), 2037–2043 (2007) (electronic). MR MR2299477 (2008h:35379)

[Muc72] Muckenhoupt, B.: Weighted norm inequalities for the Hardy maximal function. Trans. Am. Math. Soc. **165**, 207–226 (1972). MR0293384 (45 #2461)

[Pal65] Palais, R.: Seminar on the Atiyah–Singer Index Theorem. Annals of Mathematics Studies, vol. 57. Princeton Univerity Press, Princeton (1965)

[SS81] Selvaggi, R., Sisto, I.: Boundary value problems for first-order symmetric elliptic systems with constant coefficients in open sets of class $C^1$. Note Mat. **1**(2), 155–185 (1983). MR MR695147 (85h:35079)

[SS97] Selvaggi, R., Sisto, I.: A Dirichlet problem for harmonic forms in $C^1$-domains. Integ. Equ. Oper. Theory **28**(3), 343–357 (1997). MR MR1456561 (98g:58178)

[Uch78] Uchiyama, A.: On the compactness of operators of Hankel type. Tôhoku Math. J. (2) **30**(1), 163–171 (1978). MR0467384 (57 #7243)

# Towards a Unified Theory of Sobolev Inequalities

**Joaquim Martín and Mario Milman**

**Summary.** We discuss our work on pointwise inequalities for the gradient which are connected with the isoperimetric profile associated with a given geometry. We show how they can be used to unify certain aspects of the theory of Sobolev inequalities. In particular, we discuss our recent papers on fractional order inequalities, Coulhon type inequalities, transference and dimensionless inequalities and our forthcoming work on sharp higher order Sobolev inequalities that can be obtained by iteration.

## 1 Introduction

In this expository note we survey some of our previous work towards a unified theory of Sobolev inequalities using pointwise rearrangement inequalities. The presentation includes our recent results [81] on Coulhon inequalities (cf. [5, 28]), integral transference and dimensionless inequalities [82], fractional inequalities [80], extrapolation and self-improvement [79]. We also include a brief section on how iterations of our estimates lead to sharp higher order pointwise inequalities [83]. We hope that this attempt to summarize and organize some of the material, together with the inclusion of our motivation and details on the methods used, could be of some interest.

In conclusion we should mention three important lessons we learned from Gian Carlo Rota [106]:

**1.** *Publish the same results several times* **2.** *Do not worry about your mistakes and* **3.** *Write informative introductions.*

J. Martín
Department of Mathematics, Universitat Autònoma de Barcelona, Barcelona, Spain
e-mail: jmartin@mat.uab.cat

M. Milman (✉)
Department of Mathematics, Florida Atlantic University, Boca Raton, FL, USA
e-mail: mario.milman@gmail.com

 
A.M. Stokolos et al. (eds.), *Special Functions, Partial Differential Equations, and Harmonic Analysis*, Springer Proceedings in Mathematics & Statistics 108,
DOI 10.1007/978-3-319-10545-1_13

## 2 Basic Definitions and Notation

Before starting our presentation let us recall some basic definitions and set up the notation.

Let $(\Omega,\mu)$ be a measure space $(\Omega,\mu)$; for a measurable function $u:\Omega\to\mathbb{R}$, the *distribution function* of $u$ is given by

$$\mu_u(t)=\mu\{x\in\Omega:|u(x)|>t\}\quad(t\geq 0).$$

The *decreasing rearrangement* of a function $u$ is the right-continuous non-increasing function from $[0,\mu(\Omega))$ into $\mathbb{R}$ which is equimeasurable with $u$. It can be defined by the formula

$$u_\mu^*(s)=\inf\{t\geq 0:\mu_u(t)\leq s\},\quad s\in[0,\mu(\Omega)).$$

The maximal average $u_\mu^{**}(t)$ is defined by

$$u_\mu^{**}(t)=\frac{1}{t}\int_0^t u_\mu^*(s)ds.$$

When the measure is clear from the context, or when we are dealing with Lebesgue measure, we may simply write $u^*$ and $u^{**}$, etc.

Let $X=X(\Omega)$ be a Banach function space on $(\Omega,\mu)$, with the Fatou property.[1] We shall say that $X$ is a *rearrangement-invariant* (r.i.) space, if $g\in X$ implies that all $\mu-$measurable functions $f$ with $f_\mu^*=g_\mu^*$, also belong to $X$ and moreover, $\|f\|_X=\|g\|_X$. The functional $\|\cdot\|_X$ will be called a rearrangement invariant norm. Typical examples of r.i. spaces are the $L^p$-spaces, Orlicz spaces, Lorentz spaces, Marcinkiewicz spaces, etc.

Let $X(\Omega)$ be an r.i. space, then there exists an r.i. space $\bar{X}=\bar{X}(0,\mu(\Omega))$ on $((0,\mu(\Omega)),m)$, ($m$ denotes the Lebesgue measure on the interval $(0,\mu(\Omega))$) such that

$$\|f\|_{X(\Omega)}=\|f_\mu^*\|_{\bar{X}(0,\mu(\Omega))}. \tag{1}$$

$\bar{X}$ is called the *representation space* of $X(\Omega)$ which is unique if $\mu(\Omega)$ is finite. We refer the reader to [15] for a complete treatment of the theory of (r.i.) spaces.

As usual, the symbol $f\simeq g$ will indicate the existence of a universal constant $c>0$ (independent of all parameters involved) so that $(1/c)f\leq g\leq c\,f$, while the symbol $f\preceq g$ means that $f\leq c\,g$; finally given $1\leq p\leq\infty$, $p'=p/(p-1)$.

[1] This means that if $f_n\geq 0$, and $f_n\uparrow f$, then $\|f_n\|_X\uparrow\|f\|_X$ (i.e., Fatou's Lemma holds in the $X$ norm).

## 3 Two Examples

We give two simple examples that illustrate the difficulties in deriving a unified theory of Sobolev inequalities.

*Example 1.* In dimension 1 the Sobolev embedding theorem is connected with the fundamental theorem of calculus: If $f$ is a Lip function with compact support, then

$$f(y) = \int_{-\infty}^{y} f'(s)ds$$

and therefore[2]

$$\|f\|_{\infty} \leq \|f'\|_1 .$$

In dimension $n$ the corresponding estimate does not hold. In particular, there are functions $f \in W^{1,n}(\mathbb{R}^n)$ that are not bounded. Instead, we have somewhat weaker results: (Sobolev)

$$\|f\|_{p_n} \leq c_{n,p} \|\nabla f\|_p , f \in C_0^{\infty}(\mathbb{R}^n), \text{ with } \frac{1}{p_n} = \frac{1}{p} - \frac{1}{n}, \ 1 < p < n,$$

with $c_n \to \infty$, when $p \to n$. The inequality is also true when $p = 1$ (Gagliardo–Nirenberg): there exists $c_n$ such that

$$\|f\|_{n'} \leq c_n \|\nabla f\|_1 , \ f \in C_0^{\infty}(\mathbb{R}^n).$$

One way to complete the picture for $p = n$ is by extrapolation: Trudinger [112] showed by extrapolation that if $\Omega$ is an open set in $\mathbb{R}^n$ with $|\Omega| < \infty$, then

$$W_0^{1,n}(\Omega) \subset e^{L^{n'}} .$$

This example, in particular, illustrates the crucial role of dimension in the Euclidean Sobolev inequalities.

As a counterpart, consider the following.

*Example 2.* Consider $\mathbb{R}$ but now equipped with one dimensional Gaussian measure $\gamma_1$. In this case the gain of integrability is logarithmic ("logarithmic type Sobolev inequalities")

---

[2] actually using the available cancellation, we have $2f(y) = \int_{-\infty}^{y} f'(s)ds - \int_{y}^{\infty} f'(s)ds$, therefore the constant of the embedding can be improved,

$$\|f\|_{\infty} \leq \frac{1}{2} \|f'\|_1 .$$

$$\|f\|_{L(LogL)^{1/2}(\gamma_1)} \leq c(\||\nabla f|\|_{L^1(\gamma_1)} + \|f\|_{L^1(\gamma_1)}).$$

Similarly, for $p > 1$ we have

$$\|f\|_{L^p(LogL)^{p/2}(\gamma_1)} \leq c(\||\nabla f|\|_{L^p(\gamma_1)} + \|f\|_{L^p(\gamma_1)}).$$

Moreover, the gain of integrability does not change with the dimension! For example, for $\mathbb{R}^n$ equipped with Gaussian measure $\gamma_n$, we have

$$\|f\|_{L(LogL)^{1/2}(\gamma_n)} \leq c(\||\nabla f|\|_{L^1(\gamma_n)} + \|f\|_{L^1(\gamma_n)}).$$

In other words, while in the Euclidean geometry the spaces involved in the inequalities depend on the dimension, and the gain of integrability is naturally measured with powers, i.e. using $L^p$ or $L(p,q)$ spaces, in the Gaussian world the gain of integrability is logarithmic, independent of the dimension, and the right spaces one needs to measure the gain of integrability are logarithmic Orlicz type spaces. The celebrated logarithmic Sobolev inequalities of Gross (cf. [47]) are among the most prominent examples of dimensionless inequalities in the Gaussian world (cf. [47]).

Traditionally this state of affairs has led to different methods/theories to deal with Euclidean or Gaussian Sobolev inequalities, or more generally, Sobolev inequalities in other geometries, e.g. Euclidean domains with measures of the form $w(x)dx$, Riemannian manifolds, or more generally, metric measure spaces. In particular, what is the role of dimension? What are the function spaces one needs to use to measure the integrability gains?

While some of the differences are unavoidable, one wonders if it is possible to unify at least some aspects of these disparate theories and thus, maybe, provide a better understanding. Recently there has been progress in this direction from several different directions by a number of authors. The results presented in this note describe mainly our work towards a unified theory of Sobolev inequalities using pointwise inequalities on rearrangements; as a consequence, the list of references is disproportionately tilted towards our own work. Moreover, there is a huge literature on Sobolev inequalities, and while the bibliography we have compiled is rather large, we must warn the reader that we did not attempt to cover all the references, not even the important references. We apologize in advance if your favorite paper/author is not included in the bibliography, hopefully it should not be more than one (reference) iteration away. In particular, we should explicitly mention our debt to the pioneering[3] work of V. Maz'ya which has greatly influenced our viewpoint of the subject (cf. Sect. 5 below). For more information, background and more comprehensive bibliographies we refer to [1, 30, 34, 49, 59, 69, 78, 88, 101, 104, 107], and [110].

[3]For a detailed presentation of Maz'ya's remarkable early work we refer to [51].

# 4 Rearrangement Invariant Spaces and the Brezis–Wainger–Hansson Embedding

One early difficulty in trying to develop any type of unified theory is that, even in the Euclidean case, the limiting borderline case ($p = n$) of the Sobolev embedding apparently requires the use of a different scale of spaces, e.g. an "extrapolation space"... So we started our work trying to understand the issues connected with the limiting inequalities. In a convoluted way our efforts to understand the limiting cases eventually led us to a better understanding on how to approach all the cases simultaneously![4] So it seems that this is a good point where to start with the story.

An early result in this direction was obtained by Trudinger [112], who showed that for a domain $\Omega$ in $\mathbb{R}^n$, with $|\Omega| < \infty$,

$$W_0^{1,n}(\Omega) \subset e^{L^{n'}}(\Omega). \tag{2}$$

Brezis–Wainger [23] improved this result using the rearrangement inequalities of O'Neil [99], while Hansson [50] obtained similar results. Their result reads

$$W_0^{1,n}(\Omega) \subset \mathrm{BWH}(\Omega), \tag{3}$$

where if, say, $|\Omega| = 1$, then BWH $=$ (Brezis-Wainger-Hansson) is the space defined sby

$$\|f\|_{\mathrm{BWH}} = \left\{ \int_0^1 \left( f^{**}(s) \frac{1}{1 + \log \frac{1}{s}} \right)^n \frac{ds}{s} \right\}^{1/n} < \infty.$$

Maz'ya (cf. [88, (3.1.4), p. 232] and the references therein) had obtained earlier an inequality using his capacity theory that, in particular, implies the embedding (3). The improvement over (2) is given by the easily verified fact that

$$\mathrm{BWH}(\Omega) \subset e^{L^{n'}}(\Omega).$$

On the other hand, O'Neil [99] extended the original (one dimensional) results of Hardy–Littlewood, as follows

$$W^{1,p}(\mathbb{R}^n) \subset L(p_n, p),\ 1 < p < n,\ \frac{1}{p_n} = \frac{1}{p} - \frac{1}{n},$$

---

[4] This is a somewhat disappointing turn of events for the developers of general abstract theories studying limiting inequalities (e.g., [56]) but our current understanding of Sobolev inequalities shows that: (a) Sobolev inequalities self-improve (cf. [5, 48, 79, 86]) and (b) the extrapolations of Sobolev inequalities take the form: "one inequality" implies a family of inequalities and in some cases "one inequality implies all"! (cf. [5, 29, 79]). We also refer to the forthcoming [84] for a connection with extrapolation of Sobolev inequalities à la Rubio de Francia.

where the Lorentz $L(p,q)$ spaces, for $p<\infty, q\in[1,\infty]$ are defined by

$$L(p,q)=\{f:\|f\|_{L(p,q)}=\left\{\int_0^{\infty}(f^{**}(t)t^{1/p})^q\frac{dt}{t}\right\}^{1/q}<\infty\}.$$

The results of O'Neil and Brezis–Wainger–Hansson, like the original one dimensional results of Hardy–Littlewood, can be shown to be optimal within the class of rearrangement invariant spaces: if $X$ is a rearrangement invariant space, then

$$W^{1,p}\subset X\Rightarrow L(p_n,p)\subset X, \text{ if } p<n, \text{ or BWH}\subset X, \text{ if } p=n.$$

As a consequence, the $L(p.q)$ spaces are not enough to describe the optimal Sobolev inequalities (e.g., the case $p=n$ requires a different space). This led to the introduction in the theory of Sobolev embeddings of the following modification of the $L(p,q)$ spaces which, as we shall see, resolves this difficulty (cf. [10, 11] and, as it turns out, in a different way and less explicitly in [111] and [70]). For a measure space we define

$$L(\infty,\infty)=\{f:\|f\|_{L(\infty,\infty)}=\sup_t(f^{**}(t)-f^*(t))<\infty\}.$$

This space was introduced by Bennett–DeVore–Sharpley [14] who in their paper show that for functions defined on a cube, $L(\infty,\infty)$ is the rearrangement invariant hull of BMO, i.e. $L(\infty,\infty)$ is the smallest possible space that contains all the rearrangements of functions in BMO. One should note here that the usual definition of $L(\infty,\infty)$ would give

$$\|f\|_{L(\infty,\infty)}=\sup_t f^{**}(t)=\|f\|_{L^\infty},$$

while the space $L(\infty,\infty)$ that we have defined is bigger: we have BMO $\subset L(\infty,\infty)$.

More generally, note that if we formally attempt to define the $L(\infty,q)$ spaces using the classical definition the resulting spaces are trivial:

$$\int_0^{\infty}f^{**}(s)^q\frac{ds}{s}<\infty\Rightarrow f=0.$$

On the other hand, if we redefine the $L(\infty,q)$ spaces by means of replacing $f^{**}(t)$ by the oscillation $f^{**}(t)-f^*(t)$, then the spaces defined by the condition

$$\|f\|_{L(\infty,q)}=\left\{\int_0^{|\Omega|}\left(f^{**}(t)-f^*(t)\right)^q\frac{dt}{t}\right\}^{1/q}<\infty,$$

are not trivial, since the differentiation theorem provides us with a cancellation at the origin. In [10, 11] the role of the $L(\infty, q)$ spaces in the Sobolev embedding theorem was observed. It was noted there that these spaces were relevant in connection with an inequality implicit in the paper by Alvino–Trombetti–Lions [4]: If $f$ is symmetrically decreasing, then

$$f^{**}(t) - f^{*}(t) \leq c_n t^{1/n} |\nabla f|^{**}(t). \tag{4}$$

Now, if we let $f^{\circ}$ denote the symmetric rearrangement of $f$ (cf. [61, 69]), then for smooth $f$, the Pólya–Szegö principle can be formulated as (cf. [86] and the references there to earlier work by Fournier)

$$|\nabla f^{\circ}|^{**}(t) \leq |\nabla f|^{**}(t).$$

Then, since $(f^{\circ})^{*} = f^{*}$, it follows that we can eliminate the restriction for $f$ to be symmetrically decreasing, therefore (4), indeed, holds for all smooth $f$. Next, integrating (4), we see that:

$$\left\{\int_0^{|\Omega|} \left((f^{**}(t) - f^{*}(t))\, t^{1/p-1/n}\right)^q \frac{dt}{t}\right\}^{1/q} \leq c_n \left\{\int_0^{|\Omega|} \left(t^{1/p} |\nabla f|^{**}(t)\right)^q \frac{dt}{t}\right\}^{1/q}. \tag{5}$$

The left-hand side is equivalent to $\|f\|_{L(p_n,q)}$ (cf. [73]), and moreover, for $q = p$ we have

$$\left\{\int_0^{|\Omega|} \left(t^{1/p} |\nabla f|^{**}(t)\right)^p \frac{dt}{t}\right\}^{1/p} \leq p' \, \||\nabla f|\|_p .$$

Thus, for $q = p < n$ we have recovered the classical Sobolev inequality. Moreover, the inequality (5) is valid and makes sense in the limiting case $p = n$. In particular for $q = p = n$, we have

$$\left\{\int_0^{|\Omega|} \left(f^{**}(t) - f^{*}(t)\right)^n \frac{dt}{t}\right\}^{1/n} \leq c_n \|\nabla f\|_n .$$

The condition that the left-hand side of the previous inequality is finite defines the space $L(\infty, n)$, and, moreover, we have (cf. [11])

$$L(\infty, n) \subset \mathrm{BWH}.$$

Since redefining the $L(p,q)$ spaces by means of replacing $f^{**}$ by $f^{**} - f^{*}$ gives equivalent norms when the parameters are in the usual range of the classical theory,[5] we now have a unified method to prove an inequality that includes the problematic case $p = n$,

[5] For a detailed discussion on equivalences between different Lorentz *norms* we refer to [11, 57, 73].

$$\int_0^{|\Omega|} \left((f^{**}(t) - f^*(t))\, t^{1/p-1/n}\right)^q \frac{dt}{t} \le c_n \int_0^{|\Omega|} \left(t^{1/p} |\nabla f|^{**}(t)\right)^q \frac{dt}{t}. \quad (6)$$

In fact, let us note that we can do this even if the measure is infinite,

$$\left\{\int_0^{\infty} \left((f^{**}(t) - f^*(t))\right)^n \frac{dt}{t}\right\}^{1/n} \le c_n \left\||\nabla f|\right\|_n .$$

In this notation, the original Hardy–Littlewood–O'Neil program[6] has been completed[7]:

$$\|f\|_{L(p_n,p)} \le c \left\||\nabla f|\right\|_{L^p}, \; 1 \le p \le n.$$

The improvement on Brezis–Wainger–Hansson is possible due to the fact that the class $L(\infty, n)$ is not a linear space.

A posteriori, we also understood that an inequality obtained by Kolyada on the unit cube [62], that is exactly like (4), except that the oscillation condition on the left-hand side is given in terms of $f^*(t) - f^*(2t)$, could also have been used for the Euclidean inequalities above (cf. [100]). Moreover, Tartar [111] (cf. also [70]) earlier than [11], but apparently after [10], had also obtained, using a different approach based on truncations, a discrete version of a result that, with some work, can be seen to be equivalent to (6). This was not clearly understood at the time we wrote [11].

From the point of view of the development of our program, we draw the following from this section: Redefining the target spaces using oscillations, and using the pointwise inequality (4), we could treat all the cases of the classical Sobolev inequalities in an optimal unified manner.

To proceed further with our program of understanding more general geometries the question we faced next was: What would be a substitute for (4) when dealing with other geometries? Since the oscillation condition $f^{**}(t) - f^*(t)$ is a general construct and, in fact, can be understood from an approximation point of view (cf. [57]), we concentrated our efforts on understanding better the right-hand side of the inequality.

[6] The proof to deal with the case $p = 1$ is slightly different and hinges on a variant of (4), namely (cf. [86])

$$\int_0^t (f^{**}(s) - f^*(s)) s^{1/n} \frac{ds}{s} \le c \int_0^t |\nabla f|^*(s) ds.$$

[7] Actually (6) also makes sense, and gives sharp results, when $p > n$ (cf. [80, Chapter 9]).

## 5 Martin–Milman–Pustylnik Meet Maz'ya

The Gagliardo–Nirenberg inequality

$$\|f\|_{L^{n'}} \leq c \, \||\nabla f|\|_{L^1}, f \in C_0^\infty(\mathbb{R}^n), \tag{7}$$

is well known to be equivalent to the isoperimetric inequality (cf. Maz'ya [88] and the references therein.) Moreover, using the chain rule and the scale properties of the $L^p$ spaces (here scale = Hölder's inequality) one can readily see that (7) implies "all" the classical Sobolev inequalities: simply consider positive $f$ and apply (7) to $f^\alpha$ for suitable $\alpha$ combined with the chain rule and Hölder's inequality (cf. [107]).

From our experience we knew that we could also derive "all the Sobolev inequalities" from the rearrangement inequality

$$f^{**}(t) - f^*(t) \leq c_n t^{1/n} |\nabla f|^{**}(t). \tag{8}$$

So it was natural to ask what was the connection between (7) and (8)? The key for us was the method of truncation that had been devised by Maz'ya (cf. [88]) and Talenti [110], combined with one natural idea that comes from interpolation theory, or more precisely, from the work of Marcinkiewicz and Alberto Calderón [24]. The idea simply put is to do the (smooth) cut-offs using the rearrangement of the function to be truncated evaluated at a fixed point $t$. This leads to pointwise rearrangement inequalities. In this way we could show that (7) is equivalent to (8)! (cf. [86])

One advantage of (8) is that since it is a pointwise inequality it gives all the results, even for non $L^p$ norms. This was another innovation from the work of Alberto Calderón [24]: While previously to prove interpolation theorems one had to specify the spaces in advance this was not longer necessary in A.P. Calderón's theory. This was crucially important in our program since the norms that one needs to use to measure the integrability gains depend on the geometry. We should also point out that for Euclidean geometries pointwise rearrangement inequalities had been devised and applied by Talenti (cf. [110] and the references therein), and since then we have been applied by many authors working on nonlinear PDEs…(cf. [41, 104]). In connection with compactness of Sobolev embeddings connected with the developments in this section we refer to [86, 103] and the many references therein.

## 6 The Gaussian Inequality: Ledoux's Inequality

The next item on our agenda was a test case: The Gaussian world. The questions here were: How to formulate the basic inequality (8) in the Gaussian context? What type of symmetrization was needed to replace the symmetric rearrangement? What inequality would take the role of the Gagliardo–Nirenberg inequality in the

Gaussian world? The answer to the last question had been already provided by Ledoux (cf. [66]). Ledoux's inequality is connected with the use of the Maz'ya isoperimetric profile, usually referred to as isoperimetric profile[8] $I_\gamma$, associated with Gaussian measure in $\mathbb{R}^n$,

$$I_\gamma(t) = \inf\{\gamma^+(A) : \gamma(A) = t\},$$

where $\gamma^+$ is the Minkowski content associated with Gaussian measure defined for Borel sets $A$ by

$$\gamma^+(A) = \lim_{h\to 0} \frac{\gamma((A_h)) - \gamma(A)}{h},$$

where $A_h = \{x : d(x, A) < h\}$, and $d$ is the usual $\mathbb{R}^n$ distance between $\{x\}$ and the set $A$. For Lip functions $f$, combining the isoperimetric inequality

$$I_\gamma\left(\gamma(\{|f| > t\})\right) \leq \gamma^+(\{|f| > t\}),$$

with the Gaussian co-area formula,[9] we have

$$\begin{aligned}\int_0^\infty I_\gamma\left(\gamma(\{|f| > t\})\right) dt &\leq \int_0^\infty \gamma^+(\{|f| > t\})dt \\ &= \||\nabla f|\|_{L^1(\mathbb{R}^n,\gamma)}.\end{aligned} \tag{9}$$

This is exactly the same mechanism one can use in the Euclidean world. Now, in the Gaussian world, the isoperimetric profile $I_\gamma(t)$ has the following estimate (with constants independent of the dimension)

$$I_\gamma(t) \simeq t\left(\log\frac{1}{t}\right)^{1/2}, t \in (0, 1/2),$$

while for the Euclidean isoperimetric profile we have

$$I_n(t) = c_n t^{1-1/n}, t > 0.$$

Note that in the Euclidean case,

[8]The isoperimetric profile was introduced by Maz'ya in the sixties and further developed by him in a number of publications (cf. [88] and the references therein). Independently, this useful tool was developed in parallel by geometers (cf. [12] and the references therein) and probabilists (cf. [67] and the references therein), and as we shall see plays an important role in our work formulating Sobolev pointwise inequalities on rearrangements.

[9]In the general metric case it becomes the co-area inequality (cf. [18]).

$$\begin{aligned}\int_0^\infty I_n\left(|\{|f|>t\}|\right)dt &= \int_0^\infty I_n\left(\lambda_f(t)\right)dt \\ &= \int_0^\infty I_n(t)\,df^*(t) \\ &= \int_0^\infty f^*(t)dI_n(t) \\ &= c_n\frac{1}{n'}\int_0^\infty f^*(t)t^{1/n'}\frac{dt}{t}.\end{aligned}$$

In other words, the logarithmic Sobolev inequality[10] of Ledoux (9) is the analogue[11] of the Gagliardo–Nirenberg in Gaussian world. Using the method of "symmetrization by truncation" we could then derive the Gaussian version of (8)

$$f_\gamma^{**}(t) - f_\gamma^*(t) \le \frac{t}{I_\gamma(t)}|\nabla f|_\gamma^{**}(t). \tag{10}$$

The remarkable fact is that in the Euclidean world (i.e., taking rearrangements using Lebesgue measure and using the corresponding formula for the Euclidean isoperimetric profile) this inequality is exactly (8)! In fact, these inequalities are equivalent to the corresponding isoperimetric inequalities in each of these geometries!

---

[10]In this case we have

$$\begin{aligned}\int_0^\infty I\left(\gamma(\{|f|>t\})\right)dt &\succeq \int_0^{1/2} f^*(t)d\left(t\left(\log\frac{1}{t}\right)^{1/2}\right) \\ &\simeq \int_0^{1/2} f^*(t)\left(\log\frac{1}{t}\right)^{1/2}dt.\end{aligned}$$

[11]Here again we have to allow for *generalized* Lorentz spaces since the Gaussian profile although concave is not increasing. Indeed, $I(t)$ is symmetric about $1/2$. Also note that the inequality

$$\int_0^\infty I\left(\gamma(\{|f|>t\})\right)dt \le \|\nabla f\|_{L^1(R^n,\gamma)}$$

holds for functions $f$ that do not vanish at the boundary. For example, for $f=1$, the right hand is zero and

$$\begin{aligned}\int_0^\infty I\left(\gamma(\{|f|>t\})\right)dt &= \int_0^1 I\left(1\right))\,dt \\ &= \int_0^1 0dt \\ &= 0.\end{aligned}$$

Also note that the isoperimetric profile automatically selects the spaces that need to be involved! Our approach in [74] is based on these ideas. For example, in the Gaussian case, from (10) we obtain directly the following version of a logarithmic Sobolev inequality

$$\int \left(f_\gamma^{**}(t) - f_\gamma^{*}(t)\right)^2 \left(\log \frac{1}{t}\right) dt \leq \int (|\nabla f|_\gamma^{**}(t))^2 dt.$$

Moreover, since the Gaussian isoperimetric profile does not depend on the dimension[12] the inequalities in this case are dimension free! For a complete development we refer to [74] and through this paper we refer to many other important references. Some references connected with this section [6–9, 13, 17, 19–22, 25, 35–37, 47, 53, 65]....

# 7 The Metric Case

When working on the Gaussian inequalities, we realized early on that, with a suitable definition of modulus of the gradient[13] $|\nabla f|$, and having at hand an associated co-area formula,[14] we could indeed prove (10) in the general setting of metric measure spaces (this was informally first announced in [74] and more formally in [76]). Fortunately, all the tools that we need to implement this insight had already been developed by Bobkov–Houdre [18].

In this generalized setting we work with connected metric probability spaces $(\Omega, d, \mu)$. The isoperimetric profile $I_\mu = I_{(\Omega,d,\mu)}$ is defined by

$$I_{(\Omega,d,\mu)}(t) = \inf_A \{\mu^+(A) : \mu(A) = t\},$$

where $\mu^+(A)$ is the perimeter or Minkowski content of the Borel set $A \subset X$, defined by

$$\mu^+(A) = \liminf_{h \to 0} \frac{\mu(A_h) - \mu(A)}{h},$$

[12]It is well known that the sets that realize the isoperimetric inequality are always hyperspaces: i.e., all but one of the variables are free.

[13]For all Lipschitz function $f$ on $\Omega$, the modulus of the gradient is defined by

$$|\nabla f(x)| = \limsup_{d(x,y) \to 0} \frac{|f(x) - f(y)|}{d(x,y)}.$$

[14]The strong connection between the co-area formula and Sobolev embeddings had already been emphasized by Maz'ya in his pioneering fundamental work in the early sixties (cf. [88]).

where $A_h = \{x \in \Omega : d(x, A) < h\}$. We assume that $I$ is continuous, concave, symmetric about $1/2$, and zero at zero. Further we assume[15] that for each $c \in \mathbb{R}$, and each $f \in \mathrm{Lip}(\Omega)$, $|\nabla f(x)| = 0$, $\mu - a.e.$ on the set $\{x : f(x) = c\}$. The associated isoperimetric inequality can be formulated as: for all Borel sets $A$

$$I(\mu(A)) \leq \mu^+(A). \tag{11}$$

## 7.1 The Gagliardo–Nirenberg–Bobkov–Houdre Inequality

We have the following result due to Bobkov–Houdre (cf. [18]),

$$\int_0^\infty I_\mu(\mu_f(t))dt \leq \| |\nabla f| \|_{L^1(\Omega)}, \text{ for all } f \in \mathrm{Lip}(\Omega) \Leftrightarrow \text{isoperimetric inequality.} \tag{12}$$

The reason the isoperimetric inequality is necessary is that given any Borel set $A$ we can find a sequence of Lip functions $\{f_n\}_n$ such that $\| |\nabla f_n| \|_{L^1(\Omega)} \to \mu^+(A)$, while $\int_0^\infty I_\mu(\mu_{f_n}(t))dt \to I_\mu(\mu(A))$ (cf. [18]).

## 7.2 Generalized Pólya–Szegö

One big difference between the general probability metric case and the Gaussian case is the lack of symmetry. In particular, in the Gaussian world $(\mathbb{R}^n, \gamma_n)$ there is a natural choice for a distinguished rearrangement that replaces the symmetric rearrangement. Let

$$\phi_n(x) = (2\pi)^{-n/2} e^{-\frac{|x|^2}{2}}, x \in \mathbb{R}^n, \ \Phi(r) = \int_{-\infty}^r \phi_1(t)dt, \mathbb{R} \in \mathbb{R},$$

then the Gaussian profile is given by (cf. Borell [22] and Sudakov-Tsirelson [109])

$$I_\gamma(t) = \phi_1(\Phi^{-1}(t)), \ t \in [0, 1].$$

The classical Euclidean spherical decreasing rearrangement is replaced by

$$f^\circ_{\gamma_n}(x) = f^*_{\gamma_n}(\Phi(x_1)),$$

---

[15] See [52]. Using an approximation argument developed by E. Milman [92, Remark 3.3] it is possible to prove the main inequalities of this paper without this assumption (cf. [79] and the forthcoming [85]).

and we have the Erhard analogue of Pólya–Szegö (cf. [35, 37, 74])

$$\left|\nabla\left(f_{\gamma_n}^{\circ}\right)\right|_{\gamma}^{**}(t) \leq |\nabla f|_{\gamma}^{**}(t). \tag{13}$$

In general there is no apparent symmetry and thus no distinguished rearrangement. This led us to formulate the following alternative inequality which extends, to the probability metric case, both the Pólya–Szegö inequality in the Euclidean geometry, and the Erhard inequality in the Gaussian case: for all $f \in \text{Lip}$, we have (cf. [78])

$$\int_0^t \left(I_\mu(\cdot)\frac{d}{dt}(-f_\mu^*(\cdot))\right)^*(s)ds \leq \int_0^t |\nabla f|_\mu^*(s)ds, \tag{14}$$

where the rearrangement inside the integral on the left hand side is with respect to Lebesgue measure. The usual formulation of Pólya–Szegö as a norm inequality follows directly from (14) and the Calderón–Hardy–Littlewood principle. To see this result in detail let us recall that a rearrangement invariant space $X(\Omega, \mu)$ has a representation $\bar{X}(0,1)$ such that

$$\|f\|_{X(\Omega)} = \left\|f_\mu^*\right\|_{\bar{X}(0,1)}.$$

Now, since $|\nabla f|_\mu^* = \left(|\nabla f|_\mu^*\right)^*$, we see that if (14) holds then by the Calderón–Hardy–Littlewood principle,

$$\left\|I_\mu(\cdot)\frac{d}{dt}(-f_\mu^*(\cdot))\right\|_{\bar{X}(0,1)} \leq \left\||\nabla f|_\mu^*\right\|_{\bar{X}(0,1)} = \left\||\nabla f|\right\|_{X(\Omega)},$$

which by abuse of notation (since no confusion can arise) we write as

$$\left\|I_\mu(t)\frac{d}{dt}(-f_\mu^*(t))\right\|_X \leq \||\nabla f|\|_X.$$

The proof of these inequalities follows by smooth truncation. Indeed, if we apply (12) to the smooth truncations,

$$f_{t_1}^{t_2}(x) = \begin{cases} t_2 - t_1 & \text{if } |f(x)| \geq t_2 \\ |f(x)| - t_1 & \text{if } t_1 < |f(x)| \leq t_2 \\ 0 & \text{if } |f(x)| \leq t_1 \end{cases},$$

and we use the fact that

$$\int_{t_1}^{t_2} I_\mu(\mu_f(s))ds = \int_0^{t_2-t_1} I_\mu(\mu_{f_{t_1}^{t_2}}(s))ds,$$

then we find that

$$\int_{t_1}^{t_2} I_\mu(\mu_f(s))ds \leq \int_{\{t_1<|f|<t_2\}} |\nabla f|\, d\mu.$$

A careful[16] "change of variables argument $s \mapsto f_\mu^*(u)$" (cf. [78]) on the left-hand side gives

$$\int_{t_1}^{t_2} I_\mu(u)\frac{d}{dt}(-f_\mu^*)(u)du = \int_{f_\mu^*(t_2)}^{f_\mu^*(t_1)} I_\mu(\mu_f(s))ds \leq \int_{\{f_\mu^*(t_2)<|f|<f_\mu^*(t_1)\}} |\nabla f|\, d\mu.$$

Thus, for any set $E = \bigcup(a_i, b_i)$ union of disjoint intervals, with $|E| = t$, we have

$$\begin{aligned}\int_E I(u)\frac{d}{dt}(-f_\mu^*)(u)du &\leq \sum_i \int_{\{f_\mu^*(b_i)<|f|<f_\mu^*(a_i)\}} |\nabla f|\, d\mu \\ &= \int_{\bigcup_i \{f_\mu^*(b_i)<|f|<f_\mu^*(a_i)\}} |\nabla f|\,(s)d\mu \\ &\leq \int_0^{\sum \mu\{f_\mu^*(b_i)<|f|<f_\mu^*(a_i)\}} |\nabla f|_\mu^*\,(s)d\mu \\ &\leq \int_0^{\sum(b_i-a_i)} |\nabla f|_\mu^*\,(s)d\mu \\ &= \int_0^t |\nabla f|_\mu^*\,(s)d\mu.\end{aligned}$$

By approximation,

$$\begin{aligned}\int_0^t (I(\cdot)\frac{d}{dt}(-f^*)(\cdot))^*(u)du &= \sup_{|E|=t} \int_E I(u)\frac{d}{dt}(-f^*)(u)du \\ &\leq \int_0^t |\nabla f|_\mu^*\,(s)d\mu,\end{aligned}$$

and (14) follows.

[16]Under the assumption that $f$ is Lip the calculations below can be justified (cf. [69, 75, 78])

## 7.3 Maz'ya–Talenti Inequality

Likewise, with a similar direct proof, we can extend a well-known inequality apparently obtained first by Maz'ya [89] and independently, but later, by Talenti (cf. [86] and the references therein). In our setting this inequality takes the following form (cf. [78])

$$I_\mu(t)\frac{d}{dt}(-f_\mu^*(t)) \leq \frac{d}{dt}\int_{\{|f|>f^*(t)\}} |\nabla(f)|_\mu^*(s)ds. \tag{15}$$

In fact, it is also easy to see that the argument given above shows that (15) implies (14). Indeed, starting with (15), and considering first sets which are unions of disjoint intervals, we arrive at

$$\sup_{|E|=t}\int_E I_\mu(u)\frac{d}{dt}(-f_\mu^*)(u)du \leq \int_0^t |\nabla f|_\mu^*(s)d\mu.$$

## 7.4 Basic Rearrangement Inequality

Next we observe that the previous arguments with intervals also show (cf. [78]) that, integrating by parts, we obtain (recall that under our assumptions $f_\mu^*$ is absolutely continuous (cf. [69])

$$f_\mu^{**}(t) - f_\mu^*(t) = \frac{1}{t}\int_0^t s\frac{d}{dt}(-f_\mu^*)(s)ds.$$

Consequently,

$$f_\mu^{**}(t) - f_\mu^*(t) \leq \frac{1}{t}\frac{t}{I_\mu(t)}\int_0^t I_\mu(s)\frac{d}{dt}(-f_\mu^*)(s)ds \quad \left(\text{since } \frac{s}{I_\mu(s)} \text{ is increasing}\right). \tag{16}$$

By now, we have discussed several inequalities that can be used to control $\int_0^t I_\mu(s)(-f_\mu^*(s))'ds$. For example, using the definition of rearrangement and the generalized Pólya–Szegö inequality (14), we have

$$\begin{aligned}\int_0^t I_\mu(s)\frac{d}{dt}(-f_\mu^*)(s)ds &\leq \int_0^t \left(I_\mu(\cdot)\frac{d}{dt}(-f_\mu^*)(\cdot)\right)^*(s)ds \\ &\leq \int_0^t |\nabla f|_\mu^*(s)ds.\end{aligned}$$

Consequently, combining the last inequality with (16) we arrive to the familiar

$$f_{\mu}^{**}(t) - f_{\mu}^{*}(t) \leq \frac{1}{I_{\mu}(t)} \int_{0}^{t} |\nabla f|_{\mu}^{*}(s) ds$$
$$= \frac{t}{I(t)} |\nabla f|_{\mu}^{**}(t). \tag{17}$$

Using a standard approximation argument (like the one outlined above) (cf. [78,79]) we see that (17) implies the isoperimetric inequality (11).

Let us also remark that is fairly easy to give a direct proof of (17) (cf. Sect. 8 below)

## 7.5 *Self-Improvement*

Under stronger assumptions on the isoperimetric profile we can get stronger inequalities. For example, suppose that

$$\int_{t}^{1} \frac{I_{\mu}(s)}{s} \frac{ds}{s} \leq c \frac{I_{\mu}(t)}{t}, \quad t \in (0,1).$$

Then, for all $f \in \mathrm{Lip}(\Omega)$,

$$\int_{0}^{t} (f_{\mu}^{**}(s) - f_{\mu}^{*}(s)) \frac{I(s)}{s} ds \leq \int_{0}^{t} \left( \frac{I(\cdot)}{(\cdot)} \left[ f_{\mu}^{**}(\cdot) - f_{\mu}^{*}(\cdot) \right] \right)^{*} ds \leq C \int_{0}^{t} |\nabla f|_{\mu}^{*}(s) ds. \tag{18}$$

This extends the sharp form of the Gagliardo–Nirenberg inequality to metric spaces that have "Euclidean like profiles." Indeed, suppose that $\frac{I(s)}{s} \simeq s^{1/n'}$, then for $f \in \mathrm{Lip}_0(\Omega)$, we have

$$\int_{0}^{t} (f_{\mu}^{**}(s) - f_{\mu}^{*}(s)) s^{1/n'} \frac{ds}{s} \leq c_n \int_{0}^{t} |\nabla f|_{\mu}^{*}(s) ds.$$

Consequently, letting $t \to \infty$,

$$\int_{0}^{\infty} (f_{\mu}^{**}(s) - f_{\mu}^{*}(s)) s^{1/n'} \frac{ds}{s} \leq c_n \left\| |\nabla f| \right\|_{1}.$$

## 7.6 Poincaré Inequalities and Generalized Pólya–Szegö

Once again we follow [78]. To discuss Poincaré inequalities in the metric setting we introduced the Hardy isoperimetric operators (cf. [78]). Let us define

$$Q_I(f)(t) = \chi_{(0,1/2)}(t) \int_t^{1/2} f(s) \frac{ds}{I_\mu(s)}.$$

Suppose that for positive functions supported on $(0, 1/2)$

$$\|Q_I(f)\|_X \leq \|Q_I\|_{X\to X} \|f\|_X \,.$$

Then, the following Poincaré inequality holds

$$\left\| g - \int_\Omega g \right\|_X \preceq \|Q_I\|_{X\to X} \, \||\nabla g|\|_X \,.$$

Indeed, for $g \in \mathrm{Lip}(\Omega)$, $t \in (0, 1/2)$ we can write

$$\begin{aligned} g_\mu^*(t) - g_\mu^*(1/2) &= \int_t^{1/2} \left(-g_\mu^*(s)\right)' ds \\ &= \int_t^{1/2} I_\mu(s) \left(-g_\mu^*(s)\right)' \frac{ds}{I_\mu(s)} \\ &= Q_I(I_\mu(.) \left(-g_\mu^*(.)\right)'). \end{aligned} \tag{19}$$

Thus,

$$g_\mu^*(t) = Q_I(I_\mu(.) \left(-g_\mu^*(.)\right)') + g_\mu^*(1/2) \tag{20}$$

Now,

$$\left\|g_\mu^*\right\|_X \leq \left\|g_\mu^* \chi_{(0,1/2)}\right\|_X + \left\|g_\mu^* \chi_{(1/2,1)}\right\|_X,$$

and

$$\begin{aligned} \left\|g_\mu^* \chi_{(1/2,1)}\right\|_X &\leq g_\mu^*(1/2) \left\|\chi_{(1/2,1)}\right\|_X \\ &= g_\mu^*(1/2) \left\|\chi_{(0,1/2)}\right\|_X \\ &\leq \left\|g_\mu^* \chi_{(0,1/2)}\right\|_X, \end{aligned}$$

yields

$$\left\|g_\mu^*\right\|_X \preceq \left\|g_\mu^*\chi_{(0,1/2)}\right\|_X$$

which combined with (20) gives

$$\begin{aligned}\left\|g_\mu^*\right\|_X &\preceq \left\|g_\mu^*\chi_{(0,1/2)}\right\|_X \\ &\leq \left\|Q_I(I_\mu(\cdot)\frac{d}{dt}(-g_\mu^*)(\cdot)\right\|_X + \left\|g_\mu^*(1/2)\right\|_X \\ &\leq \|Q_I\|_{X\to X}\left\|(I_\mu(\cdot)\frac{d}{dt}(-g_\mu^*))\right\|_X + \left\|g_\mu^*(1/2)\right\|_X \\ &\leq \|Q_I\|_{X\to X}\,\||\nabla g|\|_X + g_\mu^*(1/2)\,\|1\|_X\,.\end{aligned}$$

Now, by Chebyshev's inequality $g_\mu^*(1/2) \leq 2\,\|g\|_1$. Therefore,

$$\begin{aligned}\|g\|_X &= \left\|g_\mu^*\right\|_X \\ &\preceq \|Q_I\|_{X\to X}\,(\||\nabla g|\|_X + \|g\|_1).\end{aligned}$$

Applying the previous inequality to $g - \int_\Omega g$,

$$\left\|g - \int_\Omega g\right\|_X \preceq \|Q_I\|_{X\to X}\left(\||\nabla g|\|_X + \left\|g - \int_\Omega g\right\|_1\right),$$

and combining with Cheeger's inequality yields

$$\begin{aligned}\left\|g - \int_\Omega g\right\|_1 &\preceq \||\nabla g|\|_1 \\ &\leq \||\nabla g|\|_X\,.\end{aligned}$$

Thus we finally arrive at

$$\left\|g - \int_\Omega g\right\|_X \preceq \|Q_I\|_{X\to X}\,\||\nabla g|\|_X\,.$$

## 7.7 *Identity Associated with Poincaré*

For the sake of completeness we note that the previous discussion also shows that for $g \in \text{Lip}(\Omega)$ we have (cf. 19)

$$g_\mu^*(t) - g_\mu^*(1/2) = Q_I(I_\mu(\cdot)\frac{d}{dt}(-g_\mu^*)(\cdot)).$$

This section is based entirely on [78] but builds on a long list of important contributors we refer to [78] for a more complete list of references. For versions of Pólya–Szegö in the classical format (norm inequality) we refer to the books [61,69] as well as Talenti [110], Almgren–Lieb [3], and the references therein. We should also mention the work of the French geometer Gallot [42] on symmetrization inequalities on manifolds, which as we discovered a posteriori, is close in spirit to our development here. The important paper by D. Bakry, T. Coulhon, M. Ledoux, and L. Saloff-Coste [5] opened our eyes early on to the possibilities afforded by the masterful use of cut-offs.

## 8 Coulhon Inequalities

The previous section dealt with Sobolev inequalities in probability metric spaces. The next natural questions are: how to deal with more general metric measure spaces? Secondly: what is the rearrangement version of $L^p$ Sobolev inequalities for $p > 1$? In this situation we need to replace the isoperimetric profile by a suitable capacitary profile and the formulation of the inequalities is somewhat more complicated (cf. [79]). However, one can take a different tack (cf. [81]) and work with an equivalent formulation of the Sobolev inequalities due to Coulhon (cf. [26–28]) and Bakry–Coulhon–Ledoux [5].

Let $(\Omega, d, \mu)$ be a metric measure space, let $p \in [1, \infty]$, and let $\phi$ be an increasing function on the positive half line. We consider Coulhon inequalities of the form

$$(S_\phi^p) \quad \|f\|_p \leq \phi(\|f\|_0) \, \||\nabla f|\|_p \, , \; f \in \mathrm{Lip}_0(\Omega),$$

where $\|f\|_0 = \mu\{\mathrm{support}(f)\}$.[17]

An important point to keep in mind is that the function $\phi$ must then be connected to the geometry; but now we do not specify what this connection is in advance! In [79] we find rearrangement inequalities that characterize $(S_\phi^p)$. As was to be expected the rearrangement inequalities incorporate in their formulation the function $\phi$.

To see the connection with the previous discussions let us consider the case $p = 1$. We shall show that this case is connected with the isoperimetric inequality. Before doing so, let us mention that our approach in this section is independent and, indeed, can be seen as an alternative route to rearrangement inequalities discussed in previous sections. Let us then consider the connection between $(S_\phi^1)$ and a rearrangement inequality of the form

$$f_\mu^{**}(t) - f_\mu^*(t) \leq \phi(t) \, |\nabla f|_\mu^{**}(t), \; f \in \mathrm{Lip}_0(\Omega). \tag{21}$$

[17] In particular we require that for $f \in Lip_0(\Omega)$ we have $f^*(\mu(\Omega)^-) = 0$.

Suppose (21) holds where $\phi$ is a given increasing continuous function. Let $t > 0$; multiplying both sides of (21) by $t$, we obtain

$$t\left(f_{\mu}^{**}(t) - f_{\mu}^{*}(t)\right) \leq \phi(t) \int_0^t |\nabla f|_{\mu}^{*}(s)ds.$$

Since formally $f_{\mu}^{*}(t) = \mu_f^{-1}(t)$, drawing a diagram it is easy to see that

$$\begin{aligned} t\left(f_{\mu}^{**}(t) - f_{\mu}^{*}(t)\right) &= \int_0^t f_{\mu}^{*}(s)ds - t f_{\mu}^{*}(t) \\ &= \int_{f^*(t)}^{\infty} \mu_f(s)ds. \end{aligned}$$

Consequently, if we let $t = \|f\|_0$, we see that $f_{\mu}^{*}(\|f\|_0) = 0$, $\int_{f_{\mu}^{*}(\|f\|_0)}^{\infty} \mu_f(s)ds = \|f\|_1$, and $\int_0^{\|f\|_0} |\nabla f|_{\mu}^{*}(s)ds = \||\nabla f|\|_1$. Thus,

$$\|f\|_1 \leq \phi(\|f\|_0) \, \||\nabla f|\|_1 . \tag{22}$$

Suppose, on the other hand, that an $(S_{\phi}^1)$ condition holds. For $f \in \mathrm{Lip}_0(\Omega)$, and for $t > 0$, let us apply the $(S_{\phi}^1)$ condition to $\left[f - f_{\mu}^{*}(t)\right]_+$. We compute

$$\left\|\left[f - f_{\mu}^{*}(t)\right]_+\right\|_0 = \mu\{f > f_{\mu}^{*}(t)\} \leq t,$$

moreover, since $\int_{\{f=f^*(t)\}} \left|\nabla\left[f(x) - f_{\mu}^{*}(t)\right]\right| dx = 0$, we have

$$\left\|\nabla\left[f - f_{\mu}^{*}(t)\right]_+\right\|_{L^1} = \int_{\{f>f^*(t)\}} |\nabla f|_{\mu}^{*}(s)ds.$$

We also have

$$\begin{aligned} \left\|\left[f - f_{\mu}^{*}(t)\right]_+\right\|_1 &= \int_{\{f>f^*(t)\}} \left[f(s) - f_{\mu}^{*}(t)\right]_+ d\mu(s) \\ &= \int_0^{\infty} [f_{\mu}^{*}(x) - f_{\mu}^{*}(t)]_+ \, dx \\ &= \int_0^t (f_{\mu}^{*}(x) - f_{\mu}^{*}(t)) \, dx \\ &= t(f_{\mu}^{**}(t) - f_{\mu}^{*}(t)). \end{aligned}$$

Inserting these calculations in (22), we find

$$t(f_\mu^{**}(t) - f_\mu^*(t)) \le t\phi(t)\left(\frac{1}{t}\int_0^t |\nabla f|_\mu^*(s)ds\right).$$

In other words, we have shown that

$$(S_\phi^1) \Leftrightarrow (f_\mu^{**}(t) - f_\mu^*(t)) \le \phi(t)\,|\nabla f|_\mu^{**}(t),\ f \in \mathrm{Lip}_0(\Omega).$$

In particular, if $I_\mu(t)$ is concave then $\phi(t) = \frac{t}{I_\mu(t)}$ is increasing, and we have

$$\left(S^1_{\frac{t}{I_\mu(t)}}\right) \Leftrightarrow (f_\mu^{**}(t) - f_\mu^*(t)) \le \frac{t}{I_\mu(t)}|\nabla f|_\mu^{**}(t),\ f \in \mathrm{Lip}_0(\Omega).$$

Let us now show that the function $\frac{t}{I_\mu(t)}$ is optimal. Suppose that an $(S_\phi^1)$ condition holds, then it is easy to see, by approximation, that $\frac{t}{I_\mu(t)} \le \phi(t)$ (cf. [81]). The fact that $\left(S^1_{\frac{t}{I_\mu(t)}}\right)$ itself holds is a direct consequence of the co-area inequality for Lip functions on metric spaces (cf. [18]). It is instructive to see the details. Suppose that $f \in \mathrm{Lip}_0(\Omega)$, then from $I_\mu(\mu(A)) \le \mu^+(A)$, for all Borel sets, we have the Bobkov–Houdre inequality

$$\int_0^\infty I_\mu(\mu_f(t))dt \le \||\nabla f|\|_1\,.$$

Now, since $\frac{I_\mu(t)}{t}$ decreases, and we obviously have $\mu_f(t) \le \|f\|_0$, we see that

$$\begin{aligned}\int_0^\infty I_\mu(\mu_f(t))dt &= \int_0^\infty \mu_f(t)\frac{I_\mu(\mu_f(t))}{\mu_f(t)}dt \\ &\ge \int_0^\infty \mu_f(t)dt\frac{I_\mu(\|f\|_0)}{\|f\|_0}.\end{aligned}$$

Combining these inequalities, we see that for the choice of $\phi(t) = \frac{t}{I_\mu(t)}$ we do indeed have

$$\|f\|_1 = \int_0^\infty \mu_f(t)dt \le \phi(\|f\|_0)\,\||\nabla f|\|_1\,.$$

For example, in the Euclidean space $\mathbb{R}^n$, $I(t) = d_n t^{1-1/n}$, $\phi(t) \simeq t^{1/n}$ and the best possible $(S_\phi^1)$ inequality can be written as

$$\|f\|_1 \le c_n \|f\|_0^{1/n}\,\||\nabla f|\|_1\,.$$

With the choice of $\phi(t) = t^{1/n}$, and $p = 2$, Coulhon [28] shows the equivalence of the $S_\phi^2$ inequality with the classical Nash inequality

$$\|f\|_2^{1+2/n} \leq c_n \|f\|_1^{2/n} \||\nabla f|\|_2 .$$

The general characterization of Coulhon inequalities in terms of rearrangements is given by the following (cf. [81])

**Theorem 1.** *Let $(\Omega, d, \mu)$ be a connected Borel metric measure space as described above, and let $p \in [1, \infty)$. The following statements are equivalent*

*1. $(S_\phi^p)$ holds, i.e.*

$$\|f\|_p \leq \phi(\|f\|_0) \||\nabla f|\|_p , \text{ for all } f \in \mathrm{Lip}_0(\Omega).$$

*2. Let $k \in \mathbb{N}$ be such that $k < p \leq k+1$, then for all $f \in \mathrm{Lip}_0(\Omega)$*

$$\left(\frac{f_{(p)}^{**}(t)}{\phi_{(p)}(t)}\right)^{1/p} - \left(\frac{f_{(p)}^{*}(t)}{\phi_{(p)}(t)}\right)^{1/p} \leq 2^{\frac{k+1}{p}-1} \left(|\nabla f|_{(p)}^{**}(t)\right)^{1/p}, \tag{23}$$

*where*

$$f_{(p)}^{*}(t) = \left(f^{*}(t)\right)^{p}, \; f_{(p)}^{**}(t) = \frac{1}{t}\int_0^t f_{(p)}^{*}(s)ds, \; \phi_{(p)}(t) = (\phi(t))^{p} .$$

We note that for $p = 1$ the inequality (23) of Theorem 1 coincides with (17). This new characterization for $p \geq 1$ is independent of [78], and, in fact, the proof we gave above provides a new approach to (17) as well.

For the details of the proof of Theorem 1 we must refer to [81].

We cannot resist to make a connection between Coulhon's inequalities and the theory of factorization of operators. Local operators (cf. [2, 96, 102] and the references therein) satisfy conditions of the form

$$\|Tf\|_Y \leq \phi(\|f\|_0) \|f\|_X .$$

Self-improvements in this setting are expressed via factorization and change of density. We think that the factorization of Sobolev inequalities could be an interesting line of investigation to pursue.

## 9 Connection with the Work of Emanuel Milman

In this section, following [78], we consider a connection with the work of Emanuel Milman[18] (cf. [90–95]).

For metric measure spaces $(\Omega, d, \mu)$ obtained from a $C^\infty$ complete oriented $n$-dimensional Riemannian manifold $(M, g)$, where $d$ is the induced geodesic distance and $\mu$ is absolutely continuous with respect to $\mathrm{dvol}_M$, it is known that the corresponding isoperimetric profile satisfies (cf. [12]) that $I_{(\Omega,d,\mu)}(t)$ is continuous, $I_{(\Omega,d,\mu)}(t) > 0$ for $t \in (0,1)$, and moreover,

$$I_{(\Omega,d,\mu)}(t) = I_{(\Omega,d,\mu)}(1-t), \forall t \in [0,1].$$

E. Milman further assumes some convexity conditions: $d\mu = e^{-\Psi} dvol_M$, where $\Psi$ is such that $\Psi \in C^2(M)$, and as tensor fields $\mathrm{Ric}_g + Hess_g(\Psi) \geq 0$ on $M$. In this case it then follows that $I_{(\Omega,d,\mu)}$ is also concave (cf. [90] and the extensive list of references therein). Under such conditions, E. Milman shows the equivalence of Cheeger's inequality, Poincaré's inequality, and concentration inequalities! More precisely, using a variety of different tools, including the semigroup approach of Ledoux, E. Milman has shown that (cf. also Ledoux's [68] streamlined approach to E. Milman's results in [93]).

**Theorem 2 (E. Milman).** *Let $(\Omega, d, \mu)$ be a metric probability space satisfying E. Milman's convexity conditions. Then following statements are equivalent*

*(E1) Cheeger's inequality: there exists a positive constant C such that*

$$I_{(\Omega,d,\mu)} \geq Ct, \quad t \in (0, 1/2].$$

*(E2) Poincaré's inequality: there exists a positive constant $P$ such that for all $f \in \mathrm{Lip}(\Omega)$,*

$$\|f - m_e\|_{L^2(\Omega)} \leq P \, \||\nabla f|\|_{L^2(\Omega)},$$

*where $m_e$ is a median of $f$.*

*(E3) Exponential concentration: there exist positive constants $c_1, c_2$ such that for all $f \in \mathrm{Lip}(\Omega)$ with $\|f\|_{\mathrm{Lip}(\Omega)} \leq 1$,*

$$\mu\{|f - m_e| > t\} \leq c_1 e^{-c_2 t}, \quad t \in (0,1).$$

*(E4) First moment inequality: there exists a positive constant $F$ such that for all $f \in \mathrm{Lip}(\Omega)$ with $\|f\|_{\mathrm{Lip}(\Omega)} \leq 1$,*

$$\|f - m_e\|_{L^1(\Omega)} \leq F.$$

[18] We refer to E. Milman's papers for an account of the history of the problem. Emanuel, who belongs to the Milman family of mathematicians that includes David (grandfather), Vitali (father), Pierre (uncle) (cf. [54]), has no direct relation to Mario Milman.

*Remark 1.* For the optimal relationship between the concentration profile and Cheeger's constant under suitable convexity conditions see [95].

## 9.1 Isoperimetric Hardy Type

We single out a class of metric probability spaces that are suitable for our analysis (cf. [77,78]).

**Definition 1.** We shall say that a probability metric space $(\Omega, d, \mu)$ is of isoperimetric Hardy type if for any given isoperimetric estimator[19] $I$, the following are equivalent for all r.i. spaces $X = X(\Omega)$, $Y = Y(\Omega)$.

1. There exists a constant $c = c(X, Y)$ such that for all $f \in \mathrm{Lip}(\Omega)$

$$\left\| f - \int_{\Omega} f d\mu \right\|_{Y} \leq c \, \||\nabla f|\|_{X} .$$

2. There exists a constant $c_1 = c_1(X, Y) > 0$ such that for all positive functions $f \in \bar{X}$, with $\mathrm{supp}(f) \subset (0, 1/2)$ we have

$$\|Q_I f\|_{\bar{Y}} \leq c_1 \|f\|_{\bar{X}} ,$$

where $Q_I$ is the isoperimetric Hardy operator

$$Q_I f(t) = \chi_{(0,1/2)}(t) \int_t^{1/2} f(s) \frac{ds}{I(s)}.$$

For spaces of isoperimetric type it is possible to give a very simple proof of the E. Milman's equivalences (cf. [78]).

**Theorem 3.** *Suppose that $(\Omega, d, \mu)$ is a metric probability space of isoperimetric Hardy type. Then*

$$(E1) \Leftrightarrow (E2) \Leftrightarrow (E3) \Leftrightarrow (E4).$$

*Example 3.* All the model spaces studied in [78] (including Gaussian space (cf. [74]) are of Hardy isoperimetric type.

---

[19] An isoperimetric estimator is a continous concave function $I : [0, 1] \to [0, \infty]$, with $I(0) = 0$, increasing on (0, 1/2), symmetric about the point 1/2, and such that for $t \in (0, 1/2)$,

$$I(t) \geq I_{(\Omega,d,\mu)}(t).$$

For further results connecting our work with E. Milman's work we refer to [78] and [79]. For example, the following result of E. Milman can be understood in the context of Hardy isoperimetric type (cf. [78]).

**Theorem 4.** *Let* $(\Omega, d, \mu)$ *be a space satisfying E. Milman's convexity conditions. Let* $1 \leq q < \infty$, *and let* $N$ *be a Young's function such that* $\frac{N(t)^{1/q}}{t}$ *is non-decreasing, and there exists* $\alpha > \max\{\frac{1}{q} - \frac{1}{2}, 0\}$ *such that* $\frac{N(t^{\alpha})}{t}$ *non-increasing. Then, the following statements are equivalent:*

*(E5)* $(L_N, L^q)$ *Poincaré inequality holds: there exists a positive constant* $P$ *such that for all* $f \in \mathrm{Lip}(\Omega)$

$$\|f - m_e\|_{L_N(\Omega)} \leq P \, \||\nabla f|\|_{L^q(\Omega)} .$$

*(E6) Any isoperimetric profile estimator* $I$ *satisfies: there exists a constant* $c > 0$ *such that* $I(t) \geq c \frac{t^{1-1/q}}{N^{-1}(1/t)}$, $t \in (0, 1/2]$.

E. Milman's work can also be seen as providing a program to unify Sobolev inequalities in different geometries. For more on this we refer to our paper [79] where, generalizing E. Milman's work, we in particular show why Lorentz spaces appear as optimal target spaces for Sobolev embeddings.

## 10 Transference and Dimensionless Inequalities

In this section we follow [78] and [82] to show how our pointwise inequalities can be used to transfer Sobolev inequalities from one geometry to another. The transference is of special interest when it is implemented to replace Sobolev inequalities that carry dimensional constants by weaker, but dimensionless, Sobolev inequalities. Inequalities independent of the dimension play an increasingly important role in approximation theory and its applications (cf. [46]). To see how our pointwise inequalities are relevant for this task let us recall that our typical Sobolev inequality on a metric probability space $(\Omega, d, \mu)$ takes the form

$$\left\| \left( f_{\mu}^{**}(t) - f_{\mu}^{*}(t) \right) \frac{I_{\mu}(t)}{t} \right\|_{\bar{X}} \leq c \left\| |\nabla f|_{\mu}^{**} \right\|_{\bar{X}} . \tag{24}$$

Now suppose that the metric probability space $(\Omega, d, \mu)$ is of "Gaussian isoperimetric type," that is suppose that for some universal constant independent of the dimension, it holds

$$I_{(\Omega,\mu)}(t) \succeq t \left( \log \frac{1}{t} \right)^{\frac{1}{2}}, \text{ on } \left(0, \frac{1}{2}\right);$$

then we can obviously replace $\frac{I_\mu(t)}{t}$ by $\left(\log\frac{1}{t}\right)^{\frac{1}{2}}$ in (24), and in this fashion *transfer* the Gaussian inequality to $(\Omega, d, \mu)$ :

$$\left\| \left(f_\mu^{**}(t) - f_\mu^{*}(t)\right)\left(\log\frac{1}{t}\right)^{\frac{1}{2}} \right\|_{\bar{X}} \leq c \left\| |\nabla f|_\mu^{**} \right\|_{\bar{X}}. \tag{25}$$

This argument shows how the Gaussian log Sobolev inequalities can be transferred to $(\Omega, d, \mu)$ with constants independent of the dimension. In particular, since $Q_n$, the open unit cube in $\mathbb{R}^n$, is of Gaussian type with constant equal to 1 (cf. [105]), the Gaussian Sobolev inequalities can be transferred to $Q_n$, with constants independent of the dimension.

This answered a question of Triebel [113] (cf. [63, 64, 78, 82, 114]) and the references therein). For example, the non-homogeneous form of these results take the following form

$$\|f\|_{L^q(LogL)^{q/2}(Q_n)} \leq C(q)\left(\||\nabla f|\|_{L^q(Q_n)} + \|f\|_{L^q(Q_n)}\right), \; f \in W^{1,q}(Q_n). \tag{26}$$

In turn, this result was recently improved by Krbec–Fiorenza–Schmeisser [40], using the spaces $L_{(q,q'}(Q_n)$, originally introduced by Iwaniec–Sbordone–Fiorenza (cf. [38, 55]), and characterized by Fiorenza–Karadzhov [39], using extrapolation (cf. [60]) as follows

$$\|f\|_{L_{(q,q'}(Q_n)} \simeq \int_0^1 \left(\int_0^t f^*(s)^q ds\right)^{1/q} \frac{dt}{t(\log\frac{1}{t})^{\frac{1}{2}}}.$$

The (nonhomogeneous) result in [40] yields

$$\|f\|_{L_{(q,q'}(Q_n)} \leq C(q)\left(\|\nabla f\|_{L^q(Q_n)} + \|f\|_{L^q(Q_n)}\right), \; f \in W^{1,q}(Q_n). \tag{27}$$

Consequently, since

$$L_{(q,q'}(Q_n) \subset L^q(\text{Log}L)^{q/2}(Q_n),$$

this result provides an improvement upon (26).

In [82] we showed that the inequality (27) is connected with a different transference principle. We start by reformulating (25) as

$$\left\| \left(f_\mu^{**}(t) - f_\mu^{*}(t)\right)\chi_{(0,1/2)}(t) G_\infty(t) \right\|_{\bar{X}} \leq c \left\| G_\infty(t)\frac{t}{I_\mu(t)} \right\|_{L^\infty(0,\frac{1}{2})} \left\| |\nabla f|_\mu^{**} \right\|_{\bar{X}},$$

then (25) corresponds to the choice $G_\infty(t) = (\log\frac{1}{t})^{\frac{1}{2}}$. Now, using the fact that $\frac{I_\mu(t)}{t}$ decreases we see that left-hand side of (24) can be minorized as follows:

$$\left\| \left( f_\mu^{**}(\cdot) - f_\mu^{*}(\cdot) \right) \frac{I_\mu(\cdot)}{(\cdot)} \right\|_{\bar{X}} \geq \left\| \left( f_\mu^{**}(\cdot) - f_\mu^{*}(\cdot) \right) \chi_{(0,t)}(\cdot) \frac{I_\mu(\cdot)}{(\cdot)} \right\|_{\bar{X}}$$
$$\geq \left\| \left( f_\mu^{**}(\cdot) - f_\mu^{*}(\cdot) \right) \chi_{(0,t)}(\cdot) \right\|_{\bar{X}} \frac{I_\mu(t)}{t}.$$

Therefore, we have

$$\left\| \left( f_\mu^{**}(\cdot) - f_\mu^{*}(\cdot) \right) \chi_{(0,t)}(\cdot) \right\|_{\bar{X}} \leq c \frac{t}{I_\mu(t)} \left\| |\nabla f|_\mu^{**} \right\|_{\bar{X}}. \tag{28}$$

Now, if $G_1$ is such that $\left( \int_0^1 G_1(t) \frac{t}{I_\mu(t)} dt \right) < \infty$, it follows immediately from (28) that

$$\int_0^1 \left\| \left( f_\mu^{**}(\cdot) - f_\mu^{*}(\cdot) \right) \chi_{(0,t)}(\cdot) \right\|_{\bar{X}} G(t) dt \leq C \left( \int_0^1 G_1(t) \frac{t}{I_\mu(t)} dt \right) \left\| |\nabla f|_\mu^{**} \right\|_{\bar{X}}.$$

For example, let $G_1(t) = \frac{1}{t \left( \log \frac{1}{t} \right)^{\frac{1}{2}}}$, and suppose the following (stronger) Gaussian isoperimetric transference condition is satisfied,

$$\int_0^1 \frac{dt}{I_\mu(t) (\log \frac{1}{t})^{\frac{1}{2}}} < \infty. \tag{29}$$

Then, we have (cf. [82])

$$\left\| (f_\mu^{**}(\cdot) - f_\mu^{*}(\cdot)) \left( \log(\frac{1}{\cdot}) \right)^{1/2} \right\|_{\bar{X}} \leq c \int_0^1 \left\| \left( f_\mu^{**}(s) - f_\mu^{*}(s) \right) \chi_{(0,t)}(s) \right\|_{\bar{X}} \frac{dt}{t \left( \log \frac{1}{t} \right)^{\frac{1}{2}}}$$
$$\leq c \left( \int_0^1 \frac{dt}{I_\mu(t) (\log \frac{1}{t})^{\frac{1}{2}}} \right) \left\| |\nabla f|^{**} \right\|_{\bar{X}}.$$

Let us show a concrete application. Let

$$I_n(t) = n \left( \gamma_n \right)^{1/n} t^{1-1/n},$$

where $\gamma_n = \frac{\pi^{n/2}}{\Gamma(1+n/2)}$ is the measure of the unit ball in $\mathbb{R}^n$ (i.e., $I_n(t)$ is the isoperimetric profile associated with $\mathbb{R}^n$), and consider the function

$$G_1(t) = \frac{1}{t \sqrt{\ln \left( \frac{1}{t} \right)}}, t \in (0,1).$$

Then,

$$\begin{aligned}
\int_0^1 \frac{t}{tI_n(t)} G_1(t)dt &= \frac{1}{n\,(\gamma_n)^{1/n}} \int_0^1 t^{1/n} \frac{dt}{t\,(\ln \frac{1}{t})^{\frac{1}{2}}} \\
&= \frac{1}{n\,(\gamma_n)^{1/n}} \int_0^\infty z^{-\frac{1}{2}} e^{-z/n} dz \\
&= \frac{\sqrt{\pi} n^{\frac{1}{2}}}{n\,(\gamma_n)^{1/n}} \\
&= \frac{\Gamma(1+\frac{n}{2})^{1/n}}{n^{\frac{1}{2}}}. \\
&= \left(\frac{n}{2}\right)^{1/n} \frac{\Gamma(\frac{n}{2})^{1/n}}{n^{\frac{1}{2}}} \\
&\leq \frac{1}{\sqrt{2}} \left(\frac{n}{2}\right)^{1/n} \\
&\leq c.
\end{aligned}$$

Thus,

$$\sup_n \int_0^1 \frac{dt}{I_n(t)(\log \frac{1}{t})^{\frac{1}{2}}} < \infty.$$

As a consequence the following dimensionless Sobolev inequality holds,

$$\int_0^1 \left\| \left(f^{**}(\cdot) - f^*(t)\right) \chi_{[0,t)}(\cdot) \right\|_{\bar{X}} \frac{dt}{t(\log \frac{1}{t})^{1/2}} \leq C \left\| |\nabla f|^{**} \right\|_{\bar{X}}.$$

For $X = L^q$ and for $\Omega = Q_n$ this gives the result of [40].

## 11 Rearrangement Inequalities of Garsia–Rodemich Type and Morrey's Theorem

To complement the results of previous sections we now consider fractional inequalities and the Morrey–Sobolev embedding theorem. Here we follow [80] where the reader will find a complete treatment together with many applications.

Let us describe model results that influenced our development in these directions.

*Example 4.* For all $f \in X(\mathbb{R}^n) + \dot{W}_X^1(\mathbb{R}^n)$, we have (cf. [71])

$$f^{**}(t) - f^*(t) \leq c_n \frac{\omega_X\left(t^{1/n}, f\right)}{\phi_X(t)}, \; t > 0, \tag{30}$$

where $X = X(\mathbb{R}^n)$ is a rearrangement invariant space on $\mathbb{R}^n$, $\phi_X(t) = \|\chi_A\|_X$, with $|A| = t$, is the fundamental function of $X$, and $\omega_X$ be the modulus of continuity associated with $X$:

$$\omega_X(t, g) = \sup_{|h| \leq t} \|g(\cdot + h) - g(\cdot)\|_X, \text{ for } g \in X.$$

The inequality (30) can be formulated and proved on different levels of generality on the spaces, the domains; and indeed have a long history: We refer to [15, 58, 62, 71] and the references therein.

The associated questions to Example 4 are: What is the corresponding Gaussian result? More generally: What is the metric version? What is the role of dimension? What is the connection with isoperimetry? Even to formulate metric results we need to develop suitable tools. For example: What is an appropriate replacement for the modulus of continuity?

We shall need the following definition: Consider a connected, measure metric spaces $(\Omega, d, \mu)$ equipped with a finite Borel measure $\mu$. For measurable functions $u : \Omega \to \mathbb{R}$, the signed *decreasing rearrangement* of $u$, which we denote by $u_\mu^s$, is the right-continuous non-increasing function from $[0, \mu(\Omega))$ into $\mathbb{R}$ that is equimeasurable with $u$; i.e. $u_\mu^s$ satisfies

$$\mu\{x \in \Omega : u(x) > t\} = m\left(\left\{z \in [0, \mu(\Omega)) : u_\mu^s(z) > t\right\}\right), \; t \in \mathbb{R}$$

The maximal average of $u_\mu^{ss}$ is defined by

$$u_\mu^{ss}(t) = \frac{1}{t}\int_0^t u_\mu^s(z)dz, \; (t > 0).$$

*Example 5.* In closely related work Garsia and his collaborators (cf. [43–45] and the references therein) obtained related inequalities. For example, in [44] and [43] for functions defined on the unit $n$–cube, and using signed rearrangements with respect to the Lebesgue measure

$$\left.\begin{array}{r} f^s(x) - f^s(1/2) \\ f^s(1/2) - f^s(1-x) \end{array}\right\} \leq c \int_x^1 \frac{\omega_{L^p}(t^{1/n}, f)}{t^{1/p}} \frac{dt}{t}, x \in (0, \frac{1}{2}], \tag{31}$$

where $\omega_{L^p}(t, f)$ the $L^p$ modulus of continuity. The extension from dimension 1 to dimension $n$ in these works was done through highly nontrivial combinatorial inequalities (cf. [43]).

In this context we can ask similar questions to those posed in Example 4.

*Example 6.* In the work of Garsia and his collaborators one also finds another interesting use of rearrangement inequalities to extract continuity. Through a change of scale argument, inequalities on rearrangements were used, for example, to prove versions of Morrey's Sobolev theorem. For example, in the one dimensional case (cf. [43–45]) we have

$$|f(x) - f(y)| \leq 2c \int_0^{|x-y|} \frac{\omega_{L^p}(t, f)}{t^{1/p}} \frac{dt}{t}; \; x, y \in [0, 1].$$

We now adapt the change of scale argument of Garsia et al. combined with the rearrangement inequality (17) in the context of the unit cube to prove a version of Morrey's theorem. First, let us stipulate (cf. [80]) that one can rewrite the (17) in terms of signed rearrangements (i.e., we rearrange $f$ rather than its absolute value). Suppose that $p > n$, and let $f \in W_1^1(0, 1)^n$. Starting with the fundamental theorem of calculus

$$f^{ss}(0) - f^{ss}(1) = \int_0^1 (f^{ss}(t) - f^s(t)) \frac{dt}{t}$$

and the corresponding version of (17)

$$f^{ss}(t) - f^s(t) \leq c_n \frac{t}{\min(t, 1-t)^{1-1/n}} |\nabla f|^{**}(t), \;\; 0 < t < 1,$$

we see that

$$\begin{aligned} f^{ss}(0) - f^{ss}(1) &\leq c_n \int_0^1 |\nabla f|^{**}(t) \frac{dt}{\min(t, 1-t)^{1-1/n}} \\ &\leq c_{n,p} \, \||\nabla f|\|_{L^p} \left\| \frac{1}{\min(t, 1-t)^{1-1/n}} \right\|_{L^{p'}(0,1)} \quad \text{(by Hölder's inequality)} \\ &= C_{n.p} \, \||\nabla f|\|_{L^p}. \end{aligned}$$

This works of course because for $p > n$, $\left\| \frac{1}{\min(t,1-t)^{1-1/n}} \right\|_{L^{p'}} < \infty$. Now (recall we are working with signed rearrangements) since $f^{ss}(0) = ess \sup_{x \in (0,1)^n} f$, and $f^{ss}(1) = \int_0^1 f$, it thus follows that

$$ess \sup_{x \in (0,1)^n} f - \int_0^1 f = f^{ss}(0) - f^{ss}(1) \leq C_{n.p} \, \||\nabla f|\|_{L^p}. \tag{32}$$

Applying (32) now to $-f$ yields

$$\int_0^1 f - ess \inf_{x\in(0,1)^n} f \leq C_{n.p} \left\||\nabla f|\right\|_{L^p}. \tag{33}$$

Therefore, adding (32) and (33) we obtain

$$\operatorname{Osc}(f;(0,1)^n) := ess \sup_{x\in(0,1)^n} f - ess \inf_{x\in(0,1)^n} f \leq 2C_{n.p} \left\||\nabla f|\right\|_{L^p}.$$

Now, we scale: Apply the previous argument to the restriction of $f$ to a subcube $Q$, $f\chi_Q$. We obtain

$$\operatorname{Osc}(f;Q) \leq c_{n,p} \left\| \frac{t}{\min(t,|Q|-t)^{1-1/n}} \right\|_{L^{p'}(0,|Q|)} \left\||\nabla f|\right\|_{L^p(Q)}.$$

By computation, it is now easy to see that we will have

$$|f(y) - f(z)| \leq c_{n,p} |y-z|^{(1-\frac{n}{p})} \left\||\nabla f|\right\|_p, \text{a.e. } y,z.$$

To carry out our program in metric measure spaces $(\Omega, d, \mu)$ we first need to find a replacement for the modulus of continuity. In [80] we use the Peetre[20] $K-$functional defined by

$$K(t,f;X(\Omega),S_X(\Omega)) :=$$
$$\inf\{\|f-g\|_{X(\Omega)} + t\left\||\nabla g|\right\|_{X(\Omega)} : g \in S_X(\Omega)\},$$

where for a rearrangement invariant space $X(\Omega)$, $S_X(\Omega) = \{f \in \operatorname{Lip}(\Omega) : \left\||\nabla f|\right\|_{X(\Omega)} < \infty\}$. In the classical setting we have (cf. [15, 16])

$$K(t,f;X(\mathbb{R}^n),\dot{W}_X^1(\mathbb{R}^n)) := \inf\{\|f-g\|_X + t\left\||\nabla g|\right\|_X : g \in \dot{W}_X^1(\mathbb{R}^n)\}$$
$$\simeq \omega_X(t,f).$$

We can thus think of $K(t,f;X(\Omega),S_X(\Omega))$ as generalized "modulus of continuity," and we have (cf. [80] and also [87])

**Theorem 5 (cf. [80]).** *Let $(\Omega, d, \mu)$ be a metric measure space that satisfies our standard assumptions. Then,*

*(i) For all rearrangement invariant spaces $X(\Omega)$, and for all $f \in X(\Omega)+S_X(\Omega)$,*

[20]Independently, and in parallel, A. Calderón and his student Oklander defined and studied the $K-$functional, and real interpolation, e.g. in Oklander's thesis at the University of Chicago (cf. [98]).

$$f_{\mu}^{ss}(t) - f_{\mu}^{s}(t) \leq 16 \frac{K\left(\frac{t}{I_{\mu}(t)}, f; X(\Omega), S_X(\Omega)\right)}{\phi_X(t)}, \; t \in (0, \mu(\Omega)/2), \tag{34}$$

$$(f - f_{\Omega})_{\mu}^{**}(t) - (f - f_{\Omega})_{\mu}^{*}(t) \leq 16 \frac{K\left(\frac{t}{I_{\mu}(t)}, f; X(\Omega), S_X(\Omega)\right)}{\phi_X(t)}, \; t \in (0, \mu(\Omega)), \tag{35}$$

*where*

$$f_{\Omega} = \frac{1}{\mu(\Omega)} \int_{\Omega} f d\mu. \tag{36}$$

*(ii) Conversely, suppose that $G : (0, \mu(\Omega)) \to \mathbb{R}_+$ is a continuous function, which is concave and symmetric around $\mu(\Omega)/2$, and that there exists a constant $c > 0$ such that*[21] *for all $f \in X(\Omega) + S_X(\Omega)$,*

$$f_{\mu}^{ss}(t) - f_{\mu}^{s}(t) \leq c \frac{K\left(\frac{t}{G(t)}, f; X(\Omega), S_X(\Omega)\right)}{t}, \; t \in (0, \mu(\Omega)/2).$$

*Then, there exists a constant $c > 0$ such that for all $t \in (0, \mu(\Omega))$,*

$$G(t) \leq c I_{\mu}(t).$$

Following the analysis of [80, Chapter 10], one can show that for a metric probability space $(\Omega, d, \mu)$, that satisfies our standard assumptions, the Garsia inequality (31) takes the following form,

$$\left.\begin{matrix} f^s(x) - f^s(1/2) \\ f^s(1/2) - f^s(1-x) \end{matrix}\right\} \leq c \int_x^1 \frac{K\left(\frac{t}{I_{\mu}(t)}, f; X(\Omega), S_X(\Omega)\right)}{\phi_X(t)} \frac{dt}{t}, x \in (0, \frac{1}{2}]. \tag{37}$$

For the scaling argument we outlined above we need an extra assumption. We say that an *isoperimetric inequality relative to* $G$ holds, if there exists a positive constant $C_G$ such that

$$I_G(s) \geq C_G \min(I_{\mu}(s), I_{\mu}(\mu(G) - s)).$$

We say that the metric measure space $(\Omega, d, \mu)$ has the *uniform relative isoperimetric property,* if there exist $C > 0$, $\delta > 0$, such that for any $x \in \Omega$, and any open ball $B_{\alpha}(x)$ centered on $x$, with $\mu(B_{\alpha}(x)) = \alpha$ $(0 < \alpha < \delta)$, the *relative isoperimetric profile* $I_{B_{\alpha}(x)}$ satisfies:

[21] In other words we assume that (34) holds for $X = L^1(\Omega)$, and with $\frac{t}{G(t)}$ replacing $\frac{t}{I_{\Omega}(t)}$.

$$I_{B_\alpha(x)}(s) \geq C \min(I_\mu(s), I_\mu(\alpha - s)), \quad 0 < s < \alpha.$$

Then we have (cf. [80, Chapter 4])

**Theorem 6.** *Let $(\Omega, d, \mu)$ be a metric measure space that satisfies our standard assumptions and, moreover, has the relative uniform isoperimetric property. Let $X$ be an r.i. space in $\Omega$ such that*

$$\left\| \frac{1}{I_\mu(s)} \right\|_{\bar{X}'} < \infty.$$

*Then, if $f \in X + S_X(\Omega)$ satisfies*

$$\int_0^{\mu(\Omega)} \frac{K\left(\phi_X(t) \left\| \frac{1}{I_\mu(s)} \chi_{(0,t)}(s) \right\|_{\bar{X}'}, f; X, S_X(\Omega)\right)}{\phi_X(t)} \frac{dt}{t} < \infty,$$

*it follows that $f$ is essentially bounded and essentially continuous.*

For applications we need to have explicit computations of the relevant $K-$functional. We refer to [31–33] for a treatment of $K-$functionals in one dimensional Gaussian measure. For other relevant computations of $K-$functionals see the appendix of [80].

For further developments we must refer to [80]. In connection with this section we should also mention [115] and the references therein.

## 12 Higher Order Inequalities

In this section we work with Euclidean domains $\Omega$ and measure spaces of the form $d\mu(x) = w(x)dx$, and we formulate higher order pointwise inequalities by iteration[22] that extend our previous work in [72] and [97]. The basic inequality here reads as follows. We assume that $\int d\mu(x) = 1$, and we let $I_\mu := I$, then for sufficiently smooth $f$ we have that for $k \geq 2$, (cf. [83]))

$$f^{**}(t) - f^*(t) \leq \frac{1}{(k-1)!} \frac{t}{I(t)} \int_t^{1/2} \left|d^k f\right|^{**}(u) \left( \frac{1}{I(u)} \left( \int_t^u \frac{dz}{I(z)} \right)^{k-1} \right) du$$
$$+ \frac{1}{(k-1)!} \sum_{j=1}^{k-1} \left( \frac{t}{I(t)} \left( \int_t^{1/2} \frac{dz}{I(z)} \right)^{k-j-1} \right) \left\| \left|d^{k-j} f\right| \right\|_1, \; 0 < t < 1/2.$$

[22] The iteration of Sobolev inequalities is not a new idea (cf. [88, Corollary 6.9/1 in page 379 and Theorem 7.6.5 in page 430]), the novelty here, if any, is the iteration of pointwise rearrangement inequalities

As a corollary we obtain

$$f^{**}(t) \leq c_k \int_t^{1/2} |d^k f|^{**}(s) \frac{1}{I(s)} \left( \int_t^s \frac{dz}{I(z)} \right)^k ds + \sum_{j=1}^k c_j(t) \left\| |d^{k-j} f| \right\|_1, \ 0 < t < 1/2;$$

where $c_j(t) = \left( \int_t^{1/2} \frac{dz}{I(z)} \right)^{k-j}$.

In particular, one can easily show that this result implies recent iterated inequalities that appeared in [108].

**Theorem 7.** *Let $X, Y$ r.i spaces. Suppose that the operator $T$ defined by*

$$Tf(t) = \int_t^{1/2} \frac{f(u)}{I(u)} \left( \int_t^u \frac{dz}{I(z)} \right)^{k-1} du,$$

*is bounded from $X$ to $Y$. Then*

$$\|u\|_Y \leq c \left( \left\| |d^k f|^{**} \right\|_X + \sum_{j=1}^k \left\| |d^{k-j} f| \right\|_1 \right).$$

**Acknowledgements** We are grateful to E. Milman for a number of useful comments that helped improve the presentation.

The author J. Martín was Partially supported in part by Grants MTM2010-14946, MTM-2010-16232.

The author M. Milman was partially supported by a grant from the Simons Foundation (#207929 to Mario Milman).

## References

1. Adams, D.R., Hedberg, L.I.: Function Spaces and Potential Theory. Grundlehren der Mathematischen Wissenschaften, vol. 314. Springer, Berlin (1999)
2. Allen, G.D.: Locally Continuous Operators II. Indiana Univ. Math. J. **38**, 711–743 (1989)
3. Almgren, F., Lieb, E.: Symmetric Decreasing Rearrangement is sometimes continuous. J. Am. Math. Soc. **2**, 683–773 (1989)
4. Alvino, A., Trombetti, G., Lions, P.L.: On optimization problems with prescribed rearrangements. Nonlinear Anal. **13**, 185–220 (1989)
5. Bakry, D., Coulhon, T., Ledoux, M., Saloff-Coste, L.: Sobolev inequalities in disguise. Indiana Univ. Math. J. **44**, 1033–1074 (1995)
6. Barthe, F.: Levels of concentration between exponential and Gaussian. Ann. Fac. Sci. Toulouse Math. **10**, 393–404 (2001)
7. Barthe, F.: Log-concave and spherical models in isoperimetry. Geom. Funct. Anal. **12**, 32–55 (2002)
8. F. Barthe, P. Cattiaux and C. Roberto, *Isoperimetry between exponential and Gaussian, Orlicz hyper-contractivity and isoperimetry*, Rev. Mat. Iber. **22** (2006), 993–1067.
9. Barthe, F., Cattiaux, P., Roberto, C.: Isoperimetry between exponential and Gaussian. Electron. J. Probab. **12**, 1212–1237 (2007)

10. Bastero, J., Milman, M., Ruiz, F.: On the connection between weighted norm inequalities, commutators and real interpolation. preprint, Sem A. Galdeano (1996)
11. Bastero, J., Milman, M., Ruiz, F.: A note on L(∞, q) spaces and Sobolev embeddings. Indiana Math. J. **52**, 1215–1230 (2003)
12. Bayle, V.: Propriétés de concavité du profil isopérimétrique et applications. Ph.D. thesis, Institut Joseph Fourier (2004)
13. Beckner, W., Persson, M.: On sharp Sobolev embedding and the logarithmic Sobolev inequality. Bull. Lond. Math. Soc. **30**, 80–84 (1998)
14. Bennett, C., DeVore, R., Sharpley, R.: Weak-$L^{\infty}$ and BMO. Ann. Math. **113**, 601–611 (1981)
15. Bennett, C., Sharpley, R.: Interpolation of Operators. Academic Press, Boston (1988)
16. Bergh, J., Löfström, J.: Interpolation Spaces: An Introduction. Springer, Berlin (1976)
17. Bobkov, S.G.: Isoperimetric and analytic inequalities for log-concave probability measures. Ann. Probab. **27**, 1903–1921 (1999)
18. Bobkov, S.G., Houdré, C.: Some connections between isoperimetric and Sobolev-type inequalities. Mem. Am. Math. Soc. **129** (1997).
19. Bobkov, S.G., Zegarlinski, B.: Entropy bounds and isoperimetry. Mem. Am. Math. Soc. **176** (2005)
20. Bobkov, S.G., Zegarlinski, B.: Distributions with slow tails and ergodicity of Markov semigroups in infinite dimensions. In: Laptev, A. (ed.) Around the Research of Vladimir Maz'ya I: Function Spaces, pp 13–79. Springer, New York (2010)
21. Borell, C.: The Ehrhard inequality. C. R. Math. Acad. Sci. Paris **337**, 663–666 (2003)
22. Borell, C.: The Brunn-Minkowski inequality in Gauss space. Invent. Math. **30**, 207–216 (1975)
23. Brezis, H., Wainger, S.: A note on limiting cases of Sobolev embeddings and convolution inequalities. Commun. Partial Differ. Equ. **5**, 773–789 (1980)
24. Calderón, A.P.: Spaces between $L^1$and $L^{\infty}$ and the theorem of Marcinkiewicz. Stud. Math. **26**, 273–299 (1966)
25. Cianchi, A., Pick, L.: Optimal Gaussian Sobolev embeddings. J. Funct. Anal. **256**, 3588–3642 (2009)
26. Coulhon, T.: Espaces de Lipschitz et inégalités de Poincaré. J. Funct. Anal. **136**, 81–113 (1996)
27. Coulhon, T.: Dimensions at infinity for Riemannian manifolds. Potential Anal. **4**, 335–344 (1995)
28. Coulhon, T.: Heat kernel and isoperimetry on non-compact Riemmanian manifolds. Contemp. Math. **338**, 65–99 (2003)
29. Cwikel, M., Jawerth, B., Milman, M.: A note on extrapolation of inequalities, preprint (2010)
30. Cwikel, M., Pustylnik, E.: Sobolev type embeddings in the limiting case. J. Fourier Anal. Appl. **4**, 433–446 (1998)
31. Ditzian, Z., Totik, V.: Moduli of Smoothness. Springer, New York (1987)
32. Ditzian, Z., Ivanov, K.G.: Strong converse inequalities. J. D'Analise Math. **61**, 61–111 (1993)
33. Ditzian, Z., Lubinsky, D.S.: Jackson and smoothness theorems for Freud weights in $L_p$ $(0 < p < \infty)$. Constr. Approx. **13**, 99–152 (1997)
34. Edmunds, D.E., Evans, W.D.: Hardy Operators, Function Spaces and Embeddings. Springer, Berlin (2004)
35. Ehrhard, A.: Symétrisation dans l'espace de Gauss. Math. Scand. **53**, 281–301 (1983)
36. Ehrhard, A.: Sur l'inégalité de Sobolev logarithmique de Gross. In: Séminaire de Probabilités XVII. Lecture Notes in Mathematics, vol. 1059, pp. 194–196. Springer, Heidelberg (1984)
37. Ehrhard, A.: Inégalités isopérimétriques et intégrales de Dirichlet gaussiennes. Ann. Sci. Ecole. Norm. Sup. **17**, 317–332 (1984)
38. Fiorenza, A.: Duality and reflexivity in grand Lebesgue spaces. Collect. Math. **51**, 131–148 (2000)
39. Fiorenza, A., Karadzhov, G.E.: Grand and Small Lebesgue Spaces and their analogs. Z. Anal. Anwendungen **23**, 657–681 (2004)

40. Fiorenza, A., Krbec, M., Schmeisser, H.J.: An improvement of dimension-free Sobolev imbeddings in r.i. spaces, Journal of Functional Analysis **267**, 243–261 (2014)
41. Fontana, L., Morpurgo, C.: Optimal limiting embeddings for $\Delta$-reduced Sobolev spaces in $L^1$. Ann. de l'Inst. Henri Poincaré (C) Non Linear Analysis, **31**, 217–230 (2014)
42. Gallot, S.: Inégalités isopérimétriques et analytiques sur les variétés Riemanniennes. Astérisque **163–164**, 31–91 (1988)
43. Garsia, A.M.: Combinatorial inequalities and smoothness of functions. Bull. Am. Math. Soc. **82**, 157–170 (1976)
44. Garsia, A.M.: A remarkable inequality and the uniform convergence of Fourier series. Indiana Univ. Math. J. **25**, 85–102 (1976)
45. Garsia, A., Rodemich, E.: Monotonicity of certain functionals under rearrangement. Ann. Inst. Fourier **24**, 67–116 (1974)
46. Griebel, M.: Sparse grids and related approximation schemes for higher dimensional problems. In: Pardo, L., Pinkus, A., Suli, E., Todd, M. (eds.) Foundations of Computational Mathematics (FoCM05), Santander, pp. 106–161. Cambridge University Press, Cambridge (2006)
47. Gross, L.: Logarithmic Sobolev inequalities. Am. J. Math. **97**, 1061–1083 (1975)
48. Hajlasz, P.: Sobolev inequalities, truncation method, and John domains. In: Papers in Analysis, Rep. Univ. Jyväskylä Dep. Math. Stat. **83**, Univ. Jyväskylä, Jyväskylä, pp 109–126 (2001)
49. Hajlasz, P., Koskela, P.: Sobolev met Poincaré. Mem. Am. Math. Soc. **145**, 101 pages (2000)
50. Hansson, K.: Imbedding theorems of Sobolev type in potential theory. Math Scand **45**, 77–102 (1979)
51. Hedberg, L.I.: On Maz'ya's work in potential theory and the theory of function spaces. In: The Maz'ya Anniversary Collection, vol. 1, pp. 7–16, Rostock, 1998. Operator Theory: Advances and Applications, vol. 109. Birkhäuser, Basel (1999)
52. Heinonen, J.: Lectures on Analysis on metric spaces. Lecture Notes. University of Michigan (1996)
53. Houdre, C., Ledoux, M., Milman, E., Milman, M.: Concentration, functional inequalities and isoperimetry. Contemp. Math. **545** (2011)
54. http://en.wikipedia.org/wiki/Vitali_Milman
55. Iwaniec, T., Sbordone, C.: On the integrability of the Jacobian under minimal hypotheses. Arch. Rational Mech. Anal. **119**, 129–143 (1992)
56. Jawerth, B., Milman, M.: Extrapolation theory with applications. Mem. Am. Math. Soc. **89**, 440 (1991)
57. Jawerth, B., Milman, M.: Interpolation of weak type spaces. Math. Z. **201**, 509–520 (1989)
58. Johnen, H., Scherer, K.: On the equivalence of the K-functional and moduli of continuity and some applications. In: Constructive Theory of Functions of Several Variables. Lecture Notes in Mathematics, vol. 571, pp. 119–140. Springer, Berlin (1977)
59. Karadzhov, G.E., Mehmood, Q.: Optimal regularity properties of the generalized Sobolev spaces. J. Funct. Spaces Appl. 761648, p 10 (2013)
60. Karadzhov, G.E., Milman, M.: Extrapolation theory: new results and applications. J. Approx. Theory **133**, 38–99 (2005)
61. Kesavan, S.: Symmetrization and Applications. World Scientific, Hackensack (2006)
62. Kolyada, V.I. (1989) Rearrangements of functions and embedding theorems. Uspekhi Mat. Nauk **44**, 61–95 (1989); transl. Russ. Math. Surv. **44**, 73–117 (1989)
63. Krbec, M., Schmeisser, H.J.: On dimension-free Sobolev imbeddings I. J. Math. Anal. Appl. **387**, 114–125 (2012)
64. Krbec, M., Schmeisser, H.J.: On dimension-free Sobolev imbeddings II. Rev. Mat. Complutense **25**, 247–265 (2012)
65. Ledoux, M.: Isoperimetry and gaussian analysis. Ecole d'Eté de Probabilités de Saint-Flour 1994. Springer Lecture Notes, vol. 1648, pp 165–294. Springer, Heidelberg (1996)
66. Ledoux, M.: Isopérimétrie et inégalitées de Sobolev logarithmiques gaussiennes. C. R. Acad. Sci. Paris Ser. I Math. **306**, 79–92 (1988)

67. Ledoux, M.: The Concentration of Measure Phenomenon. Mathematical Surveys and Monographs, vol. 89. American Mathematical Society, Providence (2001)
68. Ledoux, M.: From concentration to isoperimetry: Semigroup proofs. Contemp. Math. **545**, 155–166 (2011)
69. Leoni, G.: A First Course in Sobolev Spaces. Graduate Studies in Mathematics, vol. 105. American Mathematical Society, New York (2009)
70. Malý, J., Pick, L.: An elementary proof of Sharp Sobolev embeddings. Proc. Am. Math. Soc. **130**, 555–563 (2002)
71. Martín, J., Milman, M.: Symmetrization inequalities and Sobolev embeddings. Proc. Am. Math. Soc. **134**, 2335–2347 (2006)
72. Martín, J., Milman, M.: Higher-order symmetrization inequalities and applications. J. Math. Anal. Appl. **330**, 91–113 (2007)
73. Martin, J., Milman, M.: A note on Sobolev inequalities and limits of Lorentz spaces. Contemp. Math. **445**, 237–245 (2007)
74. Martin, J., Milman, M.: Isoperimetry and Symmetrization for Logarithmic Sobolev inequalities. J. Funct. Anal. **256**, 149–178 (2009)
75. Martin, J., Milman, M.: Addendum to Isoperimetry and Symmetrization for Logarithmic Sobolev inequalities. arXiv:0901.1839
76. Martin, J., Milman, M.: Isoperimetry and Symmetrization for Sobolev spaces on metric spaces. Comptes Rendus Math. **347**, 627–630 (2009)
77. Martín, J., Milman, M.: Isoperimetric Hardy type and Poincaré inequalities on metric spaces. In: Laptev, A. (ed.) Around the Research of Vladimir Maz'ya I. Function Spaces. International Mathematical Series, vol. 11, pp. 285–298. Springer, New York (2010)
78. Martin, J., Milman, M.: Pointwise symmetrization inequalities for Sobolev functions and applications. Adv. Math. **225**, 121–199 (2010)
79. Martin, J., Milman, M.: Sobolev inequalities, rearrangements, isoperimetry and interpolation spaces. Contemp. Math. **545**, 167–193 (2011)
80. Martin, J., Milman, M.: Fractional Sobolev inequalities: symmetrization, isoperimetry and interpolation, to appear in Astérisque (arXiv:1205.1584)
81. Martin, J., Milman, M.: A note on Coulhon type inequalities, to appear in Proc. Am. Math. Soc. (arXiv:1206.1584)
82. Martin, J., Milman, M.: Integral isoperimetric transference and dimensionless Sobolev inequalities, to appear in Revista Matemática Complutense (arXiv:1309.1980)
83. Martin, J., Milman, M.: A note on iterated Sobolev inequalities involving the isoperimetric profile, preprint
84. Martin, J., Milman, M.: On the Calderón-Maz'ya-Rubio de Francia extrapolation principle, preprint (2013)
85. Martin, J., Milman, M.: Symmetrization methods in the theory of Sobolev inequalities. Lecture Notes, in preparation
86. Martin, J., Milman, M., Pustylnik, E.: Sobolev inequalities: symmetrization and self improvement via truncation. J. Funct. Anal. **252**, 677–695 (2007)
87. Mastylo, M.: The modulus of smoothness in metric spaces and related problems. Potential Anal. **35**, 301–328 (2011)
88. Maz'ya, V.G.: Sobolev Spaces with Applications to Elliptic Partial Differential Equations. Second, revised and augmented edition. Grundlehren der Mathematischen Wissenschaften [Fundamental Principles of Mathematical Sciences], vol. 342. Springer, Heidelberg (2011)
89. Maz'ya, V.G.: The p-conductivity and theorems on imbedding certain functional spaces into a C-space (Russian), Dokl. Akad. Nauk SSSR **140**, 299–302 (1961) (English translation: in Soviet Math. Dokl. 3 (1962)
90. Milman, E.: Concentration and isoperimetry are equivalent assuming curvature lower bound. C. R. Math. Acad. Sci. Paris **347**, 73–76 (2009)
91. Milman, E.: On the role of convexity in isoperimetry, spectral-gap and concentration. Invent. Math. **177**, 1–43 (2009)

92. Milman, E.: On the role of convexity in functional and isoperimetric inequalities. Proc. Lond. Math. Soc. Proc. **999**, 32–66 (2009)
93. Milman, E.: Isoperimetric and concentration inequalities - equivalence under curvature lower bound. Duke Math. J. **154**, 207–239 (2010)
94. Milman, E.: A converse to the Maz'ya inequality for capacities under curvature lower bound. In: Laptev, A. (ed.) Around the Research of Vladimir Maz'ya I: Function Spaces, pp. 321–348. Springer, New York (2010)
95. Milman, E.: Isoperimetric bounds on convex manifolds. Contemp. Math **545**, 195–208 (2011)
96. Milman, M.: Local operators vs Lorentz-Marcinkiewicz spaces. Interpolation Spaces and Related Topics (Haifa, 1990). Israel Mathematical Conference Proceedings, vol. 5, pp. 151–157 (1992)
97. Milman, M., Pustylnik, E.: On sharp higher order Sobolev embeddings. Commun. Contemp. Math. **6**, 495–511 (2004)
98. Oklander, E.: Interpolacion, espacios de Lorentz y teorema de Marcinkiewicz. In: Cursos y Seminarios 20, Univ. Buenos Aires (1965) [See also Oklander, E.: On interpolation of Banach spaces, Thesis, Univ. Chicago (1963)]
99. O'Neil, R.: Convolution operators and L(p,q) spaces. Duke Math. J. **30**, 129–142 (1963)
100. Pérez Lázaro, F.J.: A note on extreme cases of Sobolev embeddings. J. Math. Anal. Appl. **320**, 973–982 (2006)
101. Pick, L., Kufner, A., John, O., Fucik, S.: Function Spaces, vol. 1. Walter de Gruyter & Co, Berlin (2012)
102. Pisier, G.: Factorization of operators through $L_{p\infty}$ or $L_{p1}$ and non-commutative generalizations. Math. Ann. **276**, 105–136 (1986)
103. Pustylnik, E.: On compactness of Sobolev embeddings in rearrangement-invariant spaces. Forum Math. **18**, 839–852 (2006)
104. Rakotoson, J.M.: Réarrangement relatif. Un instrument d'estimations dans les problèmes aux limites. Mathematics & Applications, vol. 64. Springer, Berlin (2008)
105. Ros, A.: The isoperimetric problem. In: Global Theory of Minimal Surfaces. Clay Mathematics Proceedings, vol. 2, pp. 175–209. American Mathematical Society, Providence (2005)
106. Rota, G.C.: Ten Lessons I wish I had been Taught. http://alumni.media.mit.edu/~cahn/life/gian-carlo-rota-10-lessons.html
107. Saloff-Coste, L.: Aspects of Sobolev Inequalities. Cambridge University Press, Cambridge (2002)
108. Slavíková, L.: Compactness of higher order Sobolev embeddings, Master Thesis, Charles University (2012)
109. Sudakov, V.N., Tsirelson, B.S.: Extremal properties of half-spaces for spherically invariant measures. J. Soviet. Math. **9**, 918 (1978); translated from Zap. Nauch. Sem. L.O.M.I. **41**, 1424 (1974)
110. Talenti, G.: Inequalities in rearrangement-invariant function spaces. In: Nonlinear Analysis, Function Spaces and Applications, vol. 5, pp. 177–230. Prometheus, Prague (1995) (for a comprehensive bibliography)
111. Tartar, L.: Imbedding theorems of Sobolev spaces into Lorentz spaces. Boll. Unione Mat. Ital. Sez B Artic. Ric. Mat. **1**(8), 479–500 (1998)
112. Trudinger, N.: On imbeddings into Orlicz spaces and some applications. J. Math. Mech. **17** 473–483 (1967)
113. Triebel, H.: Tractable embeddings of Besov spaces into Zygmund spaces. Function spaces IX, vol. 92, pp. 361–377. Banach Center, Polish Acad. Sci. Inst. Math., Warsaw (2011)
114. Triebel, H.: Tractable embeddings, preprint, University of Jena (2012)
115. Xiao, J., Zhai, Z.: Fractional Sobolev, Moser-Trudinger, Morrey-Sobolev inequalities under Lorentz norms. J. Math. Sci. **166**, 357–376 (2010)

# Transference of Fractional Laplacian Regularity

**Luz Roncal and Pablo Raúl Stinga**

**Summary.** In this note we show how to obtain regularity estimates for the fractional Laplacian on the multidimensional torus $\mathbb{T}^n$ from the fractional Laplacian on $\mathbb{R}^n$. Though at first glance this may seem quite natural, it must be carefully precised. A reason for that is the simple fact that $L^2$ functions on the torus cannot be identified with $L^2$ functions on $\mathbb{R}^n$. The transference is achieved through a formula that holds in the distributional sense. Such an identity allows us to transfer Harnack inequalities, to relate the extension problems, and to obtain pointwise formulas and Hölder regularity estimates.

## 1 The Transference Formula

For $0 < \sigma < 1$ and $u : \mathbb{R}^n \to \mathbb{R}$, the fractional Laplacian of order $\sigma$ in $\mathbb{R}^n$ is defined using the Fourier transform as

$$(-\Delta_{\mathbb{R}^n})^\sigma u(x) = \int_{\mathbb{R}^n} |\xi|^{2\sigma} \hat{u}(\xi) e^{ix\cdot\xi}\, d\xi, \quad x \in \mathbb{R}^n.$$

Similarly, the fractional Laplacian on $\mathbb{T}^n \equiv \mathbb{R}^n/(2\pi\mathbb{Z}^n)$ is defined via the multiple Fourier series

$$(-\Delta_{\mathbb{T}^n})^\sigma v(z) = \sum_{k\in\mathbb{Z}^n} |k|^{2\sigma} c_k(v) e^{ik\cdot z},$$

L. Roncal
Departamento de Matemáticas y Computación, Universidad de La Rioja, 26004 Logroño, Spain
e-mail: luz.roncal@unirioja.es

P.R. Stinga (✉)
Department of Mathematics, The University of Texas at Austin,
1 University Station C1200, Austin, TX 78712-1202, USA
e-mail: stinga@math.utexas.edu



A.M. Stokolos et al. (eds.), *Special Functions, Partial Differential Equations, and Harmonic Analysis*, Springer Proceedings in Mathematics & Statistics 108,
DOI 10.1007/978-3-319-10545-1_14

where $c_k(v)$ is the Fourier coefficient of $v : \mathbb{T}^n \to \mathbb{R}$. In our notation, the point $(e^{iz_1}, \ldots, e^{iz_n}) \in \mathbb{T}^n$ is uniquely identified with $z = (z_1, \ldots, z_n) \in Q_n := (-\pi, \pi]^n$, so $v(z)$ in fact means $v(e^{iz_1}, \ldots, e^{iz_n})$. In order to avoid a rather cumbersome notation, we will just write $z \in \mathbb{T}^n$.

It is clear that the fractional Laplacian on $\mathbb{R}^n$ does not preserve the Schwartz class $\mathcal{S}$. Instead,

$$(-\Delta_{\mathbb{R}^n})^\sigma : \mathcal{S} \to \mathcal{S}_\sigma,$$

where

$$\mathcal{S}_\sigma := \{\varphi \in C^\infty(\mathbb{R}^n) : (1 + |x|^2)^{\frac{n+2\sigma}{2}} D^k\varphi(x) \in L^\infty(\mathbb{R}^n), k \in \mathbb{N}_0\},$$

see [Sil07, pp. 72–73]. Observe that $\mathcal{S} \subset \mathcal{S}_\sigma$. Then the symmetry of the fractional Laplacian allows us to define $(-\Delta_{\mathbb{R}^n})^\sigma$ for $u$ in the dual space $\mathcal{S}_\sigma'$. For locally integrable *functions* $u$ in $\mathcal{S}_\sigma'$ we let

$$\langle (-\Delta_{\mathbb{R}^n})^\sigma u, \varphi\rangle_{\mathcal{S}_\sigma} := \int_{\mathbb{R}^n} u(x)(-\Delta_{\mathbb{R}^n})^\sigma \varphi(x)\,dx, \quad \varphi \in \mathcal{S}.$$

Certainly, the integral above is absolutely convergent when (see also [Sil07])

$$u \in L_\sigma := L^1(\mathbb{R}^n, (1 + |x|^2)^{-\frac{n+2\sigma}{2}}\,dx).$$

The situation with the fractional Laplacian on the torus is different than the $\mathbb{R}^n$ case. We first notice that $(-\Delta_{\mathbb{T}^n})^\sigma$ preserves the class of smooth functions on $\mathbb{T}^n$. By symmetry we are able to define this operator for any *function* $v$ that is a periodic distribution on $\mathbb{T}^n$. Indeed, we let

$$\langle (-\Delta_{\mathbb{T}^n})^\sigma v, \phi\rangle_{C^\infty(\mathbb{T}^n)} := \int_{\mathbb{T}^n} v(z)(-\Delta_{\mathbb{T}^n})^\sigma \phi(z)\,dz, \quad \phi \in C^\infty(\mathbb{T}^n).$$

To relate both fractional Laplacians we define two operators. For a function $v$ on $\mathbb{T}^n$ we define its *repetition* $Rv : \mathbb{R}^n \to \mathbb{R}$ by

$$(Rv)(x) = \sum_{k\in\mathbb{Z}^n} v(x - 2\pi k)\chi_{Q_n}(x - 2\pi k), \quad x \in \mathbb{R}^n.$$

This is nothing but the $Q_n$-periodic function on $\mathbb{R}^n$ that coincides with $v$ on $\mathbb{T}^n$. Here $\mathbb{T}^n$ is identified with $Q_n$ as explained above. For a function $u : \mathbb{R}^n \to \mathbb{R}$ we define its *periodization* as the function $p_\Sigma u : \mathbb{T}^n \to \mathbb{R}$ given (formally) by

$$(p_\Sigma u)(z) = \sum_{k\in\mathbb{Z}^n} u(z + 2\pi k), \quad z \in \mathbb{T}^n. \tag{1}$$

**Theorem A (Transference Formula).** *Let v be a function on the torus such that*

$$\sum_{k\in\mathbb{Z}^n\setminus\{0\}} |c_k(v)|\frac{e^{-|k|^2}}{|k|} < \infty. \tag{2}$$

*Then its repetition Rv is a function in $L_\sigma$ which defines a distribution in $\mathcal{S}'_\sigma$ and such that*

$$\int_{\mathbb{R}^n} (Rv)(-\Delta_{\mathbb{R}^n})^\sigma \varphi \, dx = \int_{\mathbb{T}^n} v(-\Delta_{\mathbb{T}^n})^\sigma (p_\Sigma \varphi)\, dz, \quad \varphi \in \mathcal{S}. \tag{3}$$

*In other words, when evaluated in periodizations of Schwartz functions, the periodic distribution $(-\Delta_{\mathbb{T}^n})^\sigma v$ coincides with the distributional fractional Laplacian on $\mathbb{R}^n$ of the repetition Rv.*

*Proof.* We first check that $Rv \in L_\sigma$. Let us compute

$$\begin{aligned}\int_{\mathbb{R}^n} \frac{|(Rv)(x)|}{(1+|x|^2)^{\frac{n+2\sigma}{2}}}\, dx &= \textstyle\sum_{k\in\mathbb{Z}^n} \int_{Q_n} \dfrac{|(Rv)(x+2k\pi)|}{(1+|x+2k\pi|^2)^{\frac{n+2\sigma}{2}}}\, dx \\ &= \textstyle\int_{\mathbb{T}^n} |v(z)| p_\Sigma((1+|\cdot|^2)^{-\frac{n+2\sigma}{2}})(z)\, dz. \end{aligned} \tag{4}$$

Since $(1+|\cdot|^2)^{-\frac{n+2\sigma}{2}}$ is integrable, then $[p_\Sigma(1+|\cdot|^2)^{-\frac{n+2\sigma}{2}}]$ is integrable (see [SW71, Chapter VII]). Its Fourier coefficient can be computed as follows:

$$\begin{aligned}\mathcal{F}[(1+|\cdot|^2)^{-\frac{n+2\sigma}{2}}](k) &= \mathcal{F}^{-1}(\mathcal{F}(I-\Delta_{\mathbb{R}^n})^{-\frac{n+2\sigma}{2}})(k) \\ &= \frac{1}{\Gamma(\frac{n+2\sigma}{2})} \int_0^\infty e^{-t} \frac{e^{-|k|^2/(4t)}}{(4\pi t)^{n/2}} \frac{dt}{t^{1-\frac{n+2\sigma}{2}}} \\ &= \frac{|k|^{2\sigma}}{(4\pi)^{n/2} 4^\sigma \Gamma(\frac{n+2\sigma}{2})} \int_0^\infty e^{-|k|^2/(4r)} e^{-r} \frac{dr}{r^{1+\sigma}} \\ &= c_{n,\sigma} \mathcal{K}_\sigma(|k|^2). \end{aligned}$$

Here $\mathcal{K}_\sigma(z)$ is the modified Bessel function of the third kind (see [L72, p. 119]). A well-known asymptotic formula gives that $\mathcal{K}_\sigma(|k|^2) \sim |k|^{-1}e^{-|k|^2}$, as $|k| \to \infty$. Hence, by Parseval's identity on $\mathbb{T}^n$ and the hypothesis, from (4) we get

$$\begin{aligned}\int_{\mathbb{R}^n} \frac{|(Rv)(x)|}{(1+|x|^2)^{\frac{n+2\sigma}{2}}}\, dx &= c_{n,\sigma} \sum_{k\in\mathbb{Z}^n} |c_k(v)| \mathcal{K}_\sigma(|k|^2) \\ &\le C_{n,\sigma} \sum_{k\in\mathbb{Z}^n\setminus\{0\}} |c_k(v)| \frac{e^{-|k|^2}}{|k|} < \infty. \end{aligned}$$

Thus, $Rv \in L_\sigma$ and the left-hand side of (3) is absolutely convergent.

Again, $p_\Sigma\varphi$ is integrable on $\mathbb{T}^n$ and $c_k(p_\Sigma\varphi) = \hat{\varphi}(k)$, for each $k \in \mathbb{Z}^n$. Moreover, since $\varphi$ and $\hat{\varphi}$ decay at infinity as $|x|^{-n-\delta}$, $\delta > 0$, we have

$$(p_\Sigma\varphi)(z) = \sum_{k\in\mathbb{Z}^n} \hat{\varphi}(k)e^{ik\cdot z}, \tag{5}$$

where the series converges absolutely, see [SW71, Chapter VII]. From here, using the properties of the Fourier transform, it readily follows that $p_\Sigma\varphi$ is a smooth function on the torus. Hence $(-\Delta_{\mathbb{T}^n})^\sigma(p_\Sigma\varphi)$ is smooth too and the right-hand side of (3) is absolutely convergent.

Before proving (3) we compute the periodization of $(-\Delta_{\mathbb{R}^n})^\sigma\varphi$. Since $\varphi$ is in the Schwartz class, both $(-\Delta_{\mathbb{R}^n})^\sigma\varphi$ and its Fourier transform decay as $|x|^{-(n+2\sigma)}$ at infinity. Therefore, by (5),

$$\begin{aligned}
[p_\Sigma(-\Delta_{\mathbb{R}^n})^\sigma\varphi](z) &= \sum_{k\in\mathbb{Z}^n} \widehat{(-\Delta_{\mathbb{R}^n})^\sigma\varphi}(k)e^{ik\cdot z} = \sum_{k\in\mathbb{Z}^n} |k|^{2\sigma}\hat{\varphi}(k)e^{ik\cdot z} \\
&= \sum_{k\in\mathbb{Z}^n} |k|^{2\sigma}c_k(p_\Sigma\varphi)e^{ik\cdot z} = (-\Delta_{\mathbb{T}^n})^\sigma(p_\Sigma\varphi)(z),
\end{aligned}$$

for each $z \in \mathbb{T}^n$. With this, we readily obtain

$$\begin{aligned}
&\int_{\mathbb{R}^n} (Rv)(-\Delta_{\mathbb{R}^n})^\sigma\varphi\,dx \\
&= \int_{\mathbb{R}^n} \left[\sum_{k\in\mathbb{Z}^n} v(x-2\pi k)\chi_{Q_n}(x-2\pi k)\right](-\Delta_{\mathbb{R}^n})^\sigma\varphi(x)\,dx \\
&= \sum_{k\in\mathbb{Z}^n} \int_{Q_n+2\pi k} v(x-2\pi k)(-\Delta_{\mathbb{R}^n})^\sigma\varphi(x)\,dx \\
&= \sum_{k\in\mathbb{Z}^n} \int_{Q_n} v(z)(-\Delta_{\mathbb{R}^n})^\sigma\varphi(z+2\pi k)\,dz \\
&= \int_{\mathbb{T}^n} v(z)\,[p_\Sigma(-\Delta_{\mathbb{R}^n})^\sigma\varphi](z)\,dz = \int_{\mathbb{T}^n} v(-\Delta_{\mathbb{T}^n})^\sigma(p_\Sigma\varphi)\,dz.
\end{aligned}$$

Notice that the integration on the torus with respect to the Haar measure is just the integration over $Q_n$ with respect to the Lebesgue measure, so the previous to last equality is true.

*Remark 1.1.* Formula (3) is certainly valid for functions $v \in L^p(\mathbb{T}^n)$, $1 \le p \le \infty$. Indeed, $v \in L^1(\mathbb{T}^n)$ and, by the Riemann–Lebesgue Lemma, $c_k(v) \to 0$ as $|k| \to \infty$, thus (2) holds. Observe that condition (2) also holds whenever $\sum_{k\in\mathbb{Z}^n\setminus\{0\}} |k|^{2\sigma}|c_k(v)|^2 < \infty$, that is, when $v$ is in the Sobolev space $H^\sigma = \mathrm{Dom}((-\Delta_{\mathbb{T}^n})^{\sigma/2})$.

# 2 Applications

## 2.1 Harnack Inequalities

Interior and boundary Harnack estimates for the fractional Laplacian on the torus now follow from the transference formula in Theorem A.

**Theorem 2.1 (Interior Harnack Inequality).** *Let $\mathcal{O} \subseteq \mathbb{T}^n$ be an open set. For any compact subset $\mathcal{K} \subset \mathcal{O}$, there exists a constant $C > 0$, that depends only on $n$, $\sigma$ and $\mathcal{K}$, such that*

$$\sup_{\mathcal{K}} v \leq C \inf_{\mathcal{K}} v,$$

*for all solutions $v \in \mathrm{Dom}((-\Delta_{\mathbb{T}^n})^\sigma)$ to*

$$\begin{cases} (-\Delta_{\mathbb{T}^n})^\sigma v = 0, & \text{in } \mathcal{O}, \\ v \geq 0, & \text{on } \mathbb{T}^n. \end{cases}$$

*Proof.* For $v$ as in the hypothesis, its repetition $Rv$ is a nonnegative function on $\mathbb{R}^n$ which belongs to $L_\sigma$. We can identify $\mathcal{O}$ with an open subset $\tilde{\mathcal{O}} \subset Q_n$. Take any smooth function $\varphi$ with compact support in $\tilde{\mathcal{O}}$. Then $p_\Sigma \varphi$ is a smooth function on the torus with compact support in $\mathcal{O}$. Now Theorem A gives that $\langle (-\Delta_{\mathbb{R}^n})^\sigma (Rv), \varphi \rangle_{\mathcal{S}_\sigma} = \langle (-\Delta_{\mathbb{T}^n})^\sigma v, p_\Sigma \varphi \rangle_{C^\infty(\mathbb{T}^n)} = 0$. Hence $Rv$ is a nonnegative solution to $(-\Delta_{\mathbb{R}^n})^\sigma (Rv) = 0$ in $\tilde{\mathcal{O}}$. Then $Rv$ satisfies Harnack inequality (see [CSil07, Theorem 5.1]), and so does $v$.

**Theorem 2.2 (Boundary Harnack Inequality).** *Let $0 < \sigma < 1$ and $v_1, v_2 \in \mathrm{Dom}((-\Delta_{\mathbb{T}^n})^\sigma)$ be two nonnegative functions on $\mathbb{T}^n$. Suppose that $(-\Delta_{\mathbb{T}^n})^\sigma v_j = 0$ in $\mathcal{O}$, for some open set $\mathcal{O} \subseteq \mathbb{T}^n$ and $j = 1, 2$. Let $z_0 \in \partial\mathcal{O}$ and assume that $v_j = 0$ for all $z \in B_r(z_0) \cap \mathcal{O}^c$, for some sufficiently small $r > 0$. Assume also that $\partial\mathcal{O} \cap B_r(z_0)$ is a Lipschitz graph in the direction of $z_1$. Then, there is a constant $C$ depending only on $\mathcal{O}$, $z_0$, $r$, $n$ and $\sigma$, but not on $v_1$ or $v_2$, such that*

$$\sup_{\mathcal{O} \cap B_{r/2}(z_0)} \left( \frac{v_1}{v_2} \right) \leq C \inf_{\mathcal{O} \cap B_{r/2}(z_0)} \left( \frac{v_1}{v_2} \right).$$

*Moreover, $v_1/v_2$ is $\alpha$-Hölder continuous in $\overline{\mathcal{O} \cap B_{r/2}(z_0)}$, for some universal $0 < \alpha < 1$.*

*Proof.* Again we have that $Rv_i$ is in $L_\sigma$. We identify $\mathcal{O}$ with an open subset $\tilde{\mathcal{O}} \subset Q_n$. Let us also identify $z_0 \in \partial\mathcal{O}$ with $x_0 \in \partial\tilde{\mathcal{O}}$. Then the corresponding boundary portion $\partial\tilde{\mathcal{O}} \cap B_r(x_0)$ is a Lipschitz graph in the $x_1$-direction. Using the same argument as in the proof of Theorem 2.1, it follows that $Rv_i$ are nonnegative solutions to $(-\Delta_{\mathbb{R}^n})^\sigma (Rv_i) = 0$ in $\tilde{\mathcal{O}}$, and $Rv_i = 0$ in $B_r(x_0) \cap \tilde{\mathcal{O}}^c$.

Therefore, the boundary Harnack inequality holds for $Rv_i$ (see [CSil07, Theorem 5.3]), and so does for $v_i$ by restricting $Rv_i$ to $\mathcal{O}$. The Hölder continuity of $v_1/v_2$ follows from the Hölder continuity of $(Rv_1)/(Rv_2)$.

## 2.2 Extension Problem

It is known that the Caffarelli–Silvestre extension problem characterization is valid also for the fractional Laplacian on the torus, see [St10, StT10], also [RSt12, GMSt13]. Here we can derive it directly from the Caffarelli–Silvestre result of $\mathbb{R}^n$ in [CSil07] with the explicit constants computed in [StT10]. In the proof we are going to need the following simple result.

**Lemma 2.3.** *Let $\phi$ be a smooth function on $\mathbb{T}^n$. Then there exists a smooth function $\varphi$ with compact support on $\mathbb{R}^n$ such that*

$$\phi(z) = p_\Sigma \varphi(z), \quad \text{for } z \in \mathbb{T}^n.$$

*Proof.* It is easy to see that there exists a smooth function $\psi$ with compact support on $\mathbb{R}^n$ such that $\sum_{k\in\mathbb{Z}^n} \psi(x + 2\pi k) \equiv 1$, for all $x \in \mathbb{R}^n$. Indeed, $\psi$ can be constructed as the convolution of the characteristic function of $Q_n$ with a smooth bump function that has integral 1. Set $\varphi(x) = \psi(x)(R\phi)(x)$. Clearly, $\varphi$ is smooth (see the proof of Theorem A) and has compact support. Moreover,

$$\begin{aligned}(p_\Sigma\varphi)(z) &= \sum_{k\in\mathbb{Z}^n} \psi(z+2\pi k)(R\phi)(z+2\pi k) \\ &= \sum_{k\in\mathbb{Z}^n} \psi(z+2\pi k) \sum_{j\in\mathbb{Z}^n} \phi(z+2\pi k+2\pi j)\chi_{Q_n}(z+2\pi k+2\pi j) \\ &= \phi(z) \sum_{k\in\mathbb{Z}^n} \psi(z+2\pi k) = \phi(z), \quad z \in \mathbb{T}^n.\end{aligned}$$

**Theorem 2.4 (Extension Problem).** *Let $v \in \mathrm{Dom}((-\Delta_{\mathbb{T}^n})^\sigma)$. Let $V = V(z,y)$ be the solution to the boundary value problem*

$$\begin{cases} \Delta_{\mathbb{T}^n} V + \frac{1-2\sigma}{y} V_y + V_{yy} = 0, & \text{in } \mathbb{T}^n \times (0,\infty), \\ V(z,0) = v(z), & \text{on } \mathbb{T}^n. \end{cases} \tag{6}$$

*Then, for $c_\sigma = \frac{\Gamma(1-\sigma)}{4^{\sigma-1/2}\Gamma(\sigma)} > 0$, we have that*

$$-\lim_{y\to 0+} y^{1-2\sigma} V_y(z,y) = c_\sigma(-\Delta_{\mathbb{T}^n})^\sigma v(z), \quad z \in \mathbb{T}^n. \tag{7}$$

*Proof.* Consider $u = Rv \in L_\sigma$. Let $U$ be the solution to the extension problem for $u$:

$$\begin{cases} \Delta_{\mathbb{R}^n} U + \frac{1-2\sigma}{y} U_y + U_{yy} = 0, & \text{in } \mathbb{R}^n \times (0,\infty), \\ U(x,0) = u(x), & \text{on } \mathbb{R}^n. \end{cases}$$

From [CSil07] we know that $U(x,y) = P_y^\sigma * u(x)$, for a suitable Poisson kernel $P_y^\sigma(x)$. Using this Poisson formula and analogous to the proof of Theorem A, it can be checked that $U(z,y) = (v * (p_\Sigma P_y^\sigma))(z)$, $z \in \mathbb{T}^n$, where the convolution is performed on $\mathbb{T}^n$. Then $U(z,y)$ is a solution to (6). By uniqueness, it follows that $V(\cdot,y) = v * (p_\Sigma P_y^\sigma)$, for each $y > 0$. Moreover, by Theorem A and the Caffarelli–Silvestre extension result for the fractional Laplacian on $\mathbb{R}^n$ in [CSil07],

$$\begin{aligned} c_\sigma \int_{\mathbb{T}^n} v(-\Delta_{\mathbb{T}^n})^\sigma (p_\Sigma \varphi)\, dz &= c_\sigma \int_{\mathbb{R}^n} u(-\Delta_{\mathbb{R}^n})^\sigma \varphi\, dx \\ &= - \lim_{y\to 0^+} \int_{\mathbb{R}^n} y^{1-2\sigma} U_y(x,y)\varphi(x)\, dx \\ &= - \lim_{y\to 0^+} \sum_{k\in\mathbb{Z}^n} \int_{Q_n} y^{1-2\sigma} U_y(z+2\pi k, y)\varphi(z+2\pi k)\, dz \\ &= - \lim_{y\to 0^+} \int_{Q_n} y^{1-2\sigma} U_y(z,y)(p_\Sigma\varphi)(z)\, dz \\ &= - \lim_{y\to 0^+} \int_{\mathbb{T}^n} y^{1-2\sigma} V_y(z,y)(p_\Sigma\varphi)(z)\, dz. \end{aligned}$$

Now (7) follows because any smooth function $\phi$ on the torus can be expressed as $p_\Sigma\varphi$, for some $\varphi \in \mathcal{S}$, see Lemma 2.3.

## 2.3 Pointwise Formula

Let $0 < \alpha \le 1$ and $k \in \mathbb{N}_0$. A continuous real function $v$ defined on $\mathbb{T}^n$ belongs to the Hölder space $C^{k,\alpha}(\mathbb{T}^n)$, if $v \in C^k(\mathbb{T}^n)$ and

$$[D^\gamma v]_{C^\alpha(\mathbb{T}^n)} := \sup_{\substack{x,y\in\mathbb{T}^n \\ x\neq y}} \frac{|D^\gamma v(x) - D^\gamma v(y)|}{\operatorname{dist}(x,y)^\alpha} < \infty,$$

for each multi-index $\gamma \in \mathbb{N}_0^n$ such that $|\gamma| = k$. Here $\operatorname{dist}(x,y)$ is the geodesic distance from $x$ to $y$ on $\mathbb{T}^n$. We define the norm in the spaces $C^{k,\alpha}(\mathbb{T}^n)$ as usual.

**Theorem 2.5 (Pointwise Formula).** *Let $v \in C^{0,2\sigma+\varepsilon}(\mathbb{T}^n)$ if $0 < \sigma < 1/2$ (or $v \in C^{1,2\sigma+\varepsilon-1}(\mathbb{T}^n)$ if $1/2 \leq \sigma < 1$). Then $(-\Delta_{\mathbb{T}^n})^\sigma v$ coincides with the continuous function on $\mathbb{T}^n$ given by*

$$\begin{aligned}(-\Delta_{\mathbb{T}^n})^\sigma v(x) &= \mathrm{P.V.}\int_{\mathbb{T}^n}(v(x)-v(z))K_\sigma(x-z)\,dz\\ &= \lim_{\delta\to 0^+}\int_{|x-z|>\delta, z\in\mathbb{T}^n}(v(x)-v(z))K_\sigma(x-z)\,dz, \quad x\in\mathbb{T}^n,\end{aligned}$$

*where, for $x \in \mathbb{T}^n$, $x \neq 0$,*

$$K_\sigma(x) = \frac{2^\sigma\Gamma\left(\frac{n+\sigma}{2}\right)}{|\Gamma(-\sigma/2)|\pi^{n/2}}\sum_{k\in\mathbb{Z}^n}\frac{1}{|x+2\pi k|^{n+2\sigma}}.$$

*In the case $0 < \sigma < 1/2$ the integral above is absolutely convergent.*

One may think that $K_\sigma$ is just the periodization of the kernel of the fractional Laplacian on $\mathbb{R}^n$. In fact, formally, $K_\sigma(x) = c_{n,\sigma}p_\Sigma(|x|^{-(n+2\sigma)})$. But, since $|x|^{-(n+2\sigma)}$ is not integrable on $\mathbb{R}^n$, this formal identity makes no sense.

*Proof of Theorem 2.5* Notice that $K_\sigma(x)$ is well defined for $x \neq 0$. Indeed, if $k \neq 0$, then for $x \in \mathbb{T}^n$ we have $|\pi k| \leq c_n|x - 2k\pi|$, so

$$0 \leq K_\sigma(x) \leq C_{n,\sigma}\left[\frac{1}{|x|^{n+2\sigma}} + \sum_{k\in\mathbb{Z}^n\setminus\{0\}}\frac{1}{|\pi k|^{n+2\sigma}}\right], \quad x\neq 0,$$

and the series is absolutely convergent. We have to prove that

$$\langle(-\Delta_{\mathbb{T}^n})^\sigma v,\phi\rangle_{C^\infty(\mathbb{T}^n)} = \int_{\mathbb{T}^n}h(x)\phi(x)\,dx, \quad \text{for any } \phi\in C^\infty(\mathbb{T}^n), \tag{8}$$

where the continuous function $h$ is given by

$$h(x) = \mathrm{P.V.}\int_{\mathbb{T}^n}(v(x)-v(z))K_\sigma(x-z)\,dz.$$

Let $u = Rv$. Then $u$ is bounded and it belongs to $C^{0,2\sigma+\varepsilon}(\mathbb{R}^n)$ (or to $C^{1,2\sigma+\varepsilon-1}(\mathbb{R}^n)$), so $(-\Delta_{\mathbb{R}^n})^\sigma u$ is a continuous function on $\mathbb{R}^n$ (see [Sil07, Proposition 2.4]) and

$$\begin{aligned}(-\Delta_{\mathbb{R}^n})^\sigma u(x) &= c_{n,\sigma}\,\mathrm{P.V.}\int_{\mathbb{R}^n}\frac{u(x)-u(y)}{|x-y|^{n+2\sigma}}\,dy\\ &= c_{n,\sigma}\,\mathrm{P.V.}\sum_{k\in\mathbb{Z}^n}\int_{Q_n}\frac{u(x)-u(z-2\pi k)}{|x-z+2\pi k|^{n+2\sigma}}\,dz\end{aligned}$$

$$= c_{n,\sigma}\,\text{P.V.}\sum_{k\in\mathbb{Z}^n}\int_{Q_n}\frac{v(x)-v(z)}{|x-z+2\pi k|^{n+2\sigma}}\,dz$$

$$= c_{n,\sigma}\,\text{P.V.}\int_{\mathbb{T}^n}(v(x)-v(z))K_\sigma(x-z)\,dz = h(x). \tag{9}$$

With this we conclude that $h$ is a continuous function on $\mathbb{T}^n$. Observe that $(-\Delta_{\mathbb{R}^n})^\sigma u$ is a $Q_n$-periodic function. To establish (8), let $\phi$ be any smooth function on the torus. By Lemma 2.3, there exists $\varphi \in \mathcal{S}$ such that $\phi(z) = p_\Sigma\varphi(z)$, $z \in \mathbb{T}^n$. Then, by Theorem A and (9),

$$\langle(-\Delta_{\mathbb{T}^n})^\sigma v, \phi\rangle_{C^\infty(\mathbb{T}^n)} = \langle(-\Delta_{\mathbb{T}^n})^\sigma v, p_\Sigma\varphi\rangle_{C^\infty(\mathbb{T}^n)} = \langle(-\Delta_{\mathbb{R}^n})^\sigma u, \varphi\rangle_{\mathcal{S}_\sigma}$$

$$= \int_{\mathbb{R}^n}(-\Delta_{\mathbb{R}^n})^\sigma u(x)\varphi(x)\,dx$$

$$= \sum_{k\in\mathbb{Z}^n}\int_{Q_n}(-\Delta_{\mathbb{R}^n})^\sigma u(x+2\pi k)\varphi(x+2\pi k)\,dx$$

$$= \int_{Q_n}(-\Delta_{\mathbb{R}^n})^\sigma u(x)(p_\Sigma\varphi)(x)\,dx = \int_{\mathbb{T}^n}h(x)\phi(x)\,dx.$$

## 2.4 Hölder Regularity

Hölder estimates follow directly from our transference formula and the known results for the fractional Laplacian on $\mathbb{R}^n$.

**Theorem 2.6 (Hölder Estimates).** *Take $\alpha \in (0,1]$.*

(1) *Let $v \in C^{0,\alpha}(\mathbb{T}^n)$ and $0 < 2\sigma < \alpha$. Then $(-\Delta_{\mathbb{T}^n})^\sigma v \in C^{0,\alpha-2\sigma}(\mathbb{T}^n)$ and*

$$\|(-\Delta_{\mathbb{T}^n})^\sigma v\|_{C^{0,\alpha-2\sigma}(\mathbb{T}^n)} \le C\,\|v\|_{C^{0,\alpha}(\mathbb{T}^n)}.$$

(2) *Let $v \in C^{1,\alpha}(\mathbb{T}^n)$ and $0 < 2\sigma < \alpha$. Then $(-\Delta_{\mathbb{T}^n})^\sigma v \in C^{1,\alpha-2\sigma}(\mathbb{T}^n)$ and*

$$\|(-\Delta_{\mathbb{T}^n})^\sigma v\|_{C^{1,\alpha-2\sigma}(\mathbb{T}^n)} \le C\,\|v\|_{C^{1,\alpha}(\mathbb{T}^n)}.$$

(3) *Let $v \in C^{1,\alpha}(\mathbb{T}^n)$ and $2\sigma \ge \alpha$, with $\alpha - 2\sigma + 1 \ne 0$. Then $(-\Delta_{\mathbb{T}^n})^\sigma v \in C^{0,\alpha-2\sigma+1}(\mathbb{T}^n)$ and*

$$\|(-\Delta_{\mathbb{T}^n})^\sigma v\|_{C^{0,\alpha-2\sigma+1}(\mathbb{T}^n)} \le C\,\|v\|_{C^{1,\alpha}(\mathbb{T}^n)}.$$

(4) *Let $v \in C^{k,\alpha}(\mathbb{T}^n)$ and assume that $k+\alpha-2\sigma$ is not an integer. Then $(-\Delta_{\mathbb{T}^n})^\sigma v \in C^{l,\beta}(\mathbb{T}^n)$, where $l$ is the integer part of $k+\alpha-2\sigma$ and $\beta = k+\alpha-2\sigma-l$, and*

$$\|(-\Delta_{\mathbb{T}^n})^\sigma v\|_{C^{l,\beta}(\mathbb{T}^n)} \le C\,\|v\|_{C^{k,\alpha}(\mathbb{T}^n)}.$$

*Proof.* For (1), by Theorem 2.5 and [Sil07, Proposition 2.5] we readily get

$$\begin{aligned}\|(-\Delta_{\mathbb{T}^n})^\sigma v\|_{C^{0,\alpha-2\sigma}(\mathbb{T}^n)} &= \|(-\Delta_{\mathbb{R}^n})^\sigma (Rv)\|_{C^{0,\alpha-2\sigma}(\mathbb{R}^n)} \\ &\leq C\|Rv\|_{C^{0,\alpha}(\mathbb{R}^n)} = C\|v\|_{C^{0,\alpha}(\mathbb{T}^n)}.\end{aligned}$$

Parts (2), (3), and (4) follow analogously by using Theorems A, 2.5, and the known results for $\mathbb{R}^n$ [Sil07, Proposition 2.6, Proposition 2.7].

**Acknowledgements** We thank Luis Caffarelli and José L. Torrea for delightful and pleasant discussions about the results of this work. The first author was partially supported by grant MTM2012-36732-C03-02 from Spanish Government. The second author was partially supported by MTM2011-28149-C02-01 from Spanish Government.

## References

[CSil07] Caffarelli, L., Silvestre, L.: An extension problem related to the fractional Laplacian. Comm. Part. Differ. Equat. **32**, 1245–1260 (2007)

[GMSt13] Galé, J.E., Miana, P.J., Stinga, P.R.: Extension problem and fractional operators: semigroups and wave equations. J. Evol. Equat. **13**, 343–368 (2013)

[L72] Lebedev, N.N.: Special Functions and Its Applications. Dover, New York (1972)

[RSt12] Roncal, L., Stinga, P.R.: Fractional Laplacian on the torus. arXiv:1209.6104v2, preprint (2012), 25pp.

[Sil07] Silvestre, L.: Regularity of the obstacle problem for a fractional power of the Laplace operator. Comm. Pure Appl. Math. **60**, 67–112 (2007)

[SW71] Stein, E.M., Weiss, G.: Introduction to Fourier Analysis on Euclidean Spaces. Princeton Mathematical Series, vol. 32. Princeton University Press, Princeton (1971)

[St10] Stinga, P.R.: Fractional powers of second order partial differential equations: extension problem and regularity theory. PhD thesis, Universidad Autónoma de Madrid (2010)

[StT10] Stinga, P.R., Torrea, J.L.: Extension problem and Harnack's inequality for some fractional operators. Comm. Part. Differ. Equat. **35**, 2092–2122 (2010)

# Local Sharp Maximal Functions

**Alberto Torchinsky**

In considering the resistance of materials to certain types of deformations, F. John was led to the study of quasi-isometric mappings. The setting is essentially as follows. Let $f$ be a continuous function defined on a cube $Q_0 \subset \mathbb{R}^n$. Assume that to each subcube $Q$ of $Q_0$ with sides parallel to those of $Q_0$ there is assigned a constant $c_Q$ and let $\mu_Q$ be the function of the real variable $M$ given by

$$\mu_Q(M) = \frac{|\{y \in Q : |f(y) - c_Q| > M\}|}{|Q|}.$$

Let $\phi(M) = \sup_{Q \subset Q_0} \mu_Q(M)$, $0 < s < 1/2$, and $\lambda$ a number such that $\phi(\lambda) \leq s$. Then under these assumptions

$$\phi(M) \leq A e^{-BM/\lambda}$$

holds for all nonnegative $M$ where $A, B$ are universal functions of $s$ and the dimension $n$. Thus the space of functions of bounded mean oscillation ($BMO$) was introduced and the John–Nirenberg inequality established [18].

Strömberg adopted this setting when studying spaces close to $BMO$, incorporated the value $s = 1/2$ above, which corresponds to the notion of median value $m_f(Q)$ of $f$ over $Q$, and formalized the notion of local sharp maximal function [9, 35]. This note evolves from the fact that local maximal functions allow for pointwise estimates for Calderón–Zygmund singular integral operators [17, 23].

In what follows we adopt the notations of [30, 31, 35]. In particular, all cubes have sides parallel to the axes. Also, for a cube $Q \subset \mathbb{R}^n$ and $0 < t < 1$, we say that

A. Torchinsky (✉)
Department of Mathematics, Indiana University, Bloomington, IN 47405, USA
e-mail: torchins@indiana.edu



A.M. Stokolos et al. (eds.), *Special Functions, Partial Differential Equations, and Harmonic Analysis*, Springer Proceedings in Mathematics & Statistics 108,
DOI 10.1007/978-3-319-10545-1_15

$$m_f(t,Q) = \sup\{M : |\{y \in Q : f(y) < M\}| \le t|Q|\}$$

is the (maximal) median of $f$ over $Q$ with parameter $t$. For a cube $Q_0 \subset \mathbb{R}^n$ and $0 < s \le 1/2$, the local sharp maximal function restricted to $Q_0$ of a measurable function $f$ at $x \in Q_0$ is

$$\begin{aligned} &M^{\sharp}_{0,s,Q_0} f(x) \\ &\quad = \sup_{x\in Q, Q\subset Q_0} \inf_c \inf\{\alpha \ge 0 : |\{y \in Q : |f(y) - c| > \alpha\}| < s|Q|\}, \end{aligned}$$

and the local sharp maximal function of a measurable function $f$ at $x \in \mathbb{R}^n$ is

$$M^{\sharp}_{0,s} f(x) = \sup_{x\in Q} \inf_c \inf\{\alpha \ge 0 : |\{y \in Q : |f(y) - c| > \alpha\}| < s|Q|\}.$$

Then, with the notation

$$m^{\sharp}_f(1-s,Q) = \inf_c m_{|f-c|}(1-s,Q),$$

since by (4.3) of [30], $m^{\sharp}_f(1-s,Q) \sim m_{|f-m_f(1-s,Q)|}(1-s,Q)$, we have

$$\begin{aligned} M^{\sharp}_{0,s,Q_0} f(x) &\sim \sup_{x\in Q, Q\subset Q_0} m^{\sharp}_f(1-s,Q) \\ &\sim \sup_{x\in Q, Q\subset Q_0} m_{|f-m_f(1-s,Q)|}(1-s,Q). \end{aligned}$$

A word about medians. Simple examples suggest that medians rely more heavily on the distribution of the values of $f$ than do averages [30]. On the other hand, averages and medians are not always at odds. In particular, in some sense $m_f(s,Q)$ is a good substitute for the average $f_Q$ for small $Q$. In fact, the maximal medians $m_f(s,Q)$ of an arbitrary measurable function $f$ converge to $f$ a.e., a fact observed by Fujii for the case $s = 1/2$, [13].

**Theorem 1.** *Let $f$ be a real-valued, finite a.e. measurable function on $\mathbb{R}^n$, and $0 < s < 1$. Then*

$$\lim_{x\in Q,\, Q\to x} m_f(s,Q) = f(x) \text{ a.e.} \tag{1}$$

*In particular,* (1) *holds at every point of continuity of $f$.*

*Proof.* For $k \ge 1$ and an integer $j$, let $E_{k,j} = \{x \in \mathbb{R}^n : (j-1)/2^k \le f(x) < j/2^k\}$, $a_{k,j} = (j-1)/2^k$, and put $S_k(x) = \sum_{j=-\infty}^{\infty} a_{k,j}\,\chi_{E_{k,j}}(x)$. Note that since $f$ is finite a.e., $\mathbb{R}^n = \bigcup_{k,j} E_{k,j}$ except possibly for a set of measure 0, and when $f(x)$ is finite we have $0 \le f(x) - S_k(x) \le 2^{-k}$, which gives $m_{S_k}(s,Q) \le m_f(s,Q) \le$

$m_{S_k}(s, Q) + 2^{-k}$ for all cubes $Q$. Let $A_{k,j} = \{x \in E_{k,j} : x$ is a point of density for $E_{k,j}\}$, $A_k = \bigcup_{j=-\infty}^{\infty} A_{k,j}$. Since $f$ is finite a.e., $|\mathbb{R}^n \setminus A_k| = 0$ for all $k$, and if $A = \bigcup_{k=1}^{\infty} A_k$, also $|\mathbb{R}^n \setminus A| = 0$.

We claim that the limit in question exists for $x \in A$. Given $\varepsilon > 0$, pick $k$ such that $2^{-k+1} < \varepsilon$. Then $x \in A_{k,j}$ for some $j$, and

$$\lim_{x \in Q,\, Q \to x} \frac{|A_{k,j} \cap Q|}{|Q|} = 1\,.$$

Let $\delta = \max\{s, 1 - s\}$ and note that for all cubes $Q$ with small enough measure containing $x$,

$$\frac{|A_{k,j} \cap Q|}{|Q|} > \delta\,.$$

We restrict our attention to such small cubes $Q$ containing $x$. Note that for these cubes $m_{S_k}(s, Q) = a_{k,j}$. Indeed, on the one hand, since $S_k(y) = a_{k,j}$ for $y \in A_{k,j}$, $|\{y \in Q : S_k(y) < a_{k,j}\}| \le |A_{k,j}^c \cap Q| < s|Q|$, and, therefore, $a_{k,j} \le m_{S_k}(s, Q)$. And, on the other, since for $\varepsilon > 0$, $\{y \in Q : S_k(y) < a_{k,j} + \varepsilon\} \supset A_{k,j} \cap Q$, it follows that $|\{y \in Q : S_k(y) < a_{k,j} + \varepsilon\}| \ge |A_{k,j} \cap Q| \ge s|Q|$. Hence, $m_{S_k}(s, Q) \le a_{k,j} + \varepsilon$, and since $\varepsilon$ is arbitrary, $m_{S_k}(s, Q) \le a_{k,j}$ .

Then, since $a_{k,j} = m_{S_k}(s, Q) = S_k(x)$ for $x \in A_{k,j}$ ,

$$\begin{aligned} |m_f(s, Q) - f(x)| &\le |m_f(s, Q) - m_{S_k}(s, Q)| + |m_{S_k}(s, Q) - f(x)| \\ &\le 2^{-k} + (f(x) - S_k(x)) \le 2^{-k+1} < \varepsilon\,. \end{aligned}$$

In other words, $|m_f(s, Q) - f(x)| < \varepsilon$ for $x \in A$ and all $Q$ with small enough measure containing $x$.

Now, at a point of continuity $x$ of $f$, given $\varepsilon > 0$, let $\delta > 0$ be such that $|f(y) - f(x)| \le \varepsilon$ for $y \in B(x, \delta)$. Then for $y$ in a cube $Q$ containing $x$ and contained in $B(x, \delta)$ we have $-\varepsilon \le f(y) - f(x) \le \varepsilon$, and, consequently, $-\varepsilon = m_{-\varepsilon}(s, Q) \le m_{f - f(x)}(s, Q) = m_f(s, Q) - f(x) \le m_\varepsilon(s, Q) = \varepsilon$, and so $|m_f(s, Q) - f(x)| \le \varepsilon$.

The first ingredient in exploring the role of local sharp maximal functions in the study of integral operators is a decomposition of measurable functions. The one presented here was first considered in terms of averages by Carleson [3] and Garnett and Jones [15] and suggested in terms of medians by Fujii [12]. It complements Lerner's "local mean oscillation" decomposition [21, 22], which corresponds to the case $t = 1/2$, $s = 1/4$ in Theorem 2. Although the bound below is larger than his, it holds for arbitrarily small values of $s$, which are necessary for the applications of interest to us. We sketch the proof, which relies on medians and is somewhat more geometric.

First a couple of notations. For a cube $Q \subset \mathbb{R}^n$, let $\mathscr{D}(Q)$ denote the family of dyadic subcubes relative to $Q$, and put

$$m_{Q_0}^{t,\Delta} f(x) = \sup_{x \in Q, Q \in \mathscr{D}(Q_0)} |m_f(t, Q)| .$$

A related non-dyadic maximal function was introduced by A. P. Calderón in order to exploit cancellation to obtain estimates for singular integrals [2]. Finally, $\hat{Q}$ denotes the dyadic parent of a cube $Q$.

**Theorem 2.** *Let $f$ be a measurable function on a fixed cube $Q_0 \subset \mathbb{R}^n$, $0 < s < 1/2$, and $1/2 \le t < 1-s$. Then there exists a (possibly empty) collection of subcubes $\{Q_j^\nu\} \subset \mathscr{D}(Q_0)$ and a family of collections of indices $\{I_2^\nu\}_\nu$ such that*

(i) *for a.e. $x \in Q_0$,*

$$|f(x) - m_f(t, Q_0)| \le 4M_{0,s,Q_0}^{\sharp} f(x) + \sum_{\nu=1}^{\infty} \sum_{j \in I_2^\nu} a_j^\nu \mathbb{1}_{Q_j^\nu}(x) ,$$

*where*

$$a_j^\nu \le 10n \inf_{y \in Q_j^\nu} M_{0,s,\widehat{Q_j^\nu}}^{\sharp} f(y) + 2 \inf_{y \in Q_j^\nu} M_{0,s,Q_j^\nu}^{\sharp} f(y)$$

$$\le (10n + 2) \inf_{y \in Q_j^\nu} M_{0,s,\widehat{Q_j^\nu}}^{\sharp} f(y) ;$$

(ii) *for fixed $\nu$, the $\{Q_j^\nu\}_j$ are nonoverlapping;*
(iii) *if $\Omega^\nu = \bigcup_j Q_j^\nu$, then $\Omega^{\nu+1} \subset \Omega^\nu$; and,*
(iv) *for all $j$, $|\Omega^{\nu+1} \cap Q_j^\nu| \le (s/(1-t)) |Q_j^\nu|$.*

*Proof.* Let $E^1 = \{x \in Q_0 : |f(x) - m_f(t, Q_0)| > 2 \inf_{y \in Q_0} M_{0,s,Q_0}^{\sharp} f(y)\}$. If $|E^1| = 0$, the decomposition halts—trivially, for a.e. $x \in Q_0$,

$$|f(x) - m_f(t, Q_0)| \le 2 \inf_{y \in Q_0} M_{0,s,Q_0}^{\sharp} f(y) .$$

So suppose that $|E^1| > 0$. Recall that by Lemma 4.1 in [30], for $\eta > 0$,

$$|\{x \in Q_0 : |f(x) - m_f(t, Q_0)| \ge 2 \inf_{y \in Q_0} M_{0,s,Q_0}^{\sharp} f(y) + \eta\}| < s|Q_0| .$$

Thus, picking $\eta_k \to 0^+$, by continuity from below it readily follows that

$$|\{x \in Q_0 : |f(x) - m_f(t, Q_0)| > 2 \inf_{y \in Q_0} M_{0,s,Q_0}^{\sharp} f(y)\}| \le s|Q_0| . \tag{2}$$

Now let $f^0 = (f - m_f(t, Q_0))\mathbb{1}_{Q_0}$ and

$$\Omega^1 = \{x \in Q_0 : m^{t,\Delta}_{Q_0}(f^0)(x) > 2 \inf_{y\in Q_0} M^\sharp_{0,s,Q_0} f(y)\}\,.$$

Then by Theorem 1, $E^1 \subset \Omega^1$ and $|\Omega^1| > 0$ as well. Write $\Omega^1 = \bigcup_j Q^1_j$ where the $Q^1_j$ are nonoverlapping maximal dyadic subcubes of $Q_0$ such that

$$|m_{f^0}(t, Q^1_j)| > 2 \inf_{y\in Q_0} M^\sharp_{0,s,Q_0} f(y), \text{ and } |m_{f^0}(t, \widehat{Q^1_j})| \le 2 \inf_{y\in Q_0} M^\sharp_{0,s,Q_0} f(y). \tag{3}$$

Since $m_{f^0}(t, Q_0) = 0$, $Q^1_j \ne Q_0$ for any $j$.

Now since $t \ge 1/2$, from (10) in [30] it follows that

$$2 \inf_{y\in Q_0} M^\sharp_{0,s,Q_0} f(y) < |m_{f^0}(t, Q^1_j)| \le m_{|f^0|}(t, Q^1_j)\,,$$

and therefore by the definition of median

$$|\{x \in Q^1_j : |f^0(x)| > 2 \inf_{y\in Q_0} M^\sharp_{0,s,Q_0}\}| \ge (1-t)|Q^1_j|\,. \tag{4}$$

When (4) is summed over $j$, we have by (2) that

$$\begin{aligned}(1-t)\sum_j |Q^1_j| &\le \sum_j |\{x \in Q^1_j : |f^0(x)| > 2 \inf_{y\in Q_0} M^\sharp_{0,s,Q_0} f(y)\}| \\ &\le |\{x \in Q_0 : |f^0(x)| > 2 \inf_{y\in Q_0} M^\sharp_{0,s,Q_0} f(y)\}| \le s|Q_0|\,,\end{aligned}$$

so that

$$\sum_j |Q^1_j| \le \frac{s}{1-t}\,|Q_0|\,,$$

where by the choice of $s$ and $t$, $s/(1-t) < 1$.

Let $a^1_j = m_{f^0}(t, Q^1_j)$. By Lemma 4.3 in [30] we see that

$$|m_{f^0}(t, Q^1_j) - m_{f^0}(t, \widehat{Q^1_j})| \le 10n \inf_{y\in Q^1_j} M^\sharp_{0,s,\widehat{Q^1_j}} f(y)\,, \tag{5}$$

and therefore by (3) and (5)

$$\begin{aligned}|a^1_j| &\le |m_{f^0}(t, Q^1_j) - m_{f^0}(t, \widehat{Q^1_j})| + |m_{f_0}(t, \widehat{Q^1_j})| \\ &\le 10n \inf_{y\in Q^1_j} M^\sharp_{0,s,\widehat{Q^1_j}} f(y) + 2 \inf_{y\in Q_0} M^\sharp_{0,s,Q_0} f(y)\,.\end{aligned}$$

The first iteration of the local median oscillation decomposition of $f$ when $|E^1| > 0$ is then as follows: for a.e. $x \in Q_0$, with $g^1 = f^0 \mathbb{1}_{Q_0 \setminus \Omega^1}$,

$$f^0(x) = g^1(x) + \sum_j a_j^1 \mathbb{1}_{Q_j^1}(x) + \sum_j (f^0(x) - m_{f^0}(t, Q_j^1))\mathbb{1}_{Q_j^1}(x).$$

Note that $g^1$ has support off $\Omega^1$, and clearly for a.e. $x \in Q_0$,

$$|g^1(x)| \le 2 \inf_{y \in Q_0} M^\sharp_{0,s,Q_0} f(y).$$

Now focus on the second sum. Since $f^0(x) - m_{f^0}(t, Q) = f(x) - m_f(t, Q)$ for all cubes $Q$ and functions $f$ supported in $Q$, this sum equals

$$\sum_j (f(x) - m_f(t, Q_j^1))\mathbb{1}_{Q_j^1}(x).$$

The idea is now to repeat the above argument for each of the functions $f_j^1 = (f - m_f(t, Q_j^1))\mathbb{1}_{Q_j^1}$, and so on.

Next we consider the control of a weighted local mean of a function by the weighted local mean of its local maximal function. In $\mathbb{R}^n$, for the sharp maximal function, in the unweighted case this result was established by Fefferman and Stein [11], and in the weighted case by several people, including Fujii [12]. For the local sharp maximal function and $A_\infty$ weights $w$, it follows from the fact that there exists a constant $0 < s_1 < 1$ with the following property: given $0 < s \le s_1$, there exist constants $c, c_1$ such that for all cubes $Q$,

$$w(\{y \in Q : |f(y) - m_f(s, Q)| > \lambda, M^\sharp_{0,s} f(y) < \alpha\}) \le c\, e^{-c_1\lambda/\alpha}\, w(Q),$$

for all $\lambda, \alpha > 0$. This is proved in Chapter III of [36].

We are interested in the weighted local version of these results involving weights that are not necessarily $A_\infty$. We say that the weights $(w, v)$ satisfy condition $F$ if there exist positive constants $c_1, \alpha, \beta$ with $0 < \alpha < 1$, such that for any cube $Q$ and measurable subset $E$ of $Q$ with $|E| \le \alpha|Q|$,

$$\int_E w(x)\, dx \le c_1 \Big(\frac{|E|}{|Q|}\Big)^\beta \int_{Q \setminus E} v(x)\, dx. \tag{6}$$

Fujii observed that if $(w, v)$ satisfy condition $F$, then $w(x) \le c\, v(x)$ a.e., and that for $w = v$, (6) is equivalent to the $A_\infty$ condition for $w$; he also gave a simple example of a pair $(w, v)$ that satisfy condition $F$ so that neither of them is an $A_\infty$ weight and no $A_\infty$ weight can be inserted between them [12].

Now, if $w$ is in weak $A_\infty$, or more generally in the Muckenhoupt class $C_p$, then $(w, Mw)$ satisfy condition $F$. Along these lines, for a Young function $A$ let

$$\|f\|_{L^A(Q)} = \inf\Big\{\lambda > 0 : \frac{1}{|Q|}\int_Q A\Big(\frac{|f(y)|}{\lambda}\Big)\,dy \le 1\Big\},$$

and for $0 \le \gamma < 1$, let

$$M_{\gamma,A} f(x) = \sup_{x\in Q} |Q|^\gamma\, \|f\|_{L^A(Q)}\,.$$

In particular, for $A(t) = t^r$, we denote $M_{\gamma,A} = M_{\gamma,r}$, and of course, $M_\gamma = M_{\gamma,1} \le M_{\gamma,r}$ for $1 < r < \infty$. Also, when $\gamma = 0$ we drop the subscript corresponding to $\gamma$.

Now, it readily follows that for a weight $w$, $M_A w \in A_1$ if

$$\int_0^t \frac{A(s)}{s^2}\,ds \le c\,\frac{A(t)}{t}\,.$$

Hence, for any weight $w$ and $1 < r < \infty$, $(w, M_r w)$ satisfy condition $F$.

On the other hand, for an integer $k = 0, 1, \ldots$, let $A_k(t) = t\log^k(1+t)$. Then, if $M^{k+1}$ denotes the $k+1$ composition of the Hardy–Littlewood maximal function operator with itself, $M^{k+1}$ is pointwise comparable to the maximal operator $M_{A_k}$, [4]. Then for every $k$ there exists a weight $w$ such that $(w, M_{A_k} w)$ do not satisfy condition $F$; in particular, for such a weight $w$, $M_{A_k} w \notin A_\infty$, [31].

Two remarks are in order before we proceed to our next result. First, the choice of the parameters $s$ and $t$ in (7) below remains fixed throughout the note unless otherwise noted, and second, the constant in the conclusion is linear with respect to the constant $c_1$ of the weights that satisfy condition $F$, and in particular linear in the $A_\infty$ norm of $w$. Then, as proved in Theorems 3.1 and 3.2 in [31] we have

**Theorem 3.** *Let $\Phi$ satisfy condition $C$ with doubling constant $c_0$, $(w, v)$ weights on $\mathbb{R}^n$ satisfying condition $F$ with constants $\beta, c_1$, and pick $s, t$ such that $0 < s \le 1/2$, $1/2 < t < 1 - s$, and*

$$c_0\Big(\frac{s}{1-t}\Big)^\beta < 1\,. \tag{7}$$

*Then for any measurable function $f$ and a cube $Q_0 \subset \mathbb{R}^n$, with a constant $c$ independent of $\Phi$, $Q_0$, and $f$,*

$$\int_{Q_0} \Phi(|f(x) - m_f(t, Q_0)|)\, w(x)\,dx \le c\int_{Q_0} \Phi(M^\sharp_{0,s,Q_0} f(x))\, v(x)\,dx\,. \tag{8}$$

*Furthermore, if $f$ is such that $m_f(t, Q_0) \to 0$ as $Q_0 \to \mathbb{R}^n$, then*

$$\int_{\mathbb{R}^n} \Phi(|f(x)|)\, w(x)\,dx \le c\int_{\mathbb{R}^n} \Phi(M^\sharp_{0,s} f(x))\, v(x)\,dx\,. \tag{9}$$

*Moreover, if $\Phi$ is concave, including $\Phi(u) = u$, then* (8) *and* (9) *hold with $v = Mw$ on the right-hand side there.*

A word about (9). Lerner has observed that $\lim_{Q_0 \to \mathbb{R}^n} m_f(t, Q_0) = 0$ if $f^*(+\infty) = 0$, where $f^*$ denotes the nonincreasing rearrangement of $f$, which in turn holds if and only if $|\{x \in \mathbb{R}^n : |f(x)| > \alpha\}| < \infty$ for all $\alpha > 0$, [20].

The third ingredient is the local estimate relating the operators of interest to us, which include Calderón–Zygmund singular integral operators and fractional type operators, with local sharp maximal functions. And, as illustrated below in the case of operators with kernels satisfying Hörmander-type conditions, and integral operators with homogeneous kernels, our approach applies in other instances as well.

We introduce now the operators. When $\gamma = 0$ below they correspond to the Calderón–Zygmund singular integral operators [31] and for $0 < \gamma < 1$ to the fractional type operators, which include the Riesz potentials [37]. First an observation of a geometric nature: there exists a dimensional constant $c_n$ such that for every cube $Q$ in $\mathbb{R}^n$, if $x, x' \in Q$ and $y \notin 2^m Q$ for some $m \geq 1$, then

$$\frac{|x - x'|}{|x - y|} \leq c_n 2^{-m} .$$

We then have

**Theorem 4.** *Let $0 \leq \gamma < 1$ and $T$ an integral operator defined by*

$$Tf(x) = \text{p.v.} \int_{\mathbb{R}^n} k(x, y) f(y)\, dy \tag{10}$$

*such that*

(1) *for some $c > 0$, $k(x, y)$ satisfies*

$$|k(x, y) - k(x', y)| \leq c \frac{1}{|x - y|^{n(1-\gamma)}} \, \omega\Big(\frac{|x - x'|}{|x - y|}\Big)$$

*whenever $x, x' \in Q$ and $y \in (2Q)^c$ for any cube $Q$, where $\omega(t)$ is a nondecreasing function on $(0, \infty)$ such that*

$$\int_0^1 \omega(c_n t) \frac{dt}{t} < \infty ;$$

*and*

(2) *$T$ is of weak-type $(r, r/(1 - \gamma r))$, for some $1 \leq r < \infty$.*

*Then, for $0 < s \leq 1/2$, any cube $Q_0$, and $x \in Q_0$,*

$$M^{\sharp}_{0,s,Q_0}(Tf)(x) \leq c \sup_{x \in Q, Q \subset Q_0} \inf_{y \in Q} M_{\gamma,r} f(y) . \tag{11}$$

*In particular, if* $Q_0 = \mathbb{R}^n$, *then for all* $x \in \mathbb{R}^n$,

$$M^{\sharp}_{0,s}(Tf)(x) \le c\, M_{\gamma,r} f(x)\,.$$

*Proof.* Fix a cube $Q_0 \subset \mathbb{R}^n$ and take $x \in Q_0$. Let $Q \subset Q_0$ be a cube centered at $x_Q$ containing $x$. Let $f_1 = f\,\mathbb{1}_{2Q}$, and $f_2 = f - f_1$. Then by the linearity of $T$, $Tf(z) - Tf_2(x_Q) = Tf_1(z) + Tf_2(z) - Tf_2(x_Q)$ for $z \in Q$.

We claim that there exist constants $c_1, c_2 > 0$ independent of $f$ and $Q$ such that

$$|\{z \in Q : |Tf_1(z)| > c_1 \inf_{y\in Q} M_{\gamma,r} f(y)\}| < s\,|Q|\,, \tag{12}$$

and

$$\|Tf_2 - Tf_2(x_Q)\|_{L^\infty(Q)} \le c_2 \inf_{y\in Q} M_{\gamma,r} f(y)\,. \tag{13}$$

We prove (13) first. Observe that for any $z \in Q$,

$$\begin{aligned}
&|Tf_2(z) - Tf_2(x_Q)|\\
&\quad\le \int_{(2Q)^c} |k(z,y) - k(x_Q,y)|\,|f(y)|\,dy\\
&\quad\le c\sum_{m=1}^{\infty}\int_{2^{m+1}Q\setminus 2^m Q} \frac{1}{|z-y|^{n(1-\gamma)}}\,\omega\Big(\frac{|x_Q - z|}{|y-z|}\Big)\,|f(y)|\,dy\\
&\quad\le c\sum_{m=1}^{\infty}\omega(c_n/2^m)\,\frac{1}{|2^m Q|^{1-\gamma}}\int_{2^m Q}|f(y)|\,dy\\
&\quad\le c\Big(\int_0^1 \omega(c_n t)\,\frac{dt}{t}\Big)\inf_{y\in Q} M_\gamma f(y)\,,
\end{aligned} \tag{14}$$

and (13) holds.

Next, since $T$ is of weak-type $(r, r/(1-\gamma r))$, we have that for any $\lambda > 0$,

$$\begin{aligned}
&\lambda^{r/(1-\gamma r)}|\{z \in Q : |Tf_1(z)| > \lambda\}|\\
&\qquad\le c\Big(\int_{2Q}|f(y)|^r\,dy\Big)^{1/(1-\gamma r)}\\
&\qquad= c\Big(|2Q|^\gamma\Big(\frac{1}{|2Q|}\int_{2Q}|f(y)|^r\,dy\Big)^{1/r}\Big)^{r/(1-\gamma r)}|Q|\\
&\qquad\le c\,\inf_{y\in Q} M_{\gamma,r} f(y)^{r/(1-\gamma r)}\,|Q|\,,
\end{aligned} \tag{15}$$

and (12) follows by picking $\lambda = c_1 \inf_{y\in Q} M_{\gamma,r} f(y)$ for an appropriately chosen $c_1$.

Then, with $c > \max\{c_1, c_2\}$, (12) and (13) give

$$\begin{aligned} &|\{z \in Q : |Tf(z) - Tf_2(x_Q)| > 2c \inf_{y\in Q} M_{\gamma,r} f(y)\}| \\ &\qquad \le |\{z \in Q : |Tf_2(z) - Tf_2(x_Q)| \\ &\qquad > c_2 \inf_{y\in Q} M_{\gamma,r} f(y)\}| + |\{z \in Q : |Tf_1(z)| \\ &\qquad > c_1 \inf_{y\in Q} M_{\gamma,r} f(y)\}| \\ &\qquad < s|Q|. \end{aligned}$$

Whence

$$\inf_{c'} \inf\{\alpha \ge 0 : |\{z \in Q : |Tf(z) - c'| > \alpha\}| < s|Q|\} \le c \inf_{y\in Q} M_{\gamma,r} f(y),$$

and consequently, since this holds for all $Q \subset Q_0$, $x \in Q$,

$$M^{\sharp}_{0,s,Q_0} Tf(x) \le c \sup_{x\in Q, Q\subset Q_0} \inf_{y\in Q} M_{\gamma,r} f(y)\,.$$

The proof is thus complete.

Note that if also $T(1) = 0$, under appropriate conditions on $\omega$ it follows that $M^{\sharp}_{0,s,Q_0}(Tf)(x) \le c\, \sup_{x\in Q, Q\subset Q_0} \inf_{y\in Q} M^{\sharp}_{\gamma,r} f(y)$, where $M^{\sharp}_{\gamma,A} f(x)$ has the expected definition [31].

That $M_{\gamma,r}$ is relevant on the right-hand side of (11) for all $r$, $1 \le r < \infty$, is clear from (12) above, and is useful when $T$ is not known to be of weak-type $(1, 1/(1-\gamma))$. Also, there are operators of weak-type $(1, 1/(1-\gamma))$ where $M_{\gamma,r}$ is necessary on the right-hand side of (13), and hence on the right-hand side of (14), for $1 < r < \infty$. These are the convolution operators of Dini type, i.e., $k(x) = \Omega(x')/|x|^{n(1-\gamma)}$, $x \ne 0$, where $\Omega$ is a function on $S^{n-1}$ that satisfies $\int_{S^{n-1}} \Omega(x')\, dx' = 0$ and an $L^{r'}$-Dini condition for some $1 \le r' \le \infty$. When $\gamma = 0$ the Calderón–Zygmund convolution operators of Dini type are discussed in [19], and when $0 < \gamma < 1$ the convolution fractional operators of Dini type are covered in [5, 10]. Because of their similarity with the operators with kernels satisfying Hörmander-type conditions considered in Theorem 5 below, the analysis of this case is omitted.

Now, in the case of Hörmander type operators, we have

**Theorem 5.** *Let $0 \le \gamma < 1$ and $T$ be an integral operator of weak-type $(1, 1/(1-\gamma))$ such that for a Young function $A$, every cube $Q$, and $u, v \in Q$,*

$$\sum_{m=1}^{\infty} |2^{m+1}Q|^{1-\gamma} \| \mathbb{1}_{2^{m+1}Q\setminus 2^m Q}(k(u,\cdot) - k(v,\cdot))\|_{L^A(2^{m+1}Q)} \le c_A < \infty\,.$$

*Then, with $\overline{A}$ the conjugate Young function to A, c independent of x, $Q_0$, and f,*

$$M^{\sharp}_{0,s,Q_0}(Tf)(x) \le c \sup_{x\in Q\,,Q\subset Q_0} \inf_{y\in Q} M_{\gamma,\overline{A}} f(y)\,.$$

When $\gamma = 0$ the idea of the proof is essentially that of [19] and [24], where $T$ is assumed to be of convolution type. In that case, if $k$ satisfies the $L^A$-Hörmander condition for any Young function $A$, it also satisfies the usual $L^1$-Hörmander condition, and $T$ is of weak-type $(1,1)$. For $0 < \gamma < 1$ the idea of the proof is essentially that of [10, 19] and [1]. Theorem 4.3 in [31] and Theorem 2.2 in [37] are also relevant here.

Finally, we consider integral operators with homogeneous kernels defined as follows [32,33]. If $A_1,\dots,A_m$ are invertible matrices such that $A_k - A_{k'}$ is invertible for $k \ne k'$, $1 \le k, k' \le m$, and $\gamma_i > 0$ for all $i$ with $\gamma_1 + \cdots + \gamma_m = n(1-\gamma) \le n$, then

$$Tf(x) = \int_{\mathbb{R}^n} |x - A_1 y|^{-\gamma_1} \cdots |x - A_m y|^{-\gamma_m} f(y)\,dy\,. \tag{16}$$

For these operators we have

**Theorem 6.** *For T defined as in* (16) *with* $0 \le \gamma < 1$, *any cube* $Q_0 \subset \mathbb{R}^n$, *and* $x \in Q_0$, *we have*

$$M^{\sharp}_{0,s,Q_0}(Tf)(x) \le c \sum_{i=1}^{m} \sup_{x\in Q\,,Q\subset Q_0} \inf_{y\in Q} M_{\gamma} f(A_i^{-1} y)\,.$$

For $\gamma = 0$, the proof is as in Theorem 2.1 in [33], and for $0 < \gamma < 1$, as in Theorem 3.2 in [33]. The conclusion then follows as indicated in Theorem 4.2 of [31].

The local estimates established above can be used to express the local integral control of $Tf$ in terms of $M_T f$, where $M_T$ denotes the appropriate maximal function corresponding to $T$. Specifically, we have

**Theorem 7.** *Let* $T, M_T$ *be operators such that the conditions of Theorem 4 hold. Then for any* $\Phi$ *satisfying condition C, cube* $Q_0$, *and weights* $(w, v)$ *satisfying condition F,*

$$\int_{Q_0} \Phi(|Tf(x) - m_{Tf}(t, Q_0)|)\, w(x)\,dx \le c \int_{Q_0} \Phi(M_T f(x))\, v(x)\,dx\,, \tag{17}$$

*where c is independent of* $Q_0$ *and f.*

*Furthermore, if* $\lim_{Q_0\to\mathbb{R}^n} m_{Tf}(t, Q_0) = 0$,

$$\int_{\mathbb{R}^n} \Phi(|Tf(x)|)\, w(x)\,dx \le c \int_{\mathbb{R}^n} \Phi(M_T f(x))\, v(x)\,dx\,. \tag{18}$$

The proof follows immediately from Theorem 4 and Theorem 3. Thus the result holds with $M_T = M_r$ for singular integral operators, $M_T = M_{\overline{A}}$ for singular integral operators with kernels satisfying Hörmander-type conditions, and $M_T f(x) = \sum_{i=1}^m Mf(A_i^{-1}x)$ for integral operators with homogeneous kernels. It also holds with $M_T = M_{\gamma,r}$ for fractional type operators, $M_T = M_{\gamma,\overline{A}}$ for fractional type operators with kernels satisfying Hörmander-type conditions, and $M_T f(x) = \sum_{i=1}^m M_\gamma f(A_i^{-1}x)$ for fractional type operators with homogeneous kernels.

Estimates such as (17) furnish a local version of the underlying principle of the Calderón–Zygmund theory, first expressed by Cotlar for the Hilbert transform [7], that the Hardy–Littlewood maximal function controls the Calderón–Zygmund singular integral operators, as formulated in the weighted setting by Coifman [6], and also the fact that fractional maximal functions control the Riesz potentials, formulated in the weighted setting by Muckenhoupt and Wheeden [25]. As for (18), it gives both the results of [6] and [25], which are then one and the same.

Pérez addressed the question of replacing the integral involving the maximal function on the right-hand side of (18) by an integral involving $|f|$ and a possibly larger weight for the case of singular integrals in Theorem 1 in [26] and for the Riesz potentials in Theorem 1 (A) in [29], respectively. There is a vast literature of results of this nature, pioneered by Sawyer's work [34]. The work of Young [38] and Harboure et al. [16] is also of interest.

Here we consider two-weight, $L^p_v - L^q_w$ estimates with $1 < p \le q < \infty$ that apply directly to the integral operator and where the control exerted by a maximal function is not apparent. First recall the definition of the classes $B_p$ and $B_{\alpha,p}$. The latter class was introduced by Cruz-Uribe and Moen [8] and for $0 < \alpha < 1$ and $1 < p < 1/\alpha$, it consists of those Young functions $A$ such that with $1/q = 1/p - \alpha$,

$$\|A\|_{\alpha,p} = \Big(\int_c^\infty \frac{A(t)^{q/p}}{t^q}\,\frac{dt}{t}\Big)^{1/q} < \infty\,.$$

When $\alpha = 0$ this reduces to the $B_p$ class of Pérez [28]. The result of interest to us, Theorem 3.3 in [8], is that for $A \in B_{\alpha,p}$ the fractional maximal function $M_{\alpha,A}f(x)$ maps $L^p(\mathbb{R}^n)$ continuously into $L^q(\mathbb{R}^n)$ with norm not exceeding $c\,\|A\|_{\alpha,p}$. This result also holds for $\alpha = 0$, i.e., the $B_p$ classes.

We rely on the following result of Pérez, Theorem 2.11 in [27] or Theorem 3.5 in [14]. Let $p, q$ with $1 < p \le q < \infty$ when $\gamma = 0$, and $1 < p < q < \infty$ when $0 < \gamma < 1$. Let $(w, v)$ be a pair of weights such that for every cube $Q$,

$$|Q|^\gamma\,|Q|^{1/q-1/p}\,\|w^{1/q}\|_{L^q(Q)}\,\|v^{-1/p}\|_{L^B(Q)} \le c\,, \tag{19}$$

where $B$ is a Young function with $\overline{B} \in B_p$. Then, if $f \in L^p(v)$,

$$\Big(\int_{\mathbb{R}^n} M_\gamma f(x)^q\,w(x)\,dx\Big)^{1/q} \le c\Big(\int_{\mathbb{R}^n} |f(x)|^p\,v(x)\,dx\Big)^{1/p}.$$

Before we proceed to prove our next theorem, we need an extension of a property given in Lemma 4.8 in [24] and the comments that follow it.

**Lemma 1.** *Let $T$ be an integral operator defined by* (10). *If $T$ satisfies the assumptions of Theorem 4 with $1 \le r < \infty$, and in that case*

$$\lambda_m = \omega(c_n/2^m)\,, \quad m \ge 1\,,$$

*or $T$ satisfies the assumptions of Theorem 5 and in that case*

$$\lambda_m = \sup_{u,v\in Q} |2^{m+1}Q|^{1-\gamma}\, \|\mathbb{1}_{2^{m+1}Q\setminus 2^m Q}(k(u,\cdot) - k(v,\cdot))\|_{L^{r'}(2^{m+1}Q)}\,,$$

*then $\sum_m \lambda_m < \infty$, and if $Q$ is a cube of $\mathbb{R}^n$, we have*

$$m^\sharp_{Tf}(1-s,Q) \le c\sum_{m=1}^{\infty} \lambda_m\, |2^m Q|^\gamma \Big(\frac{1}{|2^m Q|}\int_{2^m Q} |f(y)|^r\, dy\Big)^{1/r}. \tag{20}$$

*Proof.* Fix $Q$, let $x \in Q$, and put $f = f_1 + f_2$ where $f_1 = f\,\mathbb{1}_{2Q}$. We claim that there exist constants $c_1, c_2 > 0$ independent of $f$ and $Q$ such that

$$|\{z \in Q : |Tf_1(z)| > c_1\, I\}| < s\,|Q|\,, \tag{21}$$

and

$$\|Tf_2 - Tf_2(x_Q)\|_{L^\infty(Q)} \le c_2\, I\,, \tag{22}$$

where

$$I = \sum_{m=1}^{\infty} \lambda_m\, |2^m Q|^\gamma \Big(\frac{1}{|2^m Q|}\int_{2^m Q} |f(y)|^r\, dy\Big)^{1/r}.$$

First, in the case of Theorem 4, $T$ is of weak-type $(r, r/(1-\gamma r))$ and as in (15) we have that for any $\lambda > 0$,

$$\begin{aligned}&\lambda^{r/(1-\gamma r)}|\{z \in Q : |Tf_1(z)| > \lambda\}|\\ &\quad\le c\,\Big(|2Q|^\gamma\Big(\frac{1}{|2Q|}\int_{2Q} |f(y)|^r\, dy\Big)^{1/r}\Big)^{r/(1-\gamma r)}|Q|\,.\end{aligned}$$

In the case of Theorem 5, $T$ is of weak-type $(1, 1/(1-\gamma))$ and as in (15) by Hölder's inequality we have that for any $\lambda > 0$,

$$\begin{aligned}&\lambda^{1/(1-\gamma)}|\{z \in Q : |Tf_1(z)| > \lambda\}|\\ &\quad\le c\,\Big(|2Q|^\gamma\Big(\frac{1}{|2Q|}\int_{2Q} |f(y)|^r\, dy\Big)^{1/r}\Big)^{1/(1-\gamma)}|Q|\,.\end{aligned}$$

Thus, in both cases (21) holds.

Next, when $T$ satisfies the assumptions of Theorem 5, (22) holds automatically. And, if $T$ satisfies the assumptions of Theorem 4, as in (14) for any $z \in Q$,

$$|Tf_2(z) - Tf_2(x_Q)| \leq c \sum_{m=1}^{\infty} \omega(c_n/2^m) \frac{1}{|2^m Q|^{1-\gamma}} \int_{2^m Q} |f(y)|\, dy$$

and, therefore, (22) holds by Hölder's inequality.

Then, in either case, with $c > \max\{c_1, c_2\}$, as in the proof of Theorem 4, (21) and (22) give

$$|\{z \in Q : |Tf(z) - Tf_2(x_Q)| > 2c\, I\}| < s|Q|,$$

and therefore for all $Q$,

$$m^{\sharp}_{Tf}(1-s, Q) = \inf_{c'} \inf\{\alpha \geq 0 : |\{z \in Q : |Tf(z) - c'| > \alpha\}| < s|Q|\}$$
$$\leq c \sum_{m=1}^{\infty} \lambda_m\, |2^m Q|^{\gamma} \Big( \frac{1}{|2^m Q|} \int_{2^m Q} |f(y)|^r\, dy \Big)^{1/r},$$

and we have finished.

Note that Lemma 1 also applies to the convolution operators of Dini type. In that case, as in Lemma 5 in [19] or Lemma 1 in [10], the $\alpha_m$ can be estimated in terms of the $\omega_{r'}$ modulus of continuity of the kernel $k$.

Now to the result.

**Theorem 8.** *Let $T$ be an integral type operator that satisfies the assumptions of Theorem 4 with $1 \leq r < \infty$ or Theorem 5 with the Young function $t^{r'}$ there and $1 \leq r < \infty$. Let $0 \leq \gamma < 1$, $\gamma r < 1$, and $r < p \leq q < \infty$, with $p < q$ when $0 < \gamma < 1$, define $0 < \alpha < 1$ by the relation $\alpha = 1/p - 1/q$, and let $\alpha_1, \alpha_2 \geq 0$ be such that $\alpha = \alpha_1 + \alpha_2$. Further, suppose that the Young functions $A$, $B$ are so that $\overline{A} \in B_{(q/r)'} \cap B^{\alpha_2}_{q'}$ and $\overline{B} \in B^{\alpha_1 r}_{p/r'}$, and $w$ and $v$ weights such that for all cubes $Q$,*

$$\sup_Q |Q|^{\gamma r} |Q|^{r/q - r/p}\, \|w^{r/q}\|_{L^A(Q)}\, \|v^{-r/p}\|_{L^B(Q)} \leq c < \infty. \tag{23}$$

*Then, if $\lambda_m$ is defined as in Lemma 1 satisfies*

$$\sum_{m=1}^{\infty} \lambda_m 2^{mn/q} < \infty,$$

*we have*

$$\Big( \int_{\mathbb{R}^n} |Tf(x)|^q\, w(x)\, dx \Big)^{1/q} \leq c \Big( \int_{\mathbb{R}^n} |f(x)|^p\, v(x)\, dx \Big)^{1/p} \tag{24}$$

*for those $f$ such that $\lim_{Q_0 \to \mathbb{R}^n} m_{Tf}(t, Q_0) = 0$.*

*Proof.* We begin by considering the local version of (24). Fix a cube $Q_0$ and note that by the local median decomposition discussed in Theorem 6.5 in [31] there exists a family $\{Q_j^\nu\}$ of dyadic subcubes of $Q_0$ such that if $\widehat{Q_j^\nu}$ denotes the dyadic parent of $Q_j^\nu$, we have

$$\begin{aligned}&|Tf(x) - m_{Tf}(t, Q_0)|\\&\quad \le 8\, M^\sharp_{0,s,Q_0}(Tf)(x) + c \sum_{\nu,j} m^\sharp_{Tf}(1-(1-t)/2^n, \widehat{Q_j^\nu})\mathbb{1}_{Q_j^\nu}(x).\end{aligned}$$

Therefore to estimate the $L^q_w(Q_0)$ norm of $Tf(x) - m_{Tf}(t, Q_0)$ it suffices to estimate the norm of each summand above separately. Since by Theorem 4 or Theorem 5 we have

$$M^\sharp_{0,s,Q_0}(Tf)(x) \le c\, M_{\gamma,r} f(x) = c\, M_{\gamma r}(|f|^r)(x)^{1/r},$$

the first term above can be estimated by

$$\|M_{\gamma r}(|f|^r)^{1/r}\|_{L^q_w} = \|M_{\gamma r}(|f|^r)\|^{1/r}_{L^{q/r}_w}. \tag{25}$$

Now, since $\overline{A} \in B_{(q/r)'}$, by (6.17) in [31],

$$\|w^{r/q}\|_{L^{q/r}(Q)} \le c\, \|w^{r/q}\|_{L^A(Q)}$$

for all cubes $Q$, and therefore (23) implies (19) with indices $p/r$ and $q/r$ there, corresponding to the value $\gamma r$. Thus $M_{\gamma r}$ maps $L^{p/r}_v$ continuously into $L^{q/r}_w$ and therefore (25) is bounded by

$$\|M_{\gamma r}(|f|^r)\|^{1/r}_{L^{q/r}_w} \le c\, \|\,|f|^r\|^{1/r}_{L^{p/r}_v} = \|f\|_{L^p_v}.$$

Hence,

$$\|M^\sharp_{0,s,Q_0}(Tf)\|_{L^q_w} \le c\, \|f\|_{L^p_v}.$$

Next note that by a purely geometric argument, if $Q$ is any of the cubes $Q_j^\nu$, there is a dimensional constant $c$ such that

$$\begin{aligned}&\sum_{m=1}^\infty \lambda_m\, |2^m \hat{Q}|^\gamma \Big(\frac{1}{|2^m \hat{Q}|}\int_{2^m \hat{Q}} |f(y)|^r\, dy\Big)^{1/r}\\&\quad \le c \sum_{m=1}^\infty \lambda_m\, |2^m Q|^\gamma \Big(\frac{1}{|2^m Q|}\int_{2^m Q} |f(y)|^r\, dy\Big)^{1/r}.\end{aligned} \tag{26}$$

To estimate the norm of the sum by duality, let $h$ be supported in $Q_0$ with $\|h\|_{L^{q'}_v(Q_0)} = 1$ and note that by (20) and (26),

$$\int_{Q_0} \Big(\sum_{\nu,j} m^\sharp_{Tf}(1-(1-t)/2^n, \widehat{Q^\nu_j})\mathbb{1}_{Q^\nu_j}(x)\Big) w(x)^{1/q} h(x)\,dx$$
$$\le c \sum_m \lambda_m \sum_{\nu,j} |2^m Q^\nu_j|^\gamma \Big(\frac{1}{|2^m Q^\nu_j|}\int_{2^m Q^\nu_j} |f(y)|^r\,dy\Big)^{1/r} \int_{Q^\nu_j} w(x)^{1/q} h(x)\,dx. \tag{27}$$

We consider each term in the inner sum of (27) separately. First, let $D$ be the Young function defined by $D(t) = \overline{B}(t^r)$, and note that since $\|g\|_{L^D(Q)} = \||g|^r\|^{1/r}_{L^{\overline{B}}(Q)}$, by Hölder's inequality for the conjugate Young functions $B, \overline{B}$,

$$\Big(\frac{1}{|2^m Q^\nu_j|}\int_{2^m Q^\nu_j} |f(y)|^r\,dy\Big)^{1/r}$$
$$\le 2\left(\||f|^r v^{r/p}\|_{L^{\overline{B}}(2^m Q^\nu_j)}\, \|v^{-r/p}\|_{L^B(2^m Q^\nu_j)}\right)^{1/r}$$
$$= 2\,\|f v^{1/p}\|_{L^D(2^m Q^\nu_j)}\, \|v^{-r/p}\|^{1/r}_{L^B(2^m Q^\nu_j)}\,.$$

Next, let $C$ be the Young function defined by $C(t) = A(t^r)$ and note that as above, by Hölder's inequality for the conjugate Young functions $C, \overline{C}$,

$$\int_{Q^\nu_j} w(x)^{1/q}\, h(x)\,dx \le 2^{mn}|Q^\nu_j|\,\frac{1}{|2^m Q^\nu_j|}\int_{2^m Q^\nu_j} w(x)^{1/q}\, h(x)\mathbb{1}_{Q^\nu_j}(x)\,dx$$
$$\le 2\cdot 2^{mn}\,\|w^{1/q}\|_{L^C(2^m Q^\nu_j)}\|h\mathbb{1}_{Q^\nu_j}\|_{L^{\overline{C}}(2^m Q^\nu_j)}\,|Q^\nu_j|$$
$$\le 2\cdot 2^{mn}\,\|w^{r/q}\|^{1/r}_{L^A(2^m Q^\nu_j)}\|h\mathbb{1}_{Q^\nu_j}\|_{L^{\overline{C}}(2^m Q^\nu_j)}\,|Q^\nu_j|\,.$$

Moreover, since for each $\lambda > 1$ and each cube $Q$ we have

$$\|g\mathbb{1}_Q\|_{L^{\overline{C}}(\lambda Q)} \le \|g\|_{L^{\overline{C}/\lambda^n}(Q)}\,,$$

it follows that

$$\int_{Q^\nu_j} w(x)^{1/q}\, h(x)\,dx \le 2\cdot 2^{mn}\,\|w^{r/q}\|^{1/r}_{L^A(2^m Q^\nu_j)}\,\|h\|_{L^{\overline{C}/2^{mn}}(Q^\nu_j)}\,|Q^\nu_j|\,.$$

Therefore, since by (23) with $1/p - 1/q = \alpha$,

$$|2^m Q^\nu_j|^\gamma\,\|w^{r/q}\|^{1/r}_{L^A(2^m Q^\nu_j)}\,\|v^{-r/p}\|^{1/r}_{L^B(2^m Q^\nu_j)} \le c\,|2^m Q^\nu_j|^\alpha\,,$$

each term in the inner sum of (27) is bounded by

$$c\,2^{mn}\,|2^m Q_j^\nu|^\alpha\,\|f\,v^{1/p}\|_{L^D(2^m Q_j^\nu)}\,\|h\|_{L^{\overline{C}/2^{mn}}(Q_j^\nu)}\,|Q_j^\nu|\,,$$

and consequently the sum itself does not exceed

$$c\sum_{m=1}^{\infty}\lambda_m 2^{mn}\sum_{\nu,j}|2^m Q_j^\nu|^\alpha\,\|f v^{1/p}\|_{L^D(2^m Q_j^\nu)}\,\|h\|_{L^{\overline{C}/2^{mn}}(Q_j^\nu)}\,|Q_j^\nu|\,. \tag{28}$$

Let $F_j^\nu = Q_j^\nu\setminus\Omega^{\nu+1}$; then the $F_j^\nu$ are pairwise disjoint and $|F_j^\nu|\ge c|Q_j^\nu|$, where $c$ depends on $s$ and $t$ but is independent of $\nu$ and $j$. Now, since $\alpha=\alpha_1+\alpha_2$, the innermost sum in (28) is bounded by

$$J = c\sum_{\nu,j}|2^m Q_j^\nu|^{\alpha_1}\,\|f v^{1/p}\|_{L^D(2^m Q_j^\nu)}\,|2^m Q_j^\nu|^{\alpha_2}\,\|h\|_{L^{\overline{C}/2^{mn}}(Q_j^\nu)}|F_j^\nu|\,,$$

and, since

$$|2^m Q_j^\nu|^{\alpha_1}\,\|f v^{1/p}\|_{L^D(2^m Q_j^\nu)} \le \inf_{x\in F_j^\nu} M_{\alpha_1,D}(f v^{1/p})(x)$$

and similarly

$$|2^m Q_j^\nu|^{\alpha_2}\,\|h\|_{L^{\overline{C}/2^{mn}}(Q_j^\nu)} \le \inf_{x\in F_j^\nu} M_{\alpha_2,\overline{C}/2^{mn}}h(x)\,,$$

we have that

$$\begin{aligned} J &\le c\sum_{\nu,j}\int_{F_j^\nu} M_{\alpha_1,D}(f v^{1/p})(x)\,M_{\alpha_2,\overline{C}/2^{mn}}\,h(x)\,dx\\ &\le c\int_{Q_0} M_{\alpha_1,D}(f v^{1/p})(x)\,M_{\alpha_2,\overline{C}/2^{mn}}\,h(x)\,dx\,. \end{aligned}$$

Now pick $s_1, s_2$ such that

$$1/p-\alpha_1 = 1/s_1\,,\quad\text{and}\quad 1/q'-\alpha_2 = 1/s_2\,.$$

Since

$$1/s_1+1/s_2 = 1/p-\alpha_1+1-1/q-\alpha_2 = 1/p-\alpha-1/q+1 = 1\,,$$

$s_1, s_2$ are conjugate exponents, and, therefore, by Hölder's inequality,

$$\int_{Q_0} M_{\alpha_1,D}(f v^{1/p})(x) M_{\alpha_2,\overline{C}/2^{mn}} h(x)\, dx$$
$$\leq \|M_{\alpha_1,D}(f v^{1/p})\|_{L^{s_1}} \|M_{\alpha_2,\overline{C}/2^{mn}} h\|_{L^{s_2}}.$$

Now, by (iii) and (iv) in Proposition 6.1 in [31] , $D \in B_p^{\alpha_1}$ and $\overline{C} \in B_{q'}^{\alpha_2}$, respectively, and, therefore, by Theorem 3.3 in [8],

$$\int_{Q_0} M_{\alpha_1,D}(f v^{1/p})(x) M_{\alpha_2,\overline{C}/2^{mn}} h(x)\, dx \leq c\, \|f v^{1/p}\|_{L^p}\, 2^{-mn/q'}\, \|h\|_{L^{q'}(Q_0)},$$

and the right-hand side of (28) is bounded by

$$c \Big(\sum_{m=1}^{\infty} \lambda_m 2^{mn(1-1/q')}\Big) \|f\|_{L_v^p} \leq c\, \|f\|_{L_v^p}.$$

Hence, combining the above estimates,

$$\|Tf - m_{Tf}(t, Q_0)\|_{L_w^q(Q_0)} \leq c\, \|f\|_{L_v^p}.$$

Finally, by Fatou's lemma, (24) follows for functions $f$ such that $m_{Tf}(t, Q_0) \to 0$ as $Q_0 \to \mathbb{R}^n$.

## References


1. Bernardis, A.L., Lorente, M., Riveros, M.S.: Weighted inequalities for fractional integral operators with kernels satisfying Hörmander type conditions. Math. Inequal. Appl. **14**(4), 881–895 (2011)
2. Calderón, A.P.: Estimates for singular integral operators in terms of maximal functions. Stud. Math. **44**, 563–582 (1972)
3. Carleson, L.: BMO–10 years development. 18th Scandinavian Congress of Mathematicians, Aarhus, 1980. Progr. Math. vol. 11, pp. 3–21. Birkhäuser, Boston (1981)
4. Carozza, M., Passarelli Di Napoli, A.: Composition of maximal operators. Publ. Mat. **40**(2), 397–409 (1996)
5. Chanillo, S., Watson, D., Wheeden, R.L.: Some integral and maximal operators related to starlike sets. Stud. Math. **107**, 223–255 (1993)
6. Coifman, R.R.: Distribution function inequalities for singular integrals. Proc. Natl. Acad. Sci. USA **69**, 2838–2839 (1972)
7. Cotlar, M.: Some generalizations of the Hardy-Littlewood maximal theorem. Rev. Mat. Cuyana **1**(1955), 85–104 (1956)
8. Cruz-Uribe, D., Moen, K.: A fractional Muckenhoupt-Wheeden theorem and its consequences. Integr. Equ. Oper. Theory **76**(3), 421–446 (2013)
9. Cwikel, M., Sagher, Y., Shvartsman, P.: A new look at the John-Nirenberg and John-Strömberg theorems for BMO. J. Funct. Anal. **263**, 129–166 (2012)

10. Ding, Y., Lu, S.: Boundedness of homogenous fractional integrals on $L^p$ for $n/\alpha \leq p \leq \infty$. Nagoya Mat. J. **167**, 17–33 (2002)
11. Fefferman, C., Stein, E.: $H^p$ spaces of several variables. Acta Math. **129**(3–4), 137–193 (1972)
12. Fujii, N.: A proof of the Fefferman-Stein-Strömberg inequality for the sharp maximal function. Proc. Am. Math. Soc. **106**(2), 371–377 (1989)
13. Fujii, N.: A condition for the two-weight norm inequality for singular integral operators. Stud Math. **98**(3), 175–190 (1991)
14. García-Cuerva, J., Martell, J.M.: Two-weight norm inequalities for maximal operators and fractional integrals on non-homogeneous Spaces. Indiana Univ. Math. J. **50**(3), 1241–1280 (2001)
15. Garnett, J.B., Jones, P.W.: BMO from dyadic BMO. Pac. J. Math. **99**(2), 351–371 (1982)
16. Harboure, E., Macías, R.A., Segovia, C.: Boundedness of fractional operators on $L^p$ spaces with different weights. Trans. Am. Math. Soc. **285**, 629–647 (1984)
17. Jawerth, B., Torchinsky, A.: Local sharp maximal functions. J. Approx. Theory **43**, 231–270 (1985)
18. John, F.: Quasi-isometric mappings. In: 1965 Seminari 1962/63 Anal. Alg. Geom. e Topol. Ist. Naz. Alta Mat., vol. 2, pp 462–473. Ediz. Cremonese, Rome
19. Kurtz, D.S., Wheeden, R.L.: Results for weighted norm inequalities for multipliers. Trans. Am. Math. Soc. **255**, 343–362 (1979)
20. Lerner, A.K.: On the John-Strömberg characterization of BMO for nondoubling measures. Real. Anal. Exch. **28**(2), 649–660 (2002/2003)
21. Lerner, A.K.: A pointwise estimate for the local sharp maximal function with applications to singular integrals. Bull. Lond. Math. Soc. **42**(5), 843–856 (2010)
22. Lerner, A.K.: A "local mean oscillation" decomposition and some of its applications. Function Spaces, Approximation, Inequalities and Lineability. Lectures of the Spring School in Analysis, pp. 71–106. Matfyzpres, Prague (2011)
23. Lerner, A.K.: A simple proof of the $A_2$ conjecture. Int. Math. Res. Not. (2012). doi:10.1093/imrn/rns145
24. Lorente, M., Riveros, M.S., de la Torre, A.: Weighted estimates for singular integral operators satisfying Hörmander's conditions of Young type. J. Fourier Anal. Appl. **11**(5), 497–509 (2005)
25. Muckenhoup, B., Wheeden, R.: Weighted norm inequaliteies for fractional integrals. Trans. Am. Math. Soc. **192**, 261–274 (1974)
26. Pérez, C.: Weighted norm inequalities for singular integral operators. J. Lond. Math. Soc. (2) **49**(2), 296–308 (1994)
27. Pérez, C.: Two weighted inequalities for potential and fractional type maximal operators. Indiana Univ. Math. J. **43**, 1–28 (1994)
28. Pérez, C.: On sufficient conditions for the boundedness of the Hardy-Littlewood maximal operator between weighted $L^p$-spaces with different weights. Proc. Lond. Math. Soc. (3) **71**(1), 135–157 (1995)
29. Pérez, C.: Sharp $L^p$–weighted Sobolev inequalities. Ann. Inst. Fourier (Grenoble) **45**(3), 809–824 (1995)
30. Poelhuis, J., Torchinsky, A.: Medians, continuity, and vanishing oscillation. Stud. Math. **213**, 227–242 (2012)
31. Poelhuis, J., Torchinsky, A.: Weighted local estimates for singular integral operators. (2013) arXiv:1308.1134v2
32. Riveros, M.S., Urciuolo, M.: Weighted inequalities for integral operators with some homogeneous kernels. Czech. Math. J. (130) **55**(2), 423–432 (2005)
33. Riveros, M.S., Urciuolo, M.: Weighted inequalities for fractional type operators with some homogenous kernels. Acta Math. Sin. (Engl. Ser.) **29**(3), 449–460 (2013)
34. Sawyer, E.: A characterization of a two-weight norm inequality for maximal operators. Stud Math. **75**(1), 1–11 (1982)

35. Strömberg, J.-O.: Bounded mean oscillation with Orlicz norms and duality of Hardy spaces. Indiana Univ. Math. J. **28**(3), 511–544 (1979)
36. Strömberg, J.-O., Torchinsky, A.: Weighted Hardy Spaces. Lecture Notes in Mathematics, vol. 1381. Springer, Berlin (1989)
37. Torchinsky, A.: Weighted local estimates for fractional type operators. (2013) arXiv:1310.2139.
38. Young, W.-S.: Weighted norm inequalities for the Hardy-Littlewood maximal function. Proc. Am. Math. Soc. **85**, 24–26 (1982)

# Weighted Norm Estimates for Singular Integrals with $L \log L$ Kernels: Regularity of Weak Solutions of Some Degenerate Quasilinear Equations

**Richard L. Wheeden**

*Dedicated to Calixto Caldéron*

**Summary.** The purpose of this paper is to summarize some known results about two different types of problems. A number of open questions related to each of the problems are indicated. The presentation is largely descriptive; no proofs are given, but they can be found in the appropriate papers listed among the references.

## 1 Singular Integrals

In 1956, A. P. Calderón and A. Zygmund [CZ] proved boundedness on $L^p(R^n)$ for $1 < p < \infty$ of singular integrals of the form

$$Hf(x) = p.v. \int_{R^n} f(x-y)\, \frac{\Omega(y)}{|y|^n}\, dy,$$

where $\Omega$ is homogeneous of degree 0 and has integral 0 on the unit sphere $\Sigma = \{x : |x| = 1\}$, assuming only that $\Omega \in L \log^+ L(|x| = 1)$. For such rough kernels, very few weighted norm estimates are known. One derived in 1971 for power weights $|x|^\beta$ in [MW] is that

$$H \text{ is bounded on } L^p_{|x|^\beta} \text{ if } -1 < \beta < p - 1.$$

In 1981, this range of $\beta$ was shown to be sharp in [KW2].

R.L. Wheeden (✉)
Department of Mathematics, Hill Center Rutgers University, Piscataway, NJ, USA
e-mail: wheeden@math.rutgers.edu



A.M. Stokolos et al. (eds.), *Special Functions, Partial Differential Equations, and Harmonic Analysis*, Springer Proceedings in Mathematics & Statistics 108,
DOI 10.1007/978-3-319-10545-1_16

These results, as well as the original unweighted one in [CZ], are corollaries of weighted estimates obtained in [WW1, WW2]. The paper [CWW] dealing instead with maximal functions and fractional integrals related to starlike sets is also relevant. The fractional integrals treated there have the form

$$I_\alpha f(x) = \int_{R^n} f(x-y)\,\frac{\Omega(y)}{|y|^{n-\alpha}}\,dy, \quad 0<\alpha<n,$$

and then $\Omega$ is not assumed to have integral 0 on the unit sphere.

All the results are inspired by unweighted estimates for starlike maximal functions obtained in 1973 by C. P. Calderón [C].

None of our general weighted results is optimal, even if $\Omega$ is continuous on the sphere $\Sigma$ except at a few points. However, the results are sharp enough to include some known ones as special cases. Besides the two already mentioned, they recapture weighted results of both D. Watson [W] and J. Duandikoetxea [Du] which assume the stronger condition $\Omega \in L^r(|x|=1)$, $r>1$. Both those authors derive boundedness of $H$ on the weighted space $L^p_w(R^n)$ under the same conditions on $w$ used in [KW1], but without the stronger assumption in [KW1] that $\Omega$ satsifies the $L^r$-Dini condition.

The results of [WW1, WW2] fall into two categories. First, there is an analogue of the Cordoba–Fefferman [CorF] estimate

$$||Hf||_{L^p_v} \le C||f||_{L^p_{Wv}} \qquad \text{for all } v \ge 0.$$

Here the operator $W$ is the worstening operator $Wv$ defined by $Wv = \{M(v^r)\}^{1/r}$, $r>1$, where $M$ is the classical Hardy–Littlewood maximal operator. In [CorF], the authors assume that the kernel of $H$ is smooth away from the origin but not necessarily homogeneous. Their result has been improved substantially, for example by M. Wilson in [Wil] and by C. Pérez in [P], but still assuming more about the kernel than $L\,log^+L(|x|=1)$.

In [WW1] and [WW2], we prove the following:

**Theorem 1.1.** *Let $\Omega \in L\log^+ L$, $1<r<\infty$ and $1<p<\infty$. Then there is a constant $C = C(\Omega, r, p)$ such that*

$$||Hf||_{L^p_v} \le C||f||_{L^p_{Wv}} \quad \textit{for all } v \ge 0,$$

*where $W$ is the much-worsening operator $Wv = \{MGM(v^r)\}^{1/r}$.*

**What is G?** We postpone the exact definition of $G$ for the moment, until we introduce some further notation. The definition is given in (4) below. $G$ is related geometrically to $\Omega$. It is big, but not very big, because $G$ turns out to be bounded on $L^p(R^n)$ for all $1<p<\infty$.

The second category of results is one-weight estimates. To describe them (and also the operator $G$), define a set $S \subset R^n$ by

$$S = \left\{ x : \frac{|\Omega(x)|}{|x|^n} \geq 1 \right\} = \left\{ x = rx' : |x'| = 1, 0 \leq r \leq |\Omega(x')|^{\frac{1}{n}} \right\}. \tag{1}$$

If $\Omega$ is smooth, then $S$ is bounded, and roughly unit-ball-like, but for general $\Omega$, $S$ can be unbounded with a nasty boundary depending on the nature of $\Omega$. For us, the key properties of $S$ are that it is starlike about the origin, and it has finite measure

$$|S| = \frac{1}{n} \, ||\Omega||_{L^1}.$$

We associate with $S$ an averaging operator which allows cancellation, namely

$$A_t f(x) = \int_S f(x - ty) \operatorname{sign} \Omega(y) \, dy = \frac{1}{t^n} \int_{tS} f(x - y) \operatorname{sign} \Omega(y) \, dy, \tag{2}$$

where $t > 0$ and $tS$ denotes dilation of $S$ by $t$,

$$tS = \{tx : x \in S\}.$$

The set $tS$ is also starlike about the origin. $\Omega$ can be complex-valued by using a suitable definition of sign $\Omega$ (see [WW1]). Clearly, $A_t f(x)$ is dominated in absolute value by a starlike average of $|f|$, namely

$$|A_t f(x)| \leq \frac{1}{t^n} \int_{tS} |f(x - y)| \, dy,$$

but for our present purposes it is best not to discard the factor sign $\Omega$ or insert absolute values.

A key fact we use is the following representation formula for $Hf$ in terms of $A_t f(x)$, assuming $\Omega \in L \log^+ L$ with mean 0:

$$Hf(x) = n \int_0^\infty A_t f(x) \frac{dt}{t} + C_\Omega f(x), \quad \text{where} \tag{3}$$

$$C_\Omega = \frac{1}{n} \int_{|x|=1} \Omega(x) \log |\Omega(x)| \, dx.$$

If $f$ is a Schwartz function, this formula is shown in [WW1] to hold for every $x \in R^n$, and the right side converges absolutely. Also, if $f \in L^2$, the formula is shown to hold in the sense of $L^2$ principal-value convergence. More can be said, but the weakest conditions on $f$ for which (3) holds have not been determined.

If one assumes only that $\Omega \in L^1(|x| = 1)$ and is an odd function, then the representation formula holds without the additive term $C_\Omega f(x)$. The proofs are not short but are fairly easy, relying for the most part on polar coordinates.

The representation formula in (3) allows us to study $Hf$ by decomposing the integral $\int_0^\infty A_t f(x)dt/t$ on the right side of the formula and using Littlewood–Paley theory. Any particular characteristics of $\Omega$, such as its size distribution, are retained in the geometry of $S$, but this geometry is not simple to deal with when $\Omega$ is rough.

We use what we call a *stratified starlike cover* of $S$. It consists of a collection of rectangles,

$$\{R_{m,k} : \quad m = 0, 1, \ldots \text{ and } 1 \le k < k_m \le \infty\},$$

which are centered at 0 with varying orientations and which efficiently cover $S$. Here, as we will explain, $m$ measures size of $\Omega$ and $k$ measures "branching of $S$" for given size. Note that the cover does not take sign $\Omega$ into account because the rectangles are centered at 0. More precisely, if $\Sigma$ denotes the unit sphere $|x'| = 1$, start by letting

$$\Sigma_0 = \left\{x' \in \Sigma : |\Omega(x')| \le 1\right\}, \text{ and}$$

$$\Sigma_m = \left\{x' \in \Sigma : 2^{m-1} \le |\Omega(x')|^{\frac{1}{n}} < 2^m\right\}, \quad m = 1, 2, \ldots$$

Next let

$$S_m = \text{ the starlike set with polar boundary } r = |\Omega(x')|^{\frac{1}{n}} \chi_{\Sigma_m}(x').$$

For $S$ as in (1), it is then easy to see that

$$S = \bigcup_0^\infty S_m \quad \text{and} \quad S_m \cap S_\ell = \{0\} \text{ if } m \ne \ell.$$

Also, as shown in [WW1], there are rectangles $\{R_{m,k}\}$ satisfying

1. $R_{m,k}$ has center 0 and longest sidelength $\approx 2^m$.
2. The cross-section of $R_{m,k}$ perpendicular to its longest side is a square.
3. $S_m \subset \bigcup_{k=1}^{k_m} R_{m,k}$ and $\sum_k |R_{m,k}| \le C(n)\,|S_m|$ (stratification)
4. $S \subset \bigcup_{m,k} R_{m,k}$
5. $|S| \approx \sum_{m,k} |R_{m,k}| < \infty \quad (\Omega \in L^1(\Sigma))$
6. $\sum_{m,k}(m+1)|R_{m,k}| < \infty$ if and only if $\Omega \in L\,log^+L$
7. $\sum_{m,k}(m+1)2^{mn(r-1)}|R_{m,k}| < \infty$ if and only if $\Omega \in L^r$, $1 < r < \infty$.

Some typical one-weight results are listed in the next three theorems. We use the standard notation

$$L^p_w(R^n) = \left\{f : ||f||_{p,w} = \left(\int_{R^n} |f(x)|^p w(x)\,dx\right)^{1/p} < \infty\right\},$$

where $1 < p < \infty$ and $w$ is a nonnegative measurable function on $R^n$.

**Theorem 1.2.** *Let $\Omega \in L\log^+ L$, $\int_\Sigma \Omega = 0$, $1 < p < \infty$, and $w \geq 0$ on $R^n$. Then $H$ is bounded on $L^p_w(R^n)$ if there is a finite constant $C$ such that for all $m, k$ and every rectangle $R$ which is a dilate or translate of $R_{m,k}$,*

$$\left(\frac{1}{|R|}\int_R w\right)^{\frac{1}{p}} \left(\frac{1}{|R|}\int_R w^{-\frac{p'}{p}}\right)^{\frac{1}{p'}} \leq C.$$

Unfortunately, this result is not sharp. In fact, in simple cases (see [WW1]), we know that replacing the right-hand side by $C\frac{1}{|R_{m,k}|}$ is a necessary condition for boundedness on $L^p_w$. Note that $C << C\frac{1}{|R_{m,k}|}$.

**Theorem 1.3.** *Let $\Omega \in L\,log^+ L$, $\int_\Sigma \Omega = 0$, and suppose that for all $m, k$ and every rectangle $R$ which is a dilate or translate of $R_{m,k}$, the weight $w$ satisfies*

$$\left(\frac{1}{|R|}\int_R w\right)\left(\frac{1}{|R|}\int_R w^{-1}\right) \leq C_{m,k}\frac{1}{|R_{m,k}|}$$

*with*

$$\sum_{m,k}(m+1)C_{m,k} < \infty.$$

*Then $H$ is bounded on $L^2_w(R^n)$.*

This is a better result than Theorem 1.2 in case $p = 2$, but its hypothesis is still very far from the necessary condition just mentioned.

When $p \neq 2$, we need to strengthen the corresponding $A_p$-type condition by adding a "bump." For example,

**Theorem 1.4.** *Let $\Omega \in L\log^+ L$, $\int_\Sigma \Omega = 0$ and $1 < p < 2$. If there exists $r > 1$ such that*

$$\left(\frac{1}{|R|}\int_R w\right)^{\frac{1}{p}} \left(\frac{1}{|R|}\int_R w^{-\frac{rp'}{p}}\right)^{\frac{1}{rp'}} \leq C_{m,k}\frac{1}{|R_{m,k}|}$$

*for all $m, k$ and every rectangle $R$ which is a dilate or translate of $R_{m,k}$, and if*

$$\sum_{m,k}(m+1)C_{m,k} < \infty,$$

*then $H$ is bounded on $L^p_w(R^n)$.*

An analogous result holds for the range $2 < p < \infty$, but then the $r$-bump is switched to the first factor above, i.e., switched to the average of $w$:

$$\left(\frac{1}{|R|}\int_R w^r\right)^{\frac{1}{rp}}\left(\frac{1}{|R|}\int_R w^{-\frac{p'}{p}}\right)^{\frac{1}{p'}} \leq C_{m,k}\frac{1}{|R_{m,k}|}$$

for some $r > 1$ and all $m, k$, and every rectangle $R$ which is a dilate or translate of $R_{m,k}$. This result is dual to the one for $1 < p < 2$. Given $m$ and $k$, either of the two $r$-bump conditions for all $R$ which are translates or dilates of that particular $R_{m,k}$ follows from the unbumped condition with $r = 1$, but unfortunately the value of $r$ may depend on $m$ and $k$, which is not allowed in Theorem 1.4 or its analogue for $p > 2$.

Requiring an $r$-bump condition in Theorem 1.4 but not in Theorem 1.3 already calls into question the sharpness of Theorem 1.4.

Having said all this, we return to answer the question, "What is $G$ in Theorem 1.1?" The answer is that $G$ is defined as follows in terms of starlike maximal functions for the sets $S_m$. For any starlike $S$, let

$$M_S f(x) = \sup_{t>0}\frac{1}{t^n}\int_{tS}|f(x-y)|\,dy.$$

In 1973, C. P. Calderón [C] obtained unweighted norm estimates for $M_S$. Unweighted estimates were also derived in 1988 by M. Christ and J. L. Rubio de Francia in [ChR]. Both unweighted and weighted estimates for $M_S$ are proved in [WW1], and some of these are refined in [WW2], but to come to the point, the operator $G$ is defined by

$$Gf(x) = \sum_m (m+1)\, M_{S_m} f(x). \tag{4}$$

## 2 Degenerate Quasilinear Equations

This section contains a brief review of recent work with C. Rios and E. Sawyer [RSW2] about $C^\infty$ regularity in $R^n$ of weak solutions $u(x)$ of some degenerate quasilinear equations with smooth data. The earlier reference [RSW1] is also relevant.

The degeneracy of the equations may be of infinite order, as explained below.

The most general form of the equations we consider is

$$\operatorname{div} A(x,u)\nabla u + \boldsymbol{\gamma}(x,u)\cdot\nabla u + f(x,u) = 0 \quad \text{in } \Omega, \tag{1}$$

where $\Omega$ is a bounded open connected set in $R^n$. The matrix $A$ as well as the vector $\boldsymbol{\gamma}$ and the scalar $f$ are assumed to be $C^\infty$ functions of $x$ and $u$.

The matrix $A$ is assumed to be nonnegative definite and is not allowed to depend on $\nabla u$, but it may vanish to infinite order. Some further structural conditions will be imposed.

A simple example in case $n = 2$ is to choose $\boldsymbol{\gamma} = \mathbf{0}$, $f = 0$, and $A$ to be the diagonal matrix

$$A(x_1, x_2, z) = \operatorname{diag}\left(1, k(x_1, x_2, z)\right), \quad (x_1, x_2) \in \Omega,\ z \in R^1,$$

where $k$ is nonnegative and smooth. Even simpler is to choose $k$ to be independent of $x_2$:

$$\partial_x^2 u + \partial_y\left\{k(x, u(x, y))\partial_y u\right\} = 0, \quad (x, y) \in \Omega.$$

This last equation is known to be closely related via the partial Legendre transform change of variables to the two-dimensional Monge–Ampère equation

$$w_{ss}w_{tt} - w_{st}^2 = k(s, t), \quad (s, t) \in \Omega;$$

see, e.g., [SW3, RSW3] and [SW1, SW2].

In order to put our results in perspective, we list a few known facts.

(1) [GT] shows that if a vector $A(x, u, \nabla u)$ is elliptic and smooth, and if $b(x, u, \nabla u)$ is a smooth scalar, then the operator $Q$ defined by

$$Qu = \operatorname{div} A(x, u, \nabla u) + b(x, u, \nabla u)$$

is hypoelliptic, i.e., any weak solution of $Qu = 0$ is smooth.

(2) It is proved in [T] that if $A(x, u, \nabla u)$ is smooth and subelliptic, i.e., ellipticity fails only to some finite order, and the operator $Q$ in (1) is linear, then $Q$ is still hypoelliptic.

(3) If $A$ is not subelliptic and $Q$ is linear, the situation is delicate. For example, a result due to Fedii [F] shows that the two-dimensional operator $\partial_x^2 + k(x)\partial_y^2$ is hypoelliptic if $k(x)$ is smooth, $k(x) \geq 0$, and $k(x) > 0$ for all $x \neq 0$, for example if

$$k(x) = \exp(-1/x^2).$$

On the other hand, in three-dimensions, by a result of Kusuoka and Stroock [KS], if $k(x)$ is smooth, nonnegative and vanishes only at $x = 0$, then the operator

$$\partial_x^2 + k(x)\partial_y^2 + \partial_z^2$$

is hypoelliptic if and only if $\lim_{x\to 0} x \log k(x) = 0$.

Our situation is in general neither subelliptic nor linear.

Before being more precise, we list two simple special cases of our results.

**Theorem 2.1 (Regularity of Solutions).** *Let $\Omega$ be a bounded open connected set in $R^n$ containing the origin. Let $k_i(x, z)$, $i = 2, \ldots, n$, be smooth nonnegative functions in $\Omega \times R$ such that*

$$k_i(x, z) = 0 \textit{ is possible only if } x = (0, \ldots, 0, x_i, 0, \ldots, 0),$$

*and*

$$\left|\frac{\partial}{\partial z}k^i(x,z)\right| = o(k^*(x,z)) \textit{ as } k^*(x,z) \to 0 \tag{2}$$

*where* $k^* = \min\{k_2, \ldots, k_n\}$. *Then any* continuous *weak solution of*

$$\frac{\partial^2}{\partial x_1^2}u(x) + \sum_{i=2}^{n} \frac{\partial}{\partial x_i}\Big[k_i(x,u(x))\frac{\partial}{\partial x_i}u(x)\Big] = 0 \textit{ in } \Omega$$

*is also a strong solution, and satisfies* $u \in C^\infty(\Omega)$. *Moreover, for any nonnegative integer* $N$ *and subdomains* $\Omega'$ *and* $\Omega''$ *satisfying* $\overline{\Omega''} \subset \Omega'$ *and* $\overline{\Omega'} \subset \Omega$, *there is a constant* $C(N, ||u||_{L^\infty(\Omega')})$ *also depending on the* $C^{N+2}$ *norm of the* $k_i$, *etc., such that*

$$\sum_{|\alpha|\le N} ||D^\alpha u||_{L^\infty(\Omega'')} \le C(N, ||u||_{L^\infty(\Omega')}).$$

**Theorem 2.2 (Dirichlet Problem).** *Let* $\Omega$ *be a bounded, strongly convex domain in* $\mathbb{R}^n$ *containing the origin. Let* $k_i(x,z)$, $i = 2, \ldots, n$, *be as in Theorem 2.1. Then for any continuous function* $\varphi$ *on* $\partial\Omega$, *there is a unique continuous strong solution* $u(x)$ *of the Dirichlet problem*

$$\frac{\partial^2}{\partial x_1^2}u(x) + \sum_{i=2}^{n} \frac{\partial}{\partial x_i}\Big[k_i(x,u(x))\frac{\partial}{\partial x_i}u(x)\Big] = 0 \textit{ in } \Omega$$

$$u(x) = \varphi(x) \textit{ on } \partial\Omega,$$

*i.e., there is a unique* $u$ *that is both a strong solution of the differential equation in* $\Omega$ *and continuous in* $\overline{\Omega}$ *with boundary values* $\varphi$. *Furthermore, this solution* $u(x)$ *belongs to* $C^\infty(\Omega)$.

*Also, for any nonnegative integer* $N$ *and any subdomain* $\Omega'$ *with* $\overline{\Omega'} \subset \Omega$, *there is a constant* $C(N, \|\varphi\|_{L^\infty(\partial\Omega)})$ *depending also on the* $C^{N+2}$ *norm of the* $k_i$, *etc., such that*

$$\sum_{|\alpha|\le N} ||D^\alpha u||_{L^\infty(\Omega')} \le C(N, ||u||_{L^\infty(\partial\Omega')}).$$

Note that condition (2), i.e., the assumption that

$$\left|\frac{\partial}{\partial z}k^i(x,z)\right| = o(k^*(x,z)) \text{ as } k^*(x,z) \to 0$$

is automatically fulfilled if the $k_i(x,z)$ are independent of $z$, which is the linear case.

The regularity result Theorem 2.1 is a corollary of Theorem 2.2 by using a maximum principle and barriers. Some further comments about the method of proof are listed below. More general statements of both theorems are given in [RSW2].

We do not know if the conclusion of Theorem 2.1 holds without assuming the solution is continuous. In particular, for this reason, the theorem does not include Fedii's result and so is not optimal. Also, we do not know whether condition (2) can be weakened to the "big O" condition

$$\left|\frac{\partial}{\partial z}k^i(x,z)\right| \leq C\, k^*(x,z)$$

for some constant $C$. However, we know that (2) cannot be replaced by the weaker condition

$$|\partial_z k_i(x,z)| \leq C\, k^*(x,z)^{1-\varepsilon}, \quad 0 < \varepsilon < 1.$$

Indeed, if $0 < \varepsilon < 1$, it is shown in [RSW2] that there is a nonnegative smooth function $k(x, y; z)$ in $\mathbb{R}^2 \times \mathbb{R}$ which is strictly positive unless both $x = z = 0$ and which satisfies

$$|\partial_z k(x,y;z)| \leq C\, k(x,y;z)^{1-\epsilon}$$

for $(x, y)$ near $(0, 0)$ and all $z \in \mathbb{R}$, and there is a continuous weak solution $u(x, y)$ of the equation

$$\partial_x^2 u + \partial_y\big[k(x,y;u(x,y))\partial_y u\big] = 0$$

in a neighborhood of the origin which is not smooth.

Although we are not able to recover Fedii's result, we can derive $n$-dimensional results which are new. An example is given in the next theorem.

**Theorem 2.3.** *Let $k_i(x_1, \dots, x_n)$, $i = 2, \dots, n$, be smooth and suppose that $k_i$ is independent of the $i^{th}$ variable, i.e.,*

$$k_i(x_1,\dots,x_n) = k_i(\hat{x}^i), \quad \text{with } \hat{x}^i = (x_1,\dots,x_{i-1},x_{i+1},\dots,x_n),$$

*and suppose that $k_i > 0$ if $\hat{x}^i \neq 0$. Then any continuous weak solution of*

$$\left\{\partial_{x_1}^2 + k_2\partial_{x_2}^2 + \cdots k_n\partial_{x_n}^2\right\}u = 0 \text{ in } \mathbb{R}^n$$

*is smooth everywhere.*

We also do not recover the hypoellipticity result of Kusuoka–Stroock [KS] for the linear operator

$$\partial_{x_1}^2 + k(x_1)\partial_{x_2}^2 + \partial_{x_3}^2 \text{ in } \mathbb{R}^3$$

since $k(x_1)$ vanishes at all points $(x_1, x_2, x_3) = (0, x_2, x_3)$, and consequently our hypotheses are not all met. However, if $k(x_1)$ here is replaced by a function $k(x_1, x_3)$ with the property that $k(x_1, x_3)$ is smooth and positive for all $(x_1, x_3) \neq (0, 0)$, we obtain that every weak solution of

$$\{\partial_{x_1}^2 + k(x_1, x_3)\partial_{x_2}^2 + \partial_{x_3}^2\}u = 0$$

is smooth in the interior of the set where $u$ is continuous.

For quasilinear equations, we derive the following simple result in case $n = 2$.

**Theorem 2.4.** *If $k(x_1, x_2; z)$ is smooth, nonnegative, and satisfies*

$$|\partial_z k| = o(k) \quad as\ k \to 0,$$

*and if $k(x_1, x_2; 0)$ is not identically zero on any horizontal line segment in $\Omega$, then any continuous weak solution of*

$$\partial_{x_1}^2 u + \partial_{x_2}\big[k(x_1, x_2; u(x_1, x_2))\, \partial_{x_2} u\big] = 0, \quad (x_1, x_2) \in \Omega,$$

*is smooth in $\Omega$.*

Next we discuss some results for non-diagonal equations. The most general equation that we study has the form

$$\text{div}\, A(x, u)\nabla u + \boldsymbol{\gamma}(x, u) \cdot \nabla u + f(x, u) = 0 \quad \text{in } \Omega$$

with $A$, $\boldsymbol{\gamma}$ and $f$ smooth. The matrix $A$ need not be diagonal, but it must satisfy the condition

$$\sum_{i=1}^{n} k_i(x, z)\xi^2 \leq \xi^t A(x, z)\xi \leq C \sum_{i=1}^{n} k_i(x, z)\xi^2, \quad (x, z) \in \Omega \times \mathbb{R},\ \xi \in \mathbb{R}^n.$$

Here the functions $k_i$ are assumed to satisfy

1. $k_i \geq 0$ and $k_i$ is $C^2$;
2. there is a constant $C > 0$ such that for every $(x, z)$, there exists some $i$ for which $k_i(x, z) \geq C$ (e.g., $k_1 \equiv 1$);
3. for every $x \in \Omega$, there is a sequence of small rectangular boxes centered at $x$, with edges parallel to the coordinate axes, which shrink to $x$ such that $k_i(y, z) > 0$ for every $y$ on every face of the box except the faces that are perpendicular to the $x_i$ axis.

The $k_i$ need not be smooth but only $C^2$. However, $A$, $\boldsymbol{\gamma}$ and $f$ are assumed to be smooth. A condition like (2) is now imposed as follows on $A$ rather than on the $k_i$:

$$|\partial_z A(x, z)\xi|^2 \leq C\, \mu(k^*(x, z))k^*(x, z)\, \xi^t A(x, z)\xi,$$

where

$$\mu : [0, \infty) \to [0, 1], \quad \mu(0) = 0, \quad \mu \text{ is continuous at } 0.$$

There are also other conditions, such as

$$|\partial_{x_i} A(x,z)\xi|^2 \le C\, \xi^t A(x,z)\xi$$

and a similar assumption about some of the second derivatives of $A$; see [RSW2] for detailed statements.

Let us now make some comments of a general nature about the method of proof. The basic idea is to first derive a priori estimates for smooth solutions and then use them together with an approximation technique in order to obtain our existence and regularity results. Step 1 is to assume $u(x)$ is a strong solution in $\Omega$ and then derive a estimate like the one at the end of Theorem 2.1 but weaker, namely

$$\sum_{|\alpha|\le N} ||D^\alpha u||_{L^\infty(\Omega'')} \le C(N, ||u||_{L^\infty(\Omega')}, ||\nabla u||_{L^\infty(\Omega')}). \tag{3}$$

Here the norms of both $u$ and $\nabla u$ are allowed to appear on the right side of the estimate. This estimate is weaker in appearance but not in actuality, because it is proved without assuming condition (2). Deriving (3) is the main goal of [RSW1].

Step 2 of the proof is to remove the dependence on the norm of $\nabla u$ from the constant $C$ in (3). Achieving this uses condition (2); the proof is in [RSW2]. Step 3 is an approximation argument using barriers and a maximum principle for degenerate equations.

Next we define what is meant by a weak solution of our degenerate equations and, in particular, what the gradient of a weak solution means. We assume that $k_1 \equiv 1$.

First, we observe that the $n$ vector fields

$$\partial_1, \quad \sqrt{k_2(x,0)}\partial_2, \quad \cdots, \quad \sqrt{k_n(x,0)}\partial_n$$

are locally Lipschitz continuous. Showing this uses the assumption that all $k_i \in C^2$, and also uses Wirtinger's inequality in the form

$$|\nabla k_i(x,0)| \le C\sqrt{k_i(x,0)}.$$

In fact, denoting $k(x) = k_i(x,0)$, we have

$$\begin{aligned} \left|\sqrt{k(x_1)} - \sqrt{k(x_2)}\right| &\le \left\|\nabla(\sqrt{k})\right\|_{L^\infty} |x_1 - x_2| \\ &= \frac{1}{2}\left\|\frac{\nabla k}{\sqrt{k}}\right\|_{L^\infty} |x_1 - x_2| \le C|x_1 - x_2|, \end{aligned}$$

where the $L^\infty$ norm and the constant $C$ are defined locally. Thus, $\sqrt{k}$ is locally Lipschitz continuous.

Let $X_i$ denote the vector field $\sqrt{k_i(x,0)}\partial_i$, and let $\mathcal{X} = \{X_i : i = 1, \dots, n\}$. Define the standard degenerate Sobolev space $H^{1,2}_{\mathcal{X}}(\Omega)$ to be the inner product space consisting of all $w \in L^2(\Omega)$ whose weak derivatives $X_i w$ belong to $L^2(\Omega)$. Here, by the weak derivative $X_i w$, we mean that we assume there is a locally integrable function $X_i w$ on $\Omega$ such that for all $\varphi \in Lip_0(\Omega)$,

$$\int_\Omega (X_i w)\, \varphi \, dx = -\int_\Omega w\, (X_i' \varphi)\, dx = -\int_\Omega w\, \partial_i\big(\sqrt{k_i(x,0)}\varphi\big)\, dx.$$

The inner product on $H^{1,2}_{\mathcal{X}}(\Omega)$ is

$$\langle w, v\rangle_{\mathcal{X}} = \int_\Omega wv\, dx + \int_\Omega \mathcal{X}w \cdot \mathcal{X}v\, dx,$$

and the norm is

$$\|w\|_{H^{1,2}_{\mathcal{X}}(\Omega)} = \left(\|w\|^2_{L^2(\Omega)} + \|\mathcal{X}w\|^2_{L^2(\Omega)}\right)^{1/2}.$$

It is known that if $w \in H^{1,2}_{\mathcal{X}}(\Omega)$, then there is a sequence $\{w_j\}$ of functions in $Lip_{loc}(\Omega)$ and a vector $\mathbf{W}(x) = \big(W_1(x), \dots, W_n(x)\big)$ such that $\sqrt{k_i(x,0)}\,W_i(x) \in L^2(\Omega)$ for all $i$ and

$$\lim_{j\to\infty}\left(\|w_j - w\|_{L^2(\Omega)} + \sum_i \left\| \sqrt{k_i(x,0)}\partial_i w_j - \sqrt{k_i(x,0)}\,W_i \right\|_{L^2(\Omega)}\right) = 0.$$

Consequently, it is not hard to see that

$$X_i w = \sqrt{k_i(x,0)}\,W_i, \quad i = 1, \dots, n.$$

Also, even though the $W_i$ may not be unique, the functions $\sqrt{k_i(x,0)}\,W_i$ are unique.

By definition, $\nabla w = \mathbf{W}$ for any such vector $\mathbf{W}$. Then, for the simple equation

$$\operatorname{div} A(x,u)\, \nabla u = 0 \text{ in } \Omega, \tag{4}$$

we say that $u$ is a weak solution if $u \in H^{1,2}_{\mathcal{X}}(\Omega) \cap L^\infty_{loc}(\Omega)$ and

$$\int_\Omega \nabla\varphi \cdot A(x,u)\nabla u\, dx = 0 \quad \text{for all } \varphi \in Lip_{loc}(\Omega).$$

The definition for the other degenerate equations that we consider is similar. Note that we have assumed a weak solution of (4) is locally bounded. One reason for making this assumption is to ensure that the integral in (4) converges, the problem being the presence of $u$ as the second variable in the matrix $A(x,u)$. A simple Gronwall argument using the hypotheses on $A$ shows that

$$\xi \cdot A(x,z)\xi \approx \xi \cdot A(x,0)\xi \quad \text{uniformly in } z,\ x \text{ and } \xi$$

provided $z$ lies in a bounded set (and $x$ lies in a compact subset of $\Omega$). Thus local boundedness of $u$ guarantees convergence of the last integral. Alternately, by assuming that $A(x,z)$ is bounded in $z$ locally in $x$, we can drop the assumption that weak solutions are locally bounded.

## References


[C] Calderón, C.P.: Differentiation through starlike sets in $R^m$. Stud. Math. **48**, 1–13 (1973)

[CZ] Calderón, A.P., Zygmund, A.: On singular integrals. Am. J. Math. **78**, 289–309 (1956)

[CWW] Chanillo, S., Watson, D.K., Wheeden, R.L.: Some integral and maximal operators related to starlike sets. Stud. Math. **107**, 223–255 (1993)

[ChR] Christ, M., Rubio de Francia, J.L.: Weak type (1,1) bounds for rough operators, II. Invent. Math. **93**, 225–237 (1988)

[CorF] Cordoba, A., Fefferman, C.L.: A weighted norm inequality for singular integrals. Stud. Math. **57**, 97–101 (1976)

[Du] Duandikoetxea, J.: Weighted norm inequalities for homogeneous singular integrals. Trans. Am. Math. Soc. **336**, 869–880 (1993)

[F] Fedii, V.S.: On a criterion for hypoellipticity. Math. USSR Sb. **14**, 15–45 (1971)

[GT] Gilbarg, D., Trudinger, N: Elliptic Partial Differential Equations of Second Order, revised 3rd printing. Springer, Berlin (1998)

[KW1] Kurtz, D.S., Wheeden, R.L.: Results on weighted norm inequalities for multipliers. Trans. Am. Math. Soc. **255**, 343–362 (1979)

[KW2] Kurtz, D.S., Wheeden, R.L.: A note on singular integrals with weights. Proc. Am. Math. Soc. **81**, 391–397 (1981)

[KS] Kusuoka, S., Stroock, D.: Applications of the Malliavin Calculus II. J. Fac. Sci. Univ. Tokyo Sect. IA Math. **32**, 1–76 (1985)

[MW] Muckenhoupt, B., Wheeden, R.L.: Weighted norm inequalities for singular and fractional integrals. Trans. Am. Math. Soc. **161**, 249–258 (1971)

[P] Pérez, C.: Weighted norm inequalities for singular integral operators. J. Lond. Math. Soc. **49**, 296–308 (1994)

[RSW1] Rios, C., Sawyer, E.T., Wheeden, R.L.: A priori estimates for infinitely degenerate quasilinear equations. Differ. Integ. Equ. **21**, 131–200 (2008)

[RSW2] Rios, C., Sawyer, E.T., Wheeden, R.L.: Hypoellipticity for infinitely degenerate quasilinear equations and the Dirichlet problem. J. d'Analyse Math. **119**, 1–62 (2013)

[RSW3] Rios, C., Sawyer, E.T., Wheeden, R.L.: A higher-dimensional partial Legendre transform, and regularity of degenerate Monge-Ampère equations. Adv. Math. **193**, 373–415 (2005)

[SW1] Sawyer, E.T., Wheeden, R.L.: Hölder continuity of weak solutions to subelliptic equations with rough coefficients Mem. Am. Math. Soc. **847**, p. 157 (2006)

[SW2] Sawyer, E.T., Wheeden, R.L.: Degenerate Sobolev spaces and regularity of subelliptic equations. Trans. Am. Math. Soc. **362**, 1869–1906 (2010)

[SW3] Sawyer, E.T., Wheeden, R.L.: Regularity of degenerate Monge-Ampère and prescribed Gaussian curvature equations in two dimensions Potential Anal. **24**, 267–301 (2006)

[T] Treves, F.: Introduction to Pseudodifferential and Fourier Integral Operators. Plenum Press, New York (1980)

[Wat] Watson, D.K.: Weighted estimates for singular integrals via Fourier transform estimates. Duke Math. J. **60**, 389–399 (1990)

[W] Watson, D.K.: Weighted estimates for singular integrals via Fourier transform estimates. Duke math. J. **60**, 389–399 (1990)

[WW1] Watson, D.K., Wheeden, R.L.: Norm estimates and representations for Calderón-Zygmund operators using averages over starlike sets. Trans. Am. Math. Soc. **351**, 4127–4171 (1999)

[WW2] Watson, D.K., Wheeden, R.L.: Averages over starlike sets, starlike maximal functions, and homogeneous singular integrals. Trans. Am. Math. Soc. **363**, 5179–5206 (2011)

[Wil] Wilson, J.M.: Weighted norm inequalities for the continuous square function. Trans. Am. Math. Soc. **314**, 661–692 (1989)

MIX
Papier aus verantwortungsvollen Quellen
Paper from responsible sources
FSC® C105338

If you have any concerns about our products,
you can contact us on
**ProductSafety@springernature.com**

In case Publisher is established outside the EU,
the EU authorized representative is:
**Springer Nature Customer Service Center GmbH**
**Europaplatz 3, 69115 Heidelberg, Germany**

Printed by Libri Plureos GmbH
in Hamburg, Germany